Pesticide Resistance

STRATEGIES AND TACTICS FOR MANAGEMENT

Committee on Strategies for the
Management of Pesticide Resistant Pest Populations

Board on Agriculture

National Research Council

NATIONAL ACADEMY PRESS
Washington, D.C. 1986

NATIONAL ACADEMY PRESS 2101 CONSTITUTION AVENUE, NW WASHINGTON, DC 20418

NOTICE: The project that is the subject of this report was approved by the Governing Board of the National Research Council, whose members are drawn from the councils of the National Academy of Sciences, the National Academy of Engineering, and the Institute of Medicine. The members of the committee responsible for the report were chosen for their special competences and with regard for appropriate balance.

This report has been reviewed by a group other than the authors according to procedures approved by a Report Review Committee consisting of members of the National Academy of Sciences, the National Academy of Engineering, and the Institute of Medicine.

The National Research Council was established by the National Academy of Sciences in 1916 to associate the broad community of science and technology with the Academy's purposes of furthering knowledge and of advising the federal government. The Council operates in accordance with general policies determined by the Academy under the authority of its congressional charter of 1863, which establishes the Academy as a private, nonprofit, self-governing membership corporation. The Council has become the principal operating agency of both the National Academy of Sciences and the National Academy of Engineering in the conduct of their services to the government, the public, and the scientific and engineering communities. It is administered jointly by both Academies and the Institute of Medicine. The National Academy of Engineering and the Institute of Medicine were established in 1964 and 1970, respectively, under the charter of the National Academy of Sciences.

This project was supported under agreements between the following agencies and the National Academy of Sciences: Grant No. DAN-1406-G-SS-3076-00 from the U.S. Agency for International Development; Grants No. 59-32R6-2-132 and 59-3159-4-33 from the U.S. Department of Agriculture; and Contract No. CR-810761-01 from the U.S. Environmental Protection Agency. Support from the following corporate sponsors is also gratefully acknowledged: American Cyanamid Company; Ciba-Geigy Corporation; E. I. du Pont de Nemours & Company; FMC Corporation; ICI Americas, Inc.; Mobay Chemical Corporation; Monsanto Agricultural Products Company; NOR-AM Chemical Company; Rohm and Haas Company; Sandoz, Inc.; and Union Carbide Agricultural Products Company, Inc.

Library of Congress Cataloging-in-Publication Data
Main entry under title:

Pesticide resistance.

Contains papers from a symposium held in Washington, Nov. 27–29, 1984.
Includes index.
1. Pesticide resistance—Congresses. I. National Research Council (U.S.). Committee on Strategies for the Management of Pesticide Resistant Pest Populations.
SB957.M36 1985 363.7′8 85-25919
ISBN 0-309-03627-5

Printed in the United States of America

COMMITTEE ON STRATEGIES FOR THE MANAGEMENT OF PESTICIDE RESISTANT PEST POPULATIONS

EDWARD H. GLASS (*Chairman*), New York State Agricultural Experiment Station, Cornell University
PERRY L. ADKISSON, Texas A&M University
GERALD A. CARLSON, North Carolina State University
BRIAN A. CROFT, Oregon State University
DONALD E. DAVIS, Auburn University
JOSEPH W. ECKERT, University of California
GEORGE P. GEORGHIOU, University of California, Riverside
WILLIAM B. JACKSON, Bowling Green State University
HOMER M. LeBARON, Ciba-Geigy Corporation
BRUCE R. LEVIN, University of Massachusetts
FREDERICK W. PLAPP, JR., Texas A&M University
RICHARD T. ROUSH, Mississippi State University
HUGH D. SISLER, University of Maryland

Staff

ELINOR C. CRUZE, *Project Officer*
GERALDINE WILLIAMS, *Secretary*
HALCYON YORKS, *Secretary*
VANESSA LEWIS, *Secretary*

Contents

Preface

THE BRIGHT FUTURE projected for crop protection and public health as a result of the introduction of synthetic organic pesticides is now open to serious question because of an alarming increase in the number of instances of resistance in insects, plant pathogens, and vertebrates, and to a lesser extent in weeds. There are no longer available any effective pesticides against some major crop pests, such as the Colorado potato beetle on Long Island and the diamondback moth on cruciferous crops in much of the tropical world. Likewise, the malaria eradication programs of many countries are in disarray, in large part because vector mosquitoes are no longer adequately controlled with available insecticides. The incidence of malaria is resurging at an alarming rate. Because of the costs of bringing new pesticides to market, there are fewer new pesticides, and those produced are targeted only for major crops and pests. Resistance to pesticides, which first involved only insecticides, now exists for fungicides, bactericides, rodenticides, nematicides, and herbicides.

Concern for the resistance problem has been expressed by the pesticide industry, farmers, crop protection scientists and practitioners, and government agencies. During the past 25 years there have been several symposia on the subject, and considerable research has been conducted on the genetic, biochemical, and physiological bases for resistance. As a result, much has been learned about the phenomenon; however, few methods have been developed to date for preventing or delaying the onset of resistance to pesticides, other than eliminating or minimizing their use. In the past, problems have been overcome by the substitution of new pesticides. This procedure is threatened, because the rate of introduction of new pesticides has slowed dramatically during the last few years.

New technologies and information have been developed in recent years that appear to have promise for application in finding ways to avoid or at least delay development of resistance. Thus, a new study was initiated, under the aegis of the Board on Agriculture.

The evolutionary process by which organisms develop strains resistant to chemicals is universal throughout the extensive range of organisms in which the problem now exists. It was decided, therefore, to enlist the assistance of basic scientists in evolution, population genetics, modeling, and biochemistry. It was also decided to make the study inclusive across pest classes and involve international experts from academia, government, and industry. Inasmuch as the application of solutions will have to take place in the field or wherever pests are found, we also enlisted crop protection practitioners. Finally, because resistance management systems may involve economics, regulations, and policy, representatives from these fields were recruited.

The objectives of this study were to (1) identify promising strategies to avoid or delay the development of pesticide-resistant strains of pest species, as well as manage established resistant pest populations; (2) establish research priorities to develop these strategies and new approaches not currently in use; (3) stimulate pertinent research, not only in those disciplines concerned with resistance of pests affecting plants and animals, but in related fields as well; and (4) analyze the impact of changes in policy that will be needed to implement these strategies.

To accomplish these objectives, the committee organized a conference held in Washington, D.C., November 27–29, 1984. The conference consisted of a two-day symposium at which invited papers were presented, followed by a one-day workshop attended by the committee, symposium speakers, and additional scientists who were asked to participate.

The conference was designed to produce this volume, which integrates a report prepared by the Committee on Strategies for the Management of Pesticide Resistant Pest Populations and the symposium papers themselves. The report is based on the committee's deliberations, the symposium papers, and the workshop discussions, while the papers represent the ideas of the individual authors. A group of papers follows each relevant section of the report. A glossary is included to communicate as broadly as possible among the disciplines and backgrounds of the many interests concerned with management of resistance to pesticides.

We hope this book will prove useful to many people, especially those involved in pest control, whether in industry, academia, government, applied pest management, or decision making.

We are grateful to our many scientific colleagues who have given generously of their knowledge and time to this study. Special thanks and appreciation are extended to Drs. Raymond E. Frisbie, Timothy Dennehy, and A. Daniel Ashton for their contributions. We also recognize and appreciate the fine support of Dr. Elinor C. Cruze, staff officer for this study, and other staff of the Board on Agriculture.

Edward H. Glass, *Chairman*
Committee on Strategies for the Management of
Pesticide Resistant Pest Populations

Executive Summary

LITERALLY HUNDREDS OF SPECIES of insects, plant pathogens, rodents, and weeds have become resistant to chemical pesticides. Indeed, resistance to pesticides is a global phenomenon. It is growing in frequency and stands as a reminder of the resiliency of nature. Public health protection efforts have been frustrated—sometimes dramatically—by resistance in populations of insects and rodents involved in the spread of disease to human populations. Substantial effects of resistance on agricultural productivity, however, have been limited so far to a few crops and locations because nonchemical tactics and alternative pesticides have generally been available for use.

Although scientists recognized resistance of insects to chemical pesticides nearly 76 years ago, the problem became widespread in the 1940s during an era of extensive use of synthetic organic insecticides and acaricides. Research on the phenomenon of resistance progressed slowly over the next three decades, despite a steadily growing list of documented cases. In the 1970s three unrelated factors converged, heightening concern around the world and lending momentum to scientific research focused on the genetic, biochemical, and ecological factors associated with resistance.

First, entire classes of once highly effective compounds became useless in many major applications because of resistance. The number and diversity of pests displaying resistance increased appreciably worldwide, as did the list of chemicals to which resistance developed. Second, clear limits began to emerge in the ability of chemists to identify and synthesize effective and safe alternative pesticides. The stock of available compounds came to be viewed as a limited resource that could—like natural resources—be depleted through poor management. Third, tremendous progress occurred within several basic scientific disciplines: scientists experimented with powerful new tools for elucidating the genetic and biochemical modes of action of pesticides; understanding of the cellular and subcellular mechanisms by which pests develop resistance grew rapidly; and progress in unraveling the genetics of resistance led to new insights into the defense systems and vulnerability of pests. Scientists began to use these new insights—with some encouraging early results—to develop more stable and effective pest-control strategies.

The combination of these three factors profoundly influenced the thinking of most pest-control researchers, practitioners, and manufacturers. Resistance is spreading at an increasing rate among pests in some crops in virtually all parts of the world. Hard lessons for pesticide manufacturers have accompanied the economic consequences of resistance. Companies now take very seriously the prospect that resistance may limit the number of years a new product will have to recover the steadily growing costs incurred in its development, testing, production, and registration. In the United States timely progress in managing resistance is a practical necessity for many farmers struggling to stay profitable in the face of growing international competition.

The committee believes that slowing or halting the spread of resistance to pesticides should become a prominent focus in both public and private sectors. A range of activities needs to be pursued, including research, field monitoring and detection programs, education, and incorporation of strategies to manage resistance into international development and health programs. Fortunately, various individuals and groups involved in pest management have pioneered the application of some promising new strategies, and more resources and attention throughout the pest-control industry are being devoted to the verification and dissemination of data on resistance and methods to manage its evolution.

The idea and impetus for this project reflect growing concern about resistance and the sense that a more systematic and scientific approach is needed to deal with this recurrent problem. In this report we take stock of what is now known about the extent and severity of resistance problems around the world, limiting the discussion primarily to pests of agricultural importance. (Resistance in disease organisms and vectors also is extremely important, but this area has already received considerable attention.) The genetic and biochemical mechanisms of resistance are assessed and emphasis is placed on some of the new biotechnological methods used to study resistance. Application of population biology to the study of resistance is also reviewed. Papers and dialogue presented at the November 27–29, 1984 conference suggest that significant advances in understanding the development of resistance can be achieved by researchers in biochemistry, genetics, and theoretical population biology collaborating with those in applied pest-management disciplines. Such synergism and multidisciplinary cooperation may prove critical in developing, refining, and validating practical management strategies that can be adopted to halt or slow down the emergence of resistance or otherwise reduce the severity of its impact.

Biotechnology is already providing critical insights into the mode of action of a few major classes of herbicides and is expected to do the same for other pesticides. These and other insights that biotechnology can offer may eventually make most conventional pesticides obsolete. Under the best of circumstances, however, such breakthroughs are a decade off for the majority of major pests and crops. In the meantime (perhaps indefinitely) pest-control strategies involving some use of chemical pesticides will need to be developed, implemented, monitored, and adjusted to sustain control that is both efficacious and affordable. The nature and properties of new pesticides will also evolve over the next several decades. Most new products will be more selective, less toxic to mammals, and effective at lower rates of application. Many will be chemical analogs of naturally occurring chemicals that control some physiological aspect of development in pest species. Nevertheless, effective management of the propensity of pest populations to develop resistance will remain a practical necessity.

A second major focus of the symposium and this report is the critical requirement for dealing with resistance now and in the foreseeable future. Resistance is a phenomenon that typically develops rapidly. A pest population just beginning to display resistance may respond favorably to a change in management tactics for only a relatively brief period after detection. Resistance can progress within just a few seasons—or even within a season—to a point at which dramatic changes in control strategies or cropping patterns become necessary. If this narrow window is not exploited, the battle can soon be lost.

Two other conclusions surfaced at the symposium and workshops: (1) pest populations that are already resistant to one or more pesticides generally develop resistance to other compounds more rapidly, especially when the compounds are related by mode of action to previously used pesticides, and (2) most pests can be expected to retain inherited resistance to pesticides for long periods. Hence primary reliance on chemical control strategies over the long run will depend on a steady stream of new compounds with different modes of action that can also meet regulatory requirements and economic expectations—an unlikely prospect in many pest-control markets.

Throughout the United States and around the world new strategies are being formulated to slow or reverse the onset of resistance during this window of time between the detection of resistance and its often rapid evolution in severity to an unmanageable state. A necessary first step, treated at length in this volume, is the development and use of rapid, reliable methods to detect low levels of resistance in pest populations. Immunology, biochemistry, and molecular genetics are expected to play a major role in developing these methods. Methods also are needed to monitor the spread and severity of a resistance episode over time and space in order to gain an accurate sense of the size of the window and how rapidly it is closing.

Data stemming from new assay methods used in resistance detection and monitoring efforts would be extremely valuable in the development of active strategies to manage pesticide resistance. The thinking underlying the use of such strategies is closely related to the philosophy and principles of integrated pest management (IPM). Put simply, management of resistance is an attempt to integrate chemical and nonchemical control practices through a range of tactics, singly or in combination, so that the frequency of resistant members of pest populations remains within a manageable, economically acceptable level.

Management of resistance offers great promise as a complementary extension of IPM. The tools and knowledge needed to structure and analyze opportunities to manage resistance are very similar to the information needs of scientists developing, applying, monitoring, and adjusting IPM strategies. Application of theoretical concepts from population biology and the use of general and specific models may provide important new capabilities in predicting the outcome of different sets of pest-management tactics. On the other hand, we see little justification in maintaining the polite fiction that pesticide resistance is solely a technical problem that can be readily overcome with the right new pesticide or an adjustment in the way conventional pesticides are used. For even a single crop or clinical situation, the design, execution, monitoring, and long-term implementation of a pesticide-use program is a major endeavor. Even with careful monitoring, timely research, and enlightened product stewardship, the efficacy of many pesticides will prove impossible to sustain except in a very limited sense and in isolated applications.

Problems loom ahead as we are forced to deal with the practical consequences of resistance episodes. These problems must be faced and will invariably command the attention of most scientists engaged in pest-control research. Experience has taught us that resistance episodes will flare up like forest fires, sometimes unexpectedly and other times not surprisingly.

As scientists and institutions gain expertise and devote additional resources to contend with threatening resistance occurrences, it is critical that steps also be taken, steadily and collectively, to develop a deeper understanding of resistance. New institutional mechanisms and a shared commitment are vitally needed so that the lessons learned in each resistance episode are not lost. Only by learning systematically from mistakes can we hope to avoid making the same mistake elsewhere, or in other crops or for different pests or pesticides. Much of the knowledge needed will be gained more quickly if new forms of collaboration, and closer ties can be forged between applied and academic biology. A concerted effort by research administrators to underwrite such collaboration—and overcome well-entrenched barriers—will be an important step toward identifying practical solutions to pesticide resistance problems.

Resistance is a potentially powerful, pervasive natural phenomenon. The development and severity of resistance to pesticides is controlled primarily by human action. Ignorance or a lack of concern in dealing with resistance can set the stage for explosions in pest populations leading to crop failure and reversals in the effectiveness of public health protection programs.

Resistance can and must be attacked in a variety of ways. Some scientists and pest-control practitioners will focus on the need for changes in farmers' pest-control practices; some will develop methods to detect and monitor resistance; and others will attempt to find improved institutions to coordinate management of resistant pest populations among various groups of farmers, other pesticide users, and pesticide manufacturers. Some scientists will pursue fundamental work on identifying the molecular and physiological bases of resistance. Progress at one level will help at other levels in understanding the ways organisms manage to overcome external threats like those posed by pesticides. To progress most swiftly and efficiently, communication and information dissemination are critical needs not adequately met either by public or private institutions.

RECOMMENDATIONS

Basic and Applied Research

Each of these research areas will require moderate or substantial increases in funding, either from new or redirected sources of funds, or both. Some of the needed research can and probably will be undertaken by the private sector. Additional public funding should be supplied through peer-reviewed programs such as USDA's Competitive Grants Program.

The following recommendations are not listed in order of priority.

RECOMMENDATION 1. **More research is needed on the biochemistry, physiology, and molecular genetics of resistance mechanisms in species representing a range of pests. Molecular biology, including recombinant DNA technology, should be helpful in isolating and characterizing specific mechanisms of resistance.**

The information provided by these investigations is essential to develop tactics to counter resistance, rapid new techniques to monitor and detect the extent of resistance, and novel pesticides (considered in more detail in Chapters 2, 3, and 5).

RECOMMENDATION 2. **The discovery and exploitation of new "target sites" for novel pesticides should be a key focus as research efforts are initiated that combine traditional research skills with the new biotechnologies.**

The number of modes of action of pesticides in current use is limited and, as a result of resistance, the number of functional pesticides is decreasing for some pests. Pesticide control will remain a necessity in many circumstances, and new compounds will be needed (Chapter 1). The methods of contemporary biotechnology should be very useful both in the identification of these target sites and for the production of new pesticides (Chapter 2).

RECOMMENDATION 3. **Standard methods to detect and monitor resistance in key pests need to be developed, validated, and then applied more widely in the field.**

Resistance detection and monitoring techniques are essential to early warning systems and in establishing the extent and severity of resistance (Chapter 4). These methods are critical for advancing and evaluating programs to manage resistance (Chapters 3 and 5). Agricultural producers, pesticide manufacturers, and applicators will benefit from better methods to monitor resistance.

RECOMMENDATION 4. **Concepts and insights stemming from population biology research on pesticide resistance should be used more effectively to develop, implement, and evaluate strategies and tactics to manage resistance.**

Population biology theory has been useful in a retrospective manner in explaining past resistance episodes. It can also be useful in a predictive manner, for the development of optimum operational schemes to manage resistance for each pest-control situation (Chapter 3).

RECOMMENDATION 5. **The development and testing of a system of resistance risk assessment needs to be pursued.**

The ability to forecast accurately the likelihood of resistance may allow for the extension of the effective life of pesticides and offer insight into how the use pattern of a pesticide should be changed to slow the development of resistance. Experts in resistance risk assessment may eventually be able to recognize previously undocumented or unforeseen resistance episodes in time to develop alternative control strategies that halt the evolution of resistance (Chapter 4).

RECOMMENDATION 6. **Increased research and development emphasis should be directed toward laboratory and field evaluation of tactics for preventing or slowing development of resistance (Chapter 5).**

RECOMMENDATION 7. **Efforts should be expanded to develop IPM systems and steps taken to encourage their use as an essential feature of all programs to manage resistance (Chapter 5).**

Implementation of Detection and Monitoring Techniques for Key Pests and Maintenance of Practices to Manage Resistance

RECOMMENDATION 8. **It is critical to determine for resistant populations the level of tolerance to the pesticide and the relative fitness of the resistant versus the susceptible portion of the pest population.**

This information is essential to the development of a sound program for managing the resistant population (Chapter 3).

RECOMMENDATION 9. **Resistance detection, monitoring, and management organizations should be formed at the local or regional level and assume greater responsibility for education, coordination, and implementation of activities to deal with resistance problems.**

Resistance monitoring activities are most effective when they are conducted by the people immediately concerned with the problem and most familiar with the specific situation of pesticide use (Chapters 4 and 6). Building wherever possible on existing initiatives (including NBIAP, the National Biological Impact Assessment Program, organized by the U.S. Department of Agriculture), new institutional mechanisms are needed to coordinate the efforts of different scientists working at the local and regional levels on specific crops or pest-control needs.

RECOMMENDATION 10. **Continuous monitoring programs should be used to evaluate the effectiveness of tactics to manage resistance.**

Information derived from monitoring programs is essential to evaluate the effectiveness of tactics to manage resistance (Chapters 3 and 4). Continuous monitoring can help protect growers from excessive losses and provide pesticide manufacturers with an early warning that product efficacy may be in jeopardy.

RECOMMENDATION 11. **Federal agencies should support and participate in the establishment and maintenance of a permanent repository of clearly documented cases of resistance.**

A bank of information on the incidence of resistance to pesticides will be needed for the rational choice of compounds by users, the planning of programs to manage resistance, and the development of new compounds by industry. This data bank should be broad-based and include information about the incidence and level of resistance for specific pests, the affected geographic regions, and cross-resistance with other pesticides (Chapter 4).

RECOMMENDATION 12. **Departments of agriculture within each state, in considering whether to request emergency use permits to respond to pest-control needs that have arisen because of resistance to another compound, should seek advice on whether the conditions governing the emergency use**

permit are consistent with validated tactics for the management of resistance. The U.S. Environmental Protection Agency, in approving such requests, should also consider the consequences for managing resistance, especially when cross-resistance is thought to be a possibility.

RECOMMENDATION 13. **After consultation with the EPA; university, state, and federal researchers; and industry trade associations, the U.S. Department of Justice should consider issuing a voluntary ruling that clarifies the antitrust implications (if any) of private sector initiatives to combat resistance.**

Such a ruling would alleviate concerns over possible antitrust prosecutions following efforts by private companies working jointly to prescribe directions for use on labels of competing pesticide products. Such jointly developed use directions are sometimes needed to slow the onset of resistance to a family of pesticides or to a single compound sold by different companies (Chapter 6).

RECOMMENDATION 14. **The public sector should become more involved in the development of required residue chemistry and other data for minor crop uses. State and federal agencies should consider applying the IR-4 program concept in developing data needed to gain registrations of pesticides with nonagricultural minor uses.**

Such efforts will help ensure availability of efficacious pesticides for use on minor crops and for nonagricultural uses such as chemical sterilants and rodenticides (Chapter 6).

RECOMMENDATION 15. **Activities to manage resistance should not override environmental health and safety responsibilities, which should remain the highest priority mission of regulatory agencies. Appropriate groups, such as the Cooperative State Research Service, the Cooperative Extension Service, the Public Health Service, and professional societies, should take leadership roles in organizing work and educational groups within state, regional, and national IPM programs to implement efforts to manage resistance (Chapter 6).**

It is necessary for some organizations to take a leadership role—including the establishment of new funding sources and mechanisms—to help galvanize research pertinent to management of resistance and to initiate new collaboration on projects essential to scientific progress on many key fronts (Chapter 6).

RECOMMENDATION 16. **A considerable effort should be put into the development of pest-control measures that do not rely on the use of chemical pesticides.**

Control of pest populations by combining in cycles the use of old and novel chemical pesticides, as they become available, is unlikely to be a viable long-term strategy. There is no biological or evolutionary justification for the proposition that pest populations will return to sensitive states in relatively short order following the termination of the use of specific pesticides that brought on resistance. Moreover, experience suggests that novel and safe new pesticides will not always appear on the market when needed to replace compounds that have lost their effectiveness due to resistance.

* * *

We are growing familiar, through unfortunate experiences, with the development of resistance. We can and should learn from these lessons. It has become apparent that the phenomenon of resistance demands clear, thoughtful, and systematic actions to prevent the loss of valuable pesticides that can contribute greatly to meeting food needs. The day is approaching when effective, affordable alternatives simply will not be available. Then, adjustments that could at times be extremely costly will have to be made in how and where we produce food. Important changes in attitude, commitment, and priority are needed now if we are to slow and eventually reverse the spread of resistance. This report offers guidance on logical steps to get the process under way.

1

Introduction

RESISTANCE IS A CONSEQUENCE of basic evolutionary processes. Populations have genetic variance, and plants and herbivores have a history, respectively, of evolving chemical defenses and overcoming them. Some individuals in a pest population may be able to survive initial applications of a chemical designed to kill them, and this survival may be due to genetic differences rather than to escape from full exposure. The breeding population that survives initial applications of pesticide is made up of an ever-increasing proportion of individuals that are able to resist the compound and to pass this characteristic on to their offspring.

Because pesticide users often assume that the survivors did not receive a lethal dose, they may react by increasing the pesticide dosage and frequency of application, which results in a further loss of susceptible pests and an increase in the proportion of resistant individuals. Often, the next step is to switch to a new product. With time, though, resistance to the new chemical also evolves.

During the early 1950s, resistance was rare, while fully susceptible populations, of insects at least, have become rare in the 1980s. Known to occur for nearly 76 years, resistance has become most serious since the discovery and widespread use of synthetic organic compounds in the last 40 years. (See Georghiou, this volume, for a fuller treatment of the magnitude of the problem.) Resistance in plant pathogens became a problem in the mid-1960s and has increased over the last 15 years along with use of systemic fungicides. Resistance is being detected with increasing frequency in weeds that have been intensively treated with herbicides. Pesticide resistance in rodents now occurs worldwide.

Resistance in insects and mites rose from 7 species resistant to DDT in 1938 to 447 species resistant to members of all the principal classes of insecticides, i.e., DDT, cyclodienes, organophosphates, carbamates, and pyrethroids, in 1984. Nearly all (97 percent) of these species are of agricultural or veterinary importance. Almost half of these species are able to resist compounds in more than one of these classes of insecticides, and 17 species can resist compounds in all five classes. Resistance occurs as well in at least 100 species of plant pathogens (primarily to the fungicide benomyl), 55 species of weeds (mainly to the triazine herbicides), 2 species of nematodes, and 5 species of rodents.

To appreciate the gravity of resistance to pesticides in agriculture and public health, though, it is necessary to look beyond lists of species known to exhibit resistance. For example, the rate of increase in species of arthropods newly reported as resistant to some pesticide has actually declined since 1980 because more of the new cases of resistance now occur in species already "counted" as resistant to some other compound. This is an even greater cause for alarm, however, since resistance to more than one compound usually means that the pest is harder to control. Furthermore, when pests are subjected to prolonged and intensive selection, frequency of resistance may stabilize at high levels over wide areas—for example, the hops aphid in England; the green rice leafhopper in Japan, the Philippines, Taiwan, and Vietnam; cattle ticks in Australia; and anopheline mosquitoes nearly worldwide. Resistance is probably the major contemporary problem in control of vectorborne diseases, particularly malaria, in most countries.

When pest organisms are resistant to one class of pesticide compounds, they may evolve resistance more rapidly to new groups of chemicals having either similar modes of action or similar metabolic pathways for detoxication. There is particular concern that the pyrethroids may have a short useful life against many pest species because of a gene identified as *kdr*. This gene played a key role in the genetic evolution of DDT resistance and appears to provide certain insects with protection against pyrethroids. Resistance to DDT is widespread, so this genetic predisposition to cross-resistance poses a potential threat to the efficacy of pyrethroids.

Pesticides remain effective in many areas where selection has been less severe. On the Atlantic coast of Central America, *Anopheles albimanus* can still be effectively controlled by organophosphates and carbamates. In the Midwest these compounds also control the Colorado potato beetle, which is resistant to every insecticide applied to control it on Long Island. Resistance to insecticides has not yet been detected in the European corn borer, but this is an exceptional case.

Nevertheless, agricultural production and public health programs can no longer rely on a steady stream of new chemicals to control resistant pest species. Resistance is spreading at an increasing rate, while development of new compounds has declined since 1970 (Georghiou, this volume). New compounds that are superior or have different modes of action are difficult to discover and are increasingly expensive to develop. Many are not pursued because of estimates that they may not return their cost of development, which is at least partly due to the potential for resistance. Pesticide costs for many agricultural and nonagricultural uses have been increasing because of resistance, which compels a switch to generally more expensive chemicals and/or more frequent applications of pesticides.

Rational pest-control strategies must be designed to manage resistance, both to prolong the effectiveness of pesticides and to reduce environmental contamination by excessive use of chemicals. These strategies should be based on integrated-pest-management (IPM) techniques. It is also vital to pursue development of new chemicals that are effective through new modes of action. Better understanding of resistance will emerge from more effective methods to detect and monitor resistance, along with better coordination of interdisciplinary research on critical areas of genetics, biochemistry, and population biology.

Many people in science and business anticipate gains in crop protection from applications of biotechnology and other new developments. Pests, however, can be expected to evolve strains that are resistant to virtually any control agent, including pest-resistant crop varieties. This is likely to hold true whether resistant plant cultivars are developed with the new tools of biotechnology or by traditional genetic methods. While it is unrealistic to expect biotechnology to eliminate the problem of pesticide resistance, emerging science does indeed offer great hope in helping reduce the impact of resistance episodes while keeping down the economic and environmental costs of pest control. For a more detailed discussion of an optimistic view of the future and data showing falling pesticide prices to farmers, see Miranowski and Carlson (this volume).

Pesticide Resistance: Strategies and Tactics for Management.
1986. National Academy Press, Washington, D.C.

The Magnitude of the Resistance Problem

GEORGE P. GEORGHIOU

The phenomenon of pest resistance to pesticides has expanded and intensified considerably in recent years. Resistance is most acute in insects and mites, among which at least 447 species—including most major pests—have been reported to be resistant to one or more classes of chemicals. At least 23 species are known to have developed resistance to pyrethroids, the most recently introduced class of insecticides. Whereas the presence of resistance was a rare phenomenon during the early 1950s, it is the fully susceptible population that is rare in the 1980s. Serious cases of resistance are also found in plant pathogens toward fungicides and bactericides and are being reported with increasing frequency in weeds toward herbicides and in rats toward rodenticides. Unquestionably the phenomenon of resistance has come to pose a serious obstacle to the efforts of many countries to increase agricultural production and to reduce the threat of vector-borne diseases. What is urgently needed is interdisciplinary research to increase our understanding of resistance and develop practical measures for its management.

INTRODUCTION

A great variety of arthropods, pathogens, and weeds compete with us for the crops that we grow for our sustenance. In turn, we attempt to control the depredation of these pests by suppressing their densities, often by the use of chemical toxicants. The use of toxicants is not a human innovation. Natural chemical defense mechanisms are present within most of our crop plants, serving to repel or kill many of the organisms that attack them.

Through the millions of years of life on earth, a continuous process of mutual evolution has taken place between plant and animal species and the various organisms that feed on them. The host plants or animals have evolved defensive mechanisms, including chemical repellents and toxins, exploiting weaknesses in the attacking organisms. In turn the attacking organisms have evolved mechanisms that enable them to detoxify or otherwise resist the defensive chemicals of their hosts. Thus, it appears that the gene pool of most of our pest species already contains genes that enable the pests to degrade enzymatically or otherwise circumvent the toxic effect of many types of chemicals that we have developed as modern pesticides. These genes may have been retained at various frequencies as part of the genetic memory of the species.

Resistance of insects to insecticides has a history of nearly 76 years, but its greatest increase and strongest impact have occurred during the last 40 years, following the discovery and extensive use of synthetic organic insecticides and acaricides. Resistance in plant pathogens is of more recent origin, the first case having been detected 44 years ago (Farkas and Aman, 1940). Numerous cases of resistance in these organisms have been reported during the last 15 years, however, coincident with the introduction of systemic fungicides (Georgopoulos and Zaracovitis, 1967; Dekker, 1972; Ogawa et al., 1983). Resistance in noxious weeds is more recent (Ryan, 1970; Radosevich, 1983), but it is now being detected with increasing frequency in species that have been intensively treated with herbicides (LeBaron and Gressel, 1982). Pesticide resistance is also manifested worldwide in rats—species that during history have come to be associated with empty granaries and the bubonic plague.

The problem of resistance to pesticides has been the subject of several recent reviews (Dekker and Georgopoulos, 1982; LeBaron and Gressel, 1982). The Board on Agriculture's symposium on "Pesticide Resistance Management" came almost exactly 33 years after a similar symposium on "Insecticide Resistance and Insect Physiology" was convened by the National Academy of Sciences on December 8–9, 1951 (NAS, 1951). That pioneering symposium, which took place only four years after the first published report of resistance to DDT (Weismann, 1947), was evidence of considerable foresight and has paid dividends during the years that followed. Attention, however, was soon directed toward more exciting goals: walking on the moon and probing the planets and beyond. Meanwhile, pests at home and in the fields have continued to evolve biologically toward greater fitness in their chemically altered environments. Whereas the presence of resistance was a rare phenomenon during the early 1950s, it is the fully susceptible population that is rare in the 1980s. Unquestionably the phenomenon of resistance poses a serious obstacle to efforts to increase agricultural production and to reduce or eliminate the threat of vector-borne diseases.

I shall attempt to discuss briefly the magnitude of the problem as it exists today, and I hope to convey the urgent need for interdisciplinary effort in the search for greater understanding of resistance to pesticides and practical measures for its management.

STATUS OF RESISTANCE

The interdisciplinary nature of the problem is evident in the variety of living organisms that have developed resistance and the many types of chemicals that are involved (Figure 1). It is also apparent that insecticides, being broad-spectrum biocides, have exceeded their intended targets and have selected for resistance not only in insects and mites but in practically every other type of organism, from bacteria to mammals. Since genetic resistance cannot be induced by any means other than lethal action, the environmental impact of such unintentional selection may be profound.

The chronological documentation of resistance that we have been maintaining at the University of California, Riverside (Figure 2), now indicates that resistance to one or more insecticides has been reported in at least 447 species of insects and mites. In addition at least 100 species of plant pathogens (J. M. Ogawa, University of California, Davis, personal communication, 1984), 48 species of weeds (LeBaron, 1984; H. M. LeBaron, Ciba-Geigy

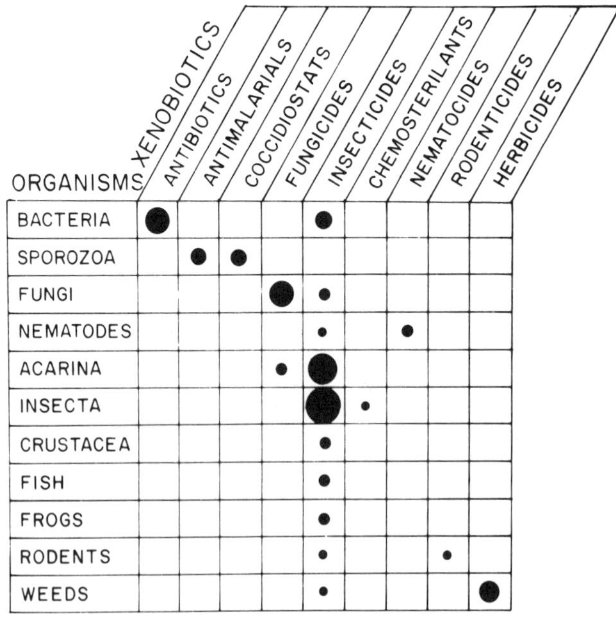

FIGURE 1 The relative frequency of resistance to xenobiotics.

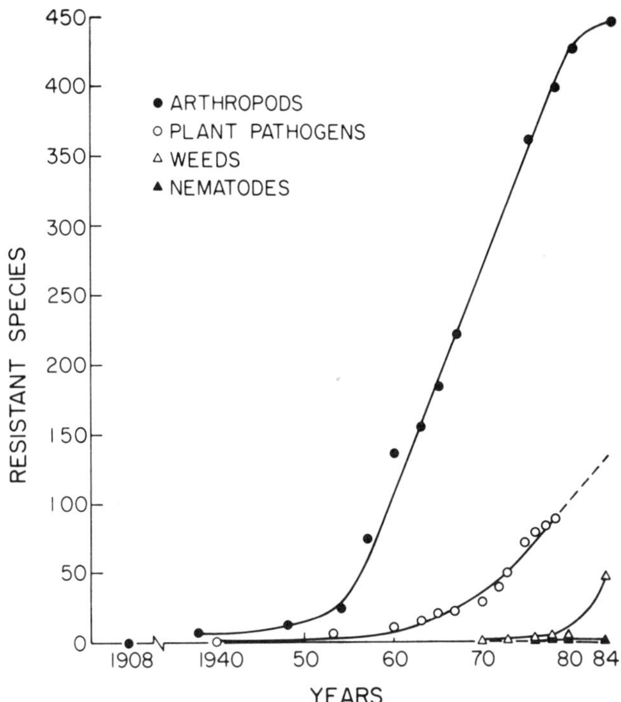

FIGURE 2 Chronological increase in number of cases of resistant species.

Corporation, personal communication, 1984), and 2 species of nematodes (Georghiou and Saito, 1983) have evolved resistance to pesticides (Figure 2). Not shown in Figure 2 are the cases of resistance in rodents, which, according to W. B. Jackson (Bowling Green State University, personal communication, 1984), now involve five species.

Resistance to the anticoagulant rodenticide warfarin was first reported in 1958 in the Norway rat (*Rattus norvegicus*) in Scotland (World Health Organization, 1976). In the United States, warfarin resistance in this species was found in North Carolina in 1970 (Jackson et al., 1971). By the mid-1970s it was detected in at least 25 percent of the sites sampled in the United States (Jackson and Ashton, 1980); at the original site in North Carolina, it occurred in essentially 100 percent of Norway rats, a truly remarkable rate of chemical selection involving a mammal.

These data concern cases of resistance that have arisen as a result of the field application of pesticides; they do not include resistance developed in laboratories through simulated selection pressure. The actual incidence of resistance must be higher than is revealed by these records, since resistance is monitored in only a few laboratories and many cases undoubtedly are not reported.

Although the rate of increase in resistant species of weeds has accelerated

TABLE 1 Increase in Cases of Resistance to Insecticides, 1980–1984[a]

	1980	1984	Percent Increase
Resistant species	428	447	4.4
Species × insecticide classes affected[b]	829	866	4.1
Species × insecticides	1,640	1,797	9.4
Species × insecticides × countries of occurrence	3,675	3,894	5.9

[a]October 1984. Data for 1980 from Georghiou, 1981.
[b]Classes: DDT, dieldrin, organophosphate, carbamate, pyrethroid, fumigant, miscellaneous.

SOURCE: Georghiou, 1981; Georghiou, unpublished.

since 1980, the rate of increase in resistant species of arthropods has declined. The reason for this decline is that an increasingly large proportion of new cases of resistance to insecticides now involves species that were recorded previously as resistant to earlier pesticides. A more realistic impression of the trend in insecticide resistance can be obtained when the increase since 1980 is viewed as the number of different insecticides to which each species is reported to be resistant. This analysis shows an increase of 9.4 percent versus a 4.4 percent rise in the number of new resistant species (Table 1).

The distribution of known cases of resistance among different orders of arthropods and the classes of chemical groups involved is indicated in Table 2. Of the 447 species concerned, 59 percent are of agricultural importance, 38 percent are of medical or veterinary importance, and 3 percent are beneficial parasites or predators.

Resistance is most frequently seen in the Diptera (156 species, or 35 percent of the total), reflecting the strong chemical selection pressure that has been applied against mosquitoes throughout the world. Substantial numbers of resistant species are also evident in such agriculturally important orders as the Lepidoptera (67 species, 15 percent), Coleoptera (66 species, 15 percent), Acarina (58 species, 13 percent), Homoptera (46 species, 10 percent), and Heteroptera (20 species, 4 percent). The resistant species include many of the major pests, since it is against these that chemical control is mainly directed.

With regard to chemical groups, cyclodiene insecticide resistance is found in 62 percent of the reported species and DDT resistance in 52 percent, followed closely by organophosphate resistance in 47 percent. Lower percentages are reported for the more recently introduced carbamate and pyrethroid insecticides. The high frequency of organophosphate resistance is undoubtedly due to the widespread use of these insecticides. It is perhaps ironic that one of the reasons organophosphates were considered more desirable than organochlorines was the prospect that these compounds, having relatively shorter persistence, would be less efficient selectors for resistance.

TABLE 2 Number of Species of Insects and Mites Resistant to Insecticides—1984[a]

Order	Chemical Group[b]							Importance[c]			
	Cyclod.	DDT	OP	Carb.	Pyr.	Fumig.	Other	Agr.	Med./Vet.	Benef.	Total (%)
Diptera	108	107	62	11	10	—	1	23	132	1	156 (35)
Lepidoptera	41	41	34	14	10	—	2	67	—	—	67 (15)
Coleoptera	57	24	26	9	4	8	5	64	—	2	66 (15)
Acarina	16	18	45	13	2	—	27	36	16	6	58 (13)
Homoptera	15	14	30	13	5	3	1	46	—	—	46 (10)
Heteroptera	16	8	6	1	—	—	—	16	4	—	20 (4)
Other	23	21	9	3	1	—	2	12	19	3	34 (8)
Total	276	233	212	64	32	11	38	264	171	12	
(%)	(62)	(52)	(47)	(14)	(7)	(2)	(9)	(59)	(38)	(3)	447

[a]Records obtained through October 1984.

[b]Cyclod. = cyclodiene, OP = organophosphate, Carb. = carbamate, Pyr. = pyrethroid, Fumig. = fumigant.

[c]Agr. = agricultural, Med./Vet. = medical/veterinary, Benef. = beneficial.

SOURCE: Georghiou, unpublished. Modified and updated from Georghiou (1981).

TABLE 3 Number of Species of Insects and Mites at Various Stages of Multiple Resistance

Year	Resistant Species	Number of Classes of Insecticides[a] that Can Be Resisted				
		1	2	3	4	5
1938[b]	7	7	0	0	0	0
1948[b]	14	13	1	0	0	0
1955[c]	25	4	18	3	0	0
1969[b]	224	155	42	23	4	0
1976[d]	364	221	70	44	22	7
1980[e]	428	245	95	53	25	10
1984[f]	447	234	119	54	23	17

[a]DDT, cyclodienes, organophosphates, carbamates, pyrethroids.
[b]Brown (1971).
[c]Metcalf (1983).
[d]Georghiou and Taylor (1976).
[e]Georghiou (1981).
[f]Records through October 1984.

SOURCE: See notes above; 1984 material new to this document.

For plant pathogens, the compilation of Ogawa et al. (1983) indicated that of the 70 species of fungi reported as resistant by 1979, 59 species (84 percent) were resistant to the systemic fungicide benomyl. Other, smaller categories involved thiophanate resistance (in 13 species of fungi) and streptomycin resistance (in 8 species of bacteria).

Among weeds most instances of resistance (41 species—28 dicots and 13 monocots) involve resistance to the triazine herbicides. In addition at least seven weed species are resistant to other herbicides, including phenoxys (e.g., 2,4-D), trifluralin, paraquat, and ureas.

Of considerable importance in exacerbating the magnitude of the resistance problem is the ability of a given population to accumulate several mechanisms of resistance. None of the present mechanisms known in field populations excludes any other mechanism from evolving. Despite the search for pairs of compounds with negatively correlated resistance, none has been discovered that would have the potential for field application. The coexistence of several resistance mechanisms (each affecting different groups of chemicals), referred to as multiresistance, has become an increasingly common phenomenon. Now almost half of the reported arthropod species can resist compounds in two, three, four, or five classes of chemicals (Table 3). Seventeen insect species can resist all five classes, including the relatively new class of pyrethroid insecticides. The species that have developed strains resistant to pyrethroids (Table 4) include some of our most important pests, such as the Colorado potato beetle (*Leptinotarsa decemlineata*) in Long Island, New

York, New Jersey, Pennsylvania, and Rhode Island; the malaria vectors *Anopheles albimanus* in Central America and *An. sacharovi* in Turkey; the house fly (*Musca domestica*) in several countries; white flies (*Bemisia tabaci*) on cotton in California; the virus vector aphid *Myzus persicae* in a number of countries; several lepidopterous pests of cotton and other crops (*Heliothis*, *Spodoptera*); and *Plutella xylostella*, a diamondback moth that is a major pest of cole crops in southeast Asia and elsewhere.

Resistance to pyrethroids has often evolved rapidly on the foundation of DDT resistance. It has been clearly demonstrated toxicologically, genetically (Omer et al., 1980; Priester and Georghiou, 1980; Malcolm, 1983), and electrophysiologically (Miller et al., 1983) that a semirecessive gene, *kdr*, often detected as one of the components of DDT resistance, is also selected by and provides protection against pyrethroid insecticides. Pyrethroid resistance that includes this gene is characteristically high, often exceeding 1,000-fold in *kdr* homozygotes, thus effectively precluding further use of pyrethroids against these resistant populations. There is valid concern that the effective life span of pyrethroids may be shorter in many developing countries, where their use directly succeeded that of DDT, than it will be in many developed countries, where the sequence after DDT has involved several years of organophosphate and carbamate use.

As in arthropods the range of compounds to which plant pathogenic fungi are resistant has expanded to include representatives of the more recently developed fungicides. Figure 3 indicates the progressive growth of fungicide resistance since 1960, with the inclusion during the last four years of cases of resistance to the dicarboximides, dichloroanilines, acylalanines, and ergosterol biosynthesis inhibitors.

FREQUENCY AND EXTENT OF RESISTANCE

When considering the magnitude of the problem, it is necessary to draw attention to the many cases of widely distributed resistance and to the high frequency of resistance genes in populations. The most frequently observed pattern of the spread of resistance is one in which isolated cases appear, initially creating a mosaic pattern that reflects the distribution and degree of selection pressure. As resistance "ages," that pattern is gradually obscured by insect dispersal and by the more widespread application of selection pressure.

In the Imperial Valley of California the pattern of resistance of the white fly *Bemisia tabaci* toward the new pyrethroid insecticides is still distinct, reflecting the number of pyrethroid treatments applied to cotton during 1984 (Figure 4). In coastal southern France the high frequency of organophosphate resistance found in *Culex pipiens* reflects the very intense chemical control

TABLE 4 Cases of Resistance to Pyrethroids[a]

Order	Species	Location	Source
Coleoptera	*Leptinotarsa decemlineata*	Ontario, Quebec, New Jersey, New York	Harris, 1984[b]; Forgash, 1981, 1984[b]
	Oryzaephilus surinamensis	New South Wales	Attia, 1984[b]
	Tribolium castaneum	Queensland	Champ and Campbell-Brown, 1970
Diptera	*Aedes aegypti*	Thailand	WHO, 1980
	Anopheles albimanus	Guatemala	Georghiou, 1980
	An. sacharovi	Turkey	Davidson, 1980
	Culex pipiens	France	Sinègre, 1984
	Haematobia irritans	Florida, Louisiana, Nebraska, Georgia, Michigan, Texas, Oklahoma, Kansas	Quisenberry et al., 1984; Keith, 1984[b], Schmidt et al., in press; Kunz, 1984[b]
		Queensland	Schnitzerling et al., 1982
	Liriomyza trifolii	California	Parrella, 1983
	Musca domestica	Europe	Sawicki et al., 1981
		Canada	Harris, 1984[b]
		California	Georghiou, 1985 (unpublished)
Homoptera	*Bemisia tabaci*	California, Arizona	Immuraju, 1984[b]
	Myzus persicae	U.K.	Sawicki and Rice, 1978
		Japan	Motoyama, 1981[b]
		Australia	Attia and Hamilton, 1978
		British Columbia	Campbell and Finlayson, 1976

	Nilaparvata lugens	Solomon Islands	Ho, 1984[b]
	Psylla pyricola	Oregon	Westigard, 1980[b]
	Trialeurodes vaporariorum	U.K.	Wardlow et al. (in press)
Lepidoptera	*Heliothis armiger*	Australia	Gunning et al., 1984
	H. virescens	Arizona, California	Martinez-Carrillo and Reynolds, 1983
	Plutella xylostella	Taiwan	Liu et al., 1981
	Scrobipalpula absoluta	Peru	Herve, 1980[b]
	Spodoptera exigua	Guatemala, El Salvador, Nicaragua	Herve, 1980[b]
	S. frugiperda	Louisiana	Wood et al., 1981
	S. littoralis	Egypt	El-Guindy et al., 1982
		Malaysia	Sudderuddin and Kok, 1978
		Singapore	Ho et al., 1983
Orthoptera	*Blatella germanica*	USSR	Smirnova et al., 1979

[a]Excluding cases of resistance to pyrethrins.

[b]Personal communications: F. I. Attia, Department of Agriculture, Rydalmere, NSW, Australia; A. J. Forgash, Rutgers University, New Brunswick, New Jersey; C. R. Harris, Agriculture Canada, London Research Center, London, Ontario; J. J. Herve, Roussel-UCL, Paris; D. T. Ho, Solrice, Honiara, Solomon Islands; J. A. Immuraju, University of California, Riverside; D. I. Keith, University of Nebraska, Lincoln; S. E. Kunz, U.S. Department of Agriculture, Kerrville, Texas; N. Motoyama, Chiba University, Matsuma, Japan, P. H. Westigard, Oregon State University, Medford.

SOURCE: See Source column and note *b* above.

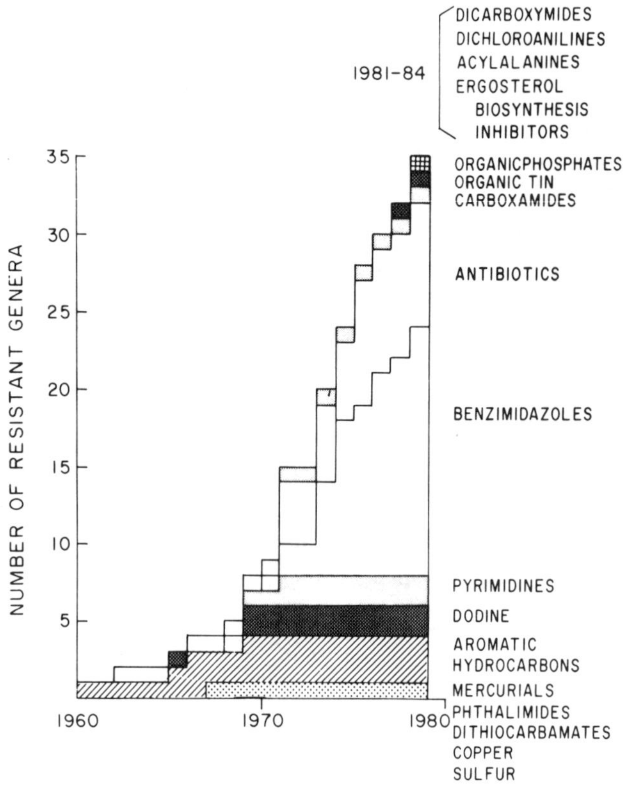

FIGURE 3 History of resistance to chemicals in plant pathogens. Source: Delp (1979), adapted from Dekker (1972), Georgopoulos (1976), and Ogawa et al. (1977); additional data from Dekker and Georgopoulos (1982) and J. M. Ogawa, University of California, Davis, personal communication, 1984.

that is being applied to protect this urbanized area. The frequency of resistance declines in the interior.

Under prolonged and intensive selection the frequency of resistance stabilizes and may show a surprising uniformity. In Great Britain, high resistance to demeton *S*-methyl was found uniformly in yearly samples of the hops aphid *Phorodon humuli* obtained from Kent during 1966–1976, compared with a susceptible population from north England during 1969–1976 (Figure 5). In another survey, involving 258 collections of the green peach aphid, only 3 collections did not contain dimethoate-resistant individuals; in 197 of the collections, more than 76 percent of the aphids were resistant (Sawicki et al., 1978).

A generally uniform pattern is evident in the distribution of resistance of

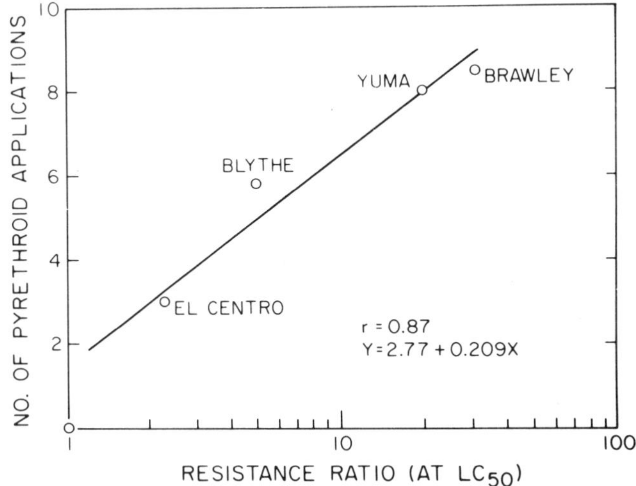

FIGURE 4 Pyrethroid resistance in *Bemisia tabaci*: relationship between resistance level and number of pyrethroid applications on cotton—1984.

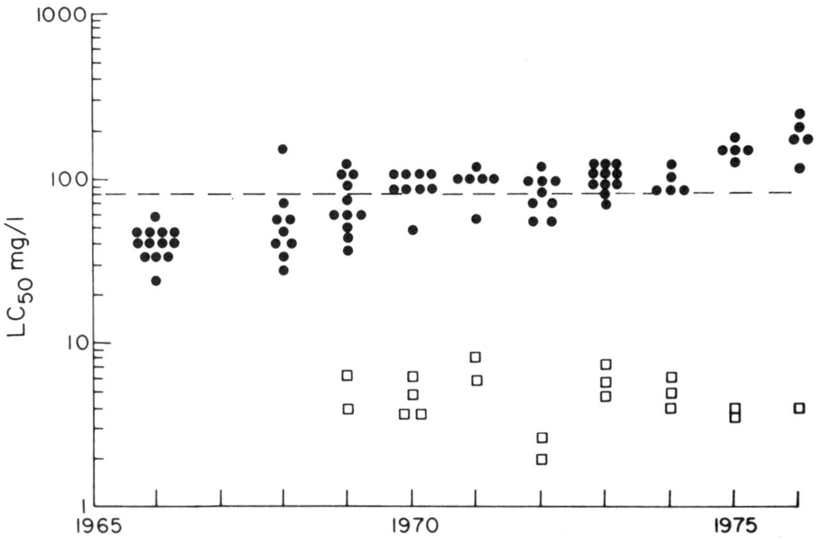

FIGURE 5 Changes in resistance to demeton *S*-methyl in stocks of *Phorodon humuli* collected from hop gardens in Kent, 1966–1976 (●), and from north England, 1969–1976 (□). Source: Muir (1979).

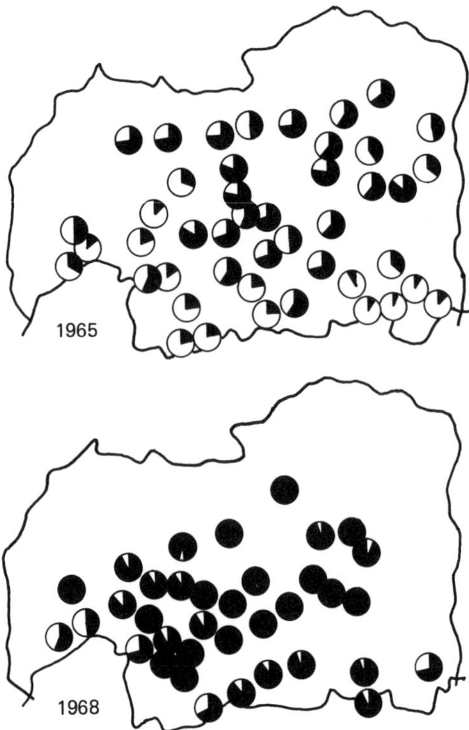

FIGURE 6 Frequency of organophosphate-resistant *Nephotettix cincticeps* in Hiroshima prefecture in 1965 and 1968. Source: Kimura and Nakazawa (1973).

the green rice leafhopper (*Nephotettix cincticeps*) in Japan (Figure 6). The frequency of resistant individuals was found to have increased rapidly from 1965 to 1968, as shown by the pattern evident in Hiroshima prefecture (Kimura and Nakazawa, 1973). Resistance of this species toward organophosphates and carbamates is now widely distributed in Japan (Figure 7), as well as in the Philippines, Taiwan, and Vietnam (Georghiou, 1981).

Likewise, resistance to organophosphates in the cattle tick (*Boophilus microplus*) in Australia is now found throughout the area of distribution of the species. In an impressive 76 percent of all sites surveyed, more than 10 percent of the ticks were resistant to organophosphates (Roulston et al., 1981). Because at this high frequency of resistance the level of control provided by organophosphate chemicals was unacceptable, tick control during the past several years has relied heavily on a group of four chemicals known collectively as amidines (Nolan, 1981). Since 1980, however, the efficacy of amidines has also declined due to resistance (J. Nolan, Commonwealth Scientific and Industrial Research Organization, Indooroopilly,

Queensland, Australia, personal communication, 1984), and emphasis is now being placed on the use of pyrethroids. Unfortunately the species has already demonstrated a low level of cross-tolerance to pyrethroids as a result of DDT resistance (Nolan et al., 1977).

Perhaps no other case of insecticide resistance has attracted as much attention as that concerning anopheline mosquitoes, vectors of malaria. The discovery of DDT enabled the launching of unprecedented programs to eradicate malaria worldwide under the guidance of the World Health Organization (WHO). These efforts have been fruitful in many areas where the disease was not endemic. But resistance in anophelines appeared soon after the program began, and it now involves 51 species, of which 47 are resistant to dieldrin, 34 to DDT, 10 to organophosphates, and 4 to carbamates (R. Pal, World Health Organization, Geneva, Switzerland, personal communication, 1984). The prospect for success of pyrethroid insecticides, which now represent the end of the line, is made uncertain by high prevailing levels of DDT resistance. Among the most critical cases, from the standpoint of frequency and intensity of multiple resistance to a variety of insecticide classes, are those of *Anopheles albimanus* in Central America, *An. sacharovi* in Turkey, and *An. stephensi* and *An. culicifacies* in the Indo-Pakistan region.

In India during 1970–1971 the frequency of genes conferring resistance to DDT in *An. culicifacies* was calculated to have been 0.34 (Georghiou and Taylor, 1976). By 1984 DDT resistance was found over much of the country, with large areas also being affected by organophosphate resistance. In *An.*

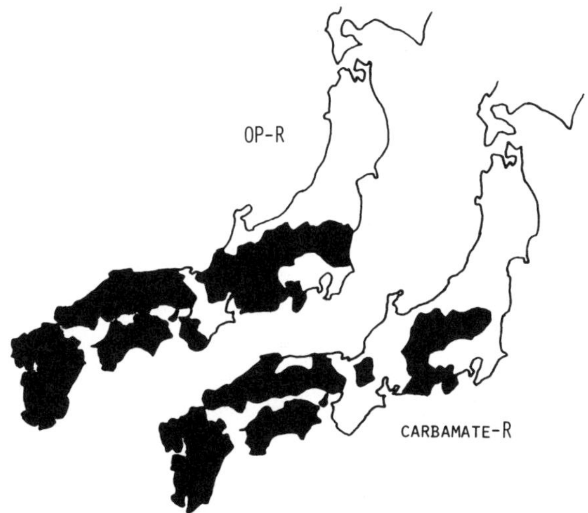

FIGURE 7 Distribution of organophosphate and carbamate-resistant *Nephotettix cincticeps* in Japan. Source: K. Ozaki, Sakaide, Japan, personal communication, 1981.

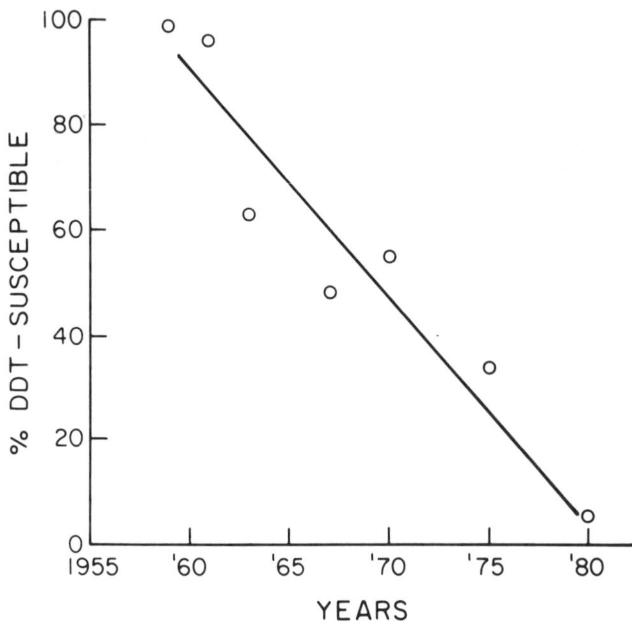

FIGURE 8 DDT susceptibility of *Anopheles albimanus* adults in Ocos, Guatemala, 1959–1980. Susceptibility determined by WHO test, 4% DDT, 1 hour. Source: H. Godoy, S.N.E.M., Guatemala, personal communication, 1981.

albimanus in Guatemala the frequency of DDT-susceptible individuals declined from nearly 100 percent in 1959 to 5 percent in 1980 (Figure 8). The propoxur susceptible genes in this species in certain areas of El Salvador had been reduced to 52 percent by 1972, leading to substantial limitation in the use of this formerly highly effective compound. The deteriorating situation of resistance in anopheline mosquitoes and its implications led the WHO Expert Committee on Insecticides to state that "it is finally becoming acknowledged that resistance is probably the largest single obstacle in the struggle against vector-borne disease and is mainly responsible for preventing successful malaria eradication in many countries" (WHO, 1976).

An important factor that exacerbates the resistance of anopheline mosquitoes in the most critical cases is widespread agricultural spraying (Georghiou, 1982). Advances in agricultural science during the past four decades have brought about the green revolution. Vast monocultures of cotton, high-yielding varieties of rice, and other crops have been developed, especially in tropical areas where the suffering from and death by malaria had previously discouraged agricultural exploitation. These areas were opened to agriculture by the malaria eradication effort. The crops in the agricultural fields became the predominant vegetation over wide areas and provided the primary resting site for adult mosquitoes. The irrigation and drainage ditches and associated ponds served as the primary breeding sites for mosquito larvae.

In these areas the agricultural pests developed resistance to one after another of the toxicants used against them, forcing applications of higher quantities of each available effective insecticide and at more frequent intervals. For example, as many as 30 insecticide treatments are applied during the six-month growing season in cotton fields in the Pacific coastal zone of Central America and southern Mexico. Records from Mexico during 1979 and 1980 show that approximately 30 liters active ingredient of a great variety of chemicals were applied per acre of cotton during the growing season (Table 5). Although these toxicants are not directed intentionally against mosquitoes, a large proportion of each generation of mosquitoes is exposed to them, often during both adult and larval stages; thus, a considerable selection for resistance genes occurs.

Insecticide resistance in *An. albimanus* in Central America is quantitatively and qualitatively correlated with the types of chemicals and the frequency of their application in cotton fields (Georghiou et al., 1973). As shown in Figure 9, resistance in *An. albimanus* in El Salvador increased in concert with the annual cotton-spraying cycle. Figure 10 illustrates the strong suppressing—and, therefore, selecting—effect of agricultural sprays on the mosquito population and the consequent increase in resistance to insecticides. Multiple resistance in these populations is now so broad as to hinder their successful control with any one of the available insecticides.

Nowhere is the end of the line of effective toxicants so clearly evident as in the Colorado potato beetle on Long Island, New York. Here, intensive chemical treatment of potato crops has resulted in the selection of a strain whose repertoire of resistance mechanisms has increased rapidly to include every insecticide that has been applied for its control (Table 6). As described recently by Forgash (1984a,b) the Colorado potato beetle "has weathered the onslaught of arsenicals . . . chlorinated hydrocarbons, organophosphorus compounds . . . carbamates and pyrethroids." This remarkable propensity for resistance, despite only two generations completed per year, is evident in the data in Table 7. The generation overwintering from 1979 had a 20-fold resistance to fenvalerate; this rose to 100-fold in the second generation of 1980, to 130-fold in 1981, and to more than 600-fold in 1982. Although combining fenvalerate with the synergist piperonyl butoxide reestablished control in 1982, this combination failed in 1983 (Forgash, 1984b). Outside Long Island a similar pattern of organophosphate-carbamate-pyrethroid resistance has been detected in several localities of the northeastern United States. As indicated in Table 6, control of the Colorado potato beetle on Long Island during 1984 was based on rotenone, a plant derivative that had been used as an insecticide for more than a century, but was superseded by

TABLE 5 Insecticides Applied on Cotton in Tapachula, Mexico, 1979–1981 (liters of active ingredient)

Insecticide Class	Compound	1979–1980	1980–1981
Organophosphates	Methyl parathion	369,626	340,800
	Parathion	60,091	50,000
	Monocrotophos	35,771	30,350
	Profenofos	30,344	30,000
	Methamidophos	14,441	21,880
	Mevinphos	7,380	15,000
	Sulprofos	7,589	14,400
	Mephosfolan	1,773	10,000
	Azinphosmethyl	2,595	4,000
	EPN	1,441	4,500
	Dicrotophos	1,687	3,496
	Dimethoate	684	
	Omethoate		500
	Total	533,422	524,926
Cyclodienes	Toxaphene	209,009	153,300
	Endrin	4,896	3,797
	Endosulfan	232	
	Total	214,137	157,097
Carbamates	Carbaryl	7,420	15,560
	Bufencarb	688	
	Total	8,108	15,560
Pyrethroids	Permethrin	2,314	5,200
	Cypermethrin	660	1,300
	Fenvalerate	529	690
	Deltamethrin	60	50
	Total	3,563	7,240
DDT	DDT	44,388	60,000
Other	Chlordimeform	24,450	25,000
GRAND TOTAL (liters)		828,068	789,823
Hectares treated		28,000	27,000
Liters a.i./HA		29.57	29.25

SOURCE: Georghiou and Mellon (1983).

DDT. Whether rotenone will continue to provide effective control remains questionable. The fact that rotenone must be combined with piperonyl butoxide to achieve control of the Colorado potato beetle indicates that metabolic enzymes capable of detoxifying rotenone are present in the population.

This somber account of critical cases of resistance does not imply that the pesticides involved are ineffective throughout the areas of distribution of the respective species. There are many examples of continued effectiveness of

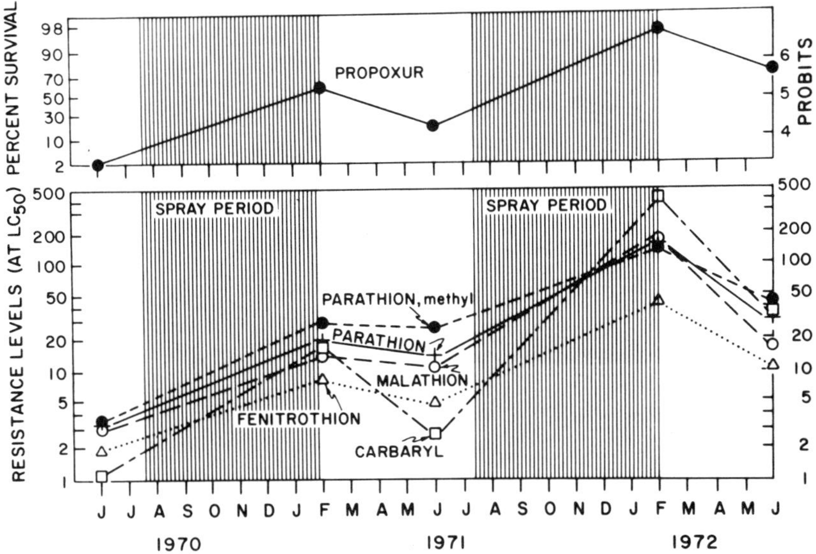

FIGURE 9 Fluctuations in resistance levels in *Anopheles albimanus* with reference to alternating agricultural spray and nonspray periods, El Salvador. Source: Georghiou et al. (1973).

the same chemical in areas where selection has been less severe. For example, organophosphates and carbamates are still effective against *An. albimanus* on the Atlantic coast of Central America; the Colorado potato beetle is still apparently susceptible to organophosphates and carbamates in the Midwest; and in the very exceptional case of the European corn borer, insecticide resistance has yet to be detected.

CONSEQUENCES OF RESISTANCE

The consequences of resistance must be immense. Farmers tend to be risk aversive (Craig et al., 1982). Thus, they have a high reliance on insurance spraying, which is probably a major cause of resistance. Usually the first response by a farmer when a pesticide is losing effectiveness is to increase the dosage applied and the frequency of application. The next step is a change to new toxicants that, typically, are more expensive than the earlier materials. The shift to new toxicants without a basic change in the philosophy and strategy of chemical control is a transient solution because, with time, resistance will probably develop to each of them. A result of these increases in dosages and frequencies of application, as well as the changes to new and invariably more expensive compounds, must be a many-fold increase in the

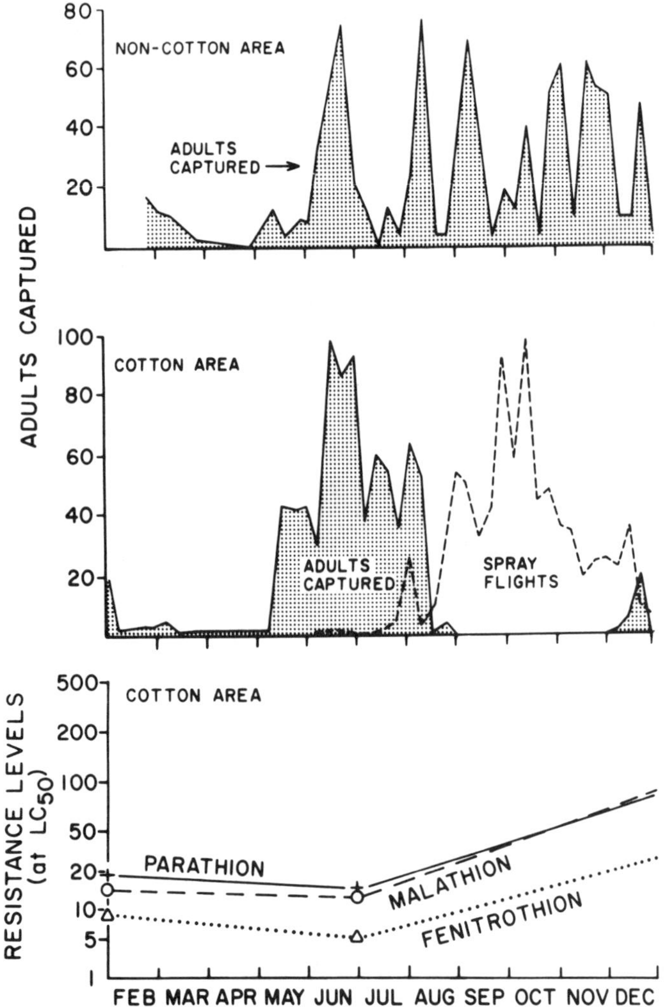

FIGURE 10 Suppression of *Anopheles albimanus* densities in cotton areas of El Salvador by agricultural sprays in 1972 and effect on resistance. Source: Hobbs (1973), Georghiou et al. (1973).

direct costs of pest control. The cost of the chemical control effort directed against the European red mite increased 5- to 8-fold as parathion was succeeded by diazinon and phenkapton and later by summer oil, omethoate, and dinocap (Figure 11) (Steiner, 1973).

In the malaria control campaigns the relative cost of insecticides for residual house spraying increased 5.3-fold when DDT was replaced by malathion and

TABLE 6 An Abbreviated Chronology of Colorado Potato Beetle Resistance to Insecticides in Long Island, New York[a]

Insecticide	Year Introduced	Year First Failure Detected
Arsenicals	1880	1940s
DDT	1945	1952
Dieldrin	1954	1957
Endrin	1957	1958
Carbaryl	1959	1963
Azinphosmethyl	1959	1964
Monocrotophos	1973	1973
Phosmet	1973	1973
Phorate	1973	1974
Disulfoton	1973	1974
Carbofuran	1974	1976
Oxamyl	1978	1978
Fenvalerate[b]	1979	1981
Permethrin[b]	1979	1981
Fenvalerate + p.b.[b]	1982	1983
Rotenone + p.b.[b]	1984	?

[a]Gauthier et al. (1981); Forgash (1984b).

[b]M. Semel, New York State Agr. Exp. Station, Riverhead, New York, personal communication, 1984; p.b. = piperonyl butoxide.

SOURCE: See notes *a* and *b* above.

15- to 20-fold when it was replaced by propoxur, fenitrothion, or deltamethrin (Table 8). Pimentel et al. (1979, 1980) estimated that the total direct costs of pesticide control measures in the United States were $2.8 billion. They also estimated that the costs due to increased resistance were $133 million (Table 9). Worldwide, excluding Russia and China, the end-user value of all pesticides purchased in 1980 was estimated at $9.7 billion (Braunholtz, 1981). If only one tenth of these pesticide applications was due to resistance (a conservative estimate), the cost of the extra chemicals alone would approximate $1 billion. Many extra applications, of course, may also be due to the suppression of natural enemies by pesticides, so the increased cost problem becomes even more intensified.

The loss of pesticide development investment must be added to the estimated cost of $1 billion. The cost of developing an agricultural chemical was estimated at $1.2 million in 1956 and at least $20 million in 1981 (Figure 12). Considering that the performance of the great majority of chemicals has been adversely affected by resistance, it may be assumed that a number of chemicals have not returned the investment involved in their development. No estimates are available of these losses, but they may be assumed to be substantial.

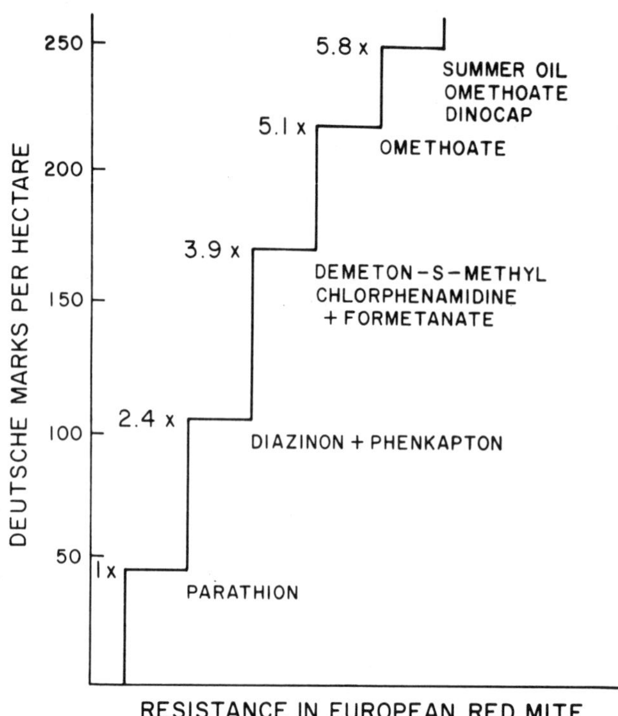

FIGURE 11 Increasing control effort and costs as pesticide resistance increases in the European red mite. Source: Steiner (1973).

TABLE 7 Development of Resistance to Aldicarb, Fenvalerate, and Synergized Fenvalerate in a Long Island Population of Colorado Potato Beetle

		Resistance Factor at LD_{50}		
Year	Generation	Aldicarb	Fenvalerate	Fenvalerate + Piperonyl butoxide
1980	Overwintering	—	20×	—
	First	13×	30×	—
	Second	22×	100×	—
1981	Overwintering	9×	30×	1.3×
	First	33×	—	—
	Second	33×	130×	4×
1982	First	—	130×	4×
	Second	60×	>600×	—
1983	Overwintering	—	>600×	200×
	First	—	>600×	200×

SOURCE: Forgash, 1984b.

TABLE 8 Relative Costs of Insecticides for Residual House Spraying

Insecticide	Dosage g/m^2 (tech.)	Approximate residual effect on mud—months	Cost per kg[a]	Cost per lb[b]	Relative cost per 6 months
DDT 75% wp	2.0	6	\$0.33	\$0.34	1.0[a]
Dieldrin 50% wp	0.5	6		2.34	1.7
Lindane 50% wp	0.5	3		3.45	5.1
Malathion 50% wp	2.0	3	0.89	1.02	5.3[a]
Propoxur 50% wp	2.0	3	3.40		20.4[a]
Fenitrothion 40% wp	2.0	3	2.63		15.9[a]
Deltamethrin 5% wp	0.1	3		~\$50.00	14.6[b]

NOTE: wp = wettable powder.

[a]World Health Organization data; Wright et al. (1972); Fontaine et al. (1978).
[b]Estimated from relative wholesale price of technical compound, Metcalf (1983).

SOURCE: Metcalf (1983).

Therefore, it is not surprising that the rate of introduction of new pesticides declined precipitously between 1970 and 1980 (Figure 13). Although several factors may have been responsible for this decline, it is strongly suspected that industry frustration with resistance has played an important role.

The question may be posed, therefore, whether we have already selected

TABLE 9 Estimated Environmental Costs Due to Loss of Natural Enemies and Insecticide Resistance in Pest Insect and Mite Populations

	Total Added Insecticide Costs (\$) Due to	
	Loss of Natural Enemies	Increased Resistance
Field crops	133,007,000	101,810,000
Vegetable crops	6,235,000	7,958,000
Fruits and nuts	14,242,000	8,312,000
Livestock and public health	>0	15,000,000
Total	153,484,000	133,080,000

SOURCE: Pimentel et al. (1979).

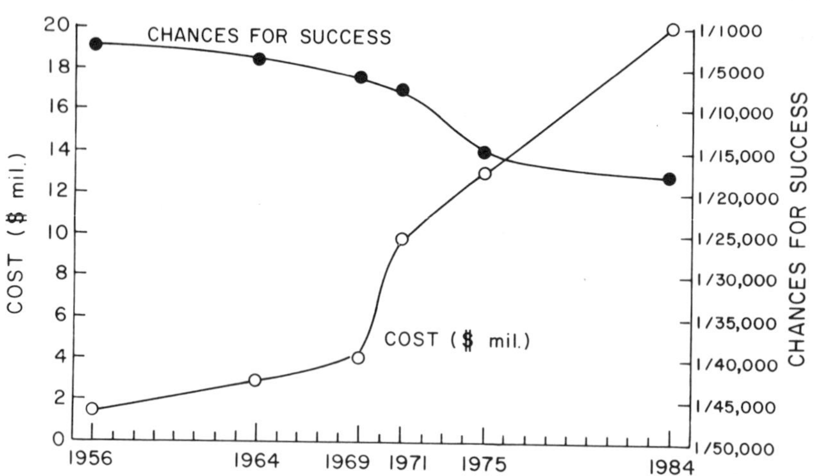

FIGURE 12 Estimated cost of developing an agricultural chemical and chance for a new chemical to become a product. Source: Mullison (1976) and others.

in pests all the various oxidases, esterases, glutathione transferases, dehydrochlorinases, and other enzyme systems that may enable them to quickly evolve resistance to practically any toxicant that may be used against them. The answer will be provided in time by the pests themselves. This concern has not deterred the search for new chemical weapons, however (Magee et al., 1984). The new emphasis is characterized by a more rational approach.

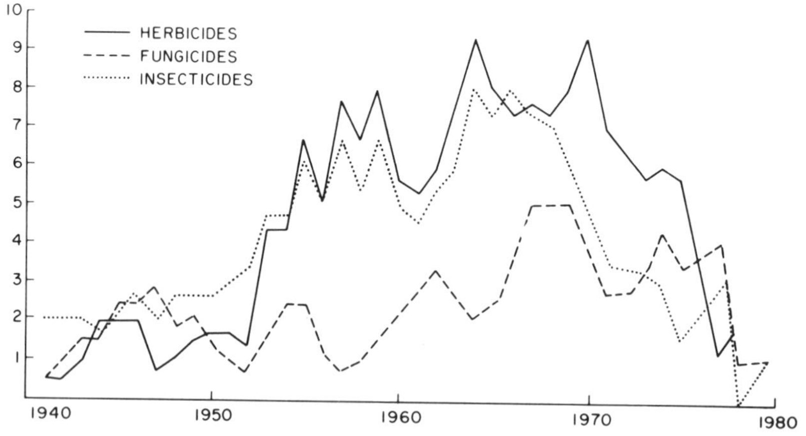

FIGURE 13 Annual introduction of new pesticides during the period 1940–1980. Source: Martin and Worthing (1977), Worthing (1979), Patton et al. (1982).

TABLE 10 Chronology of Insecticide Discoveries

Decade	Discovery
1940s	*Chlorinated hydrocarbons*: DDT, BHC, aldrin, chlordane, toxaphene *OPs*: parathion, methyl parathion *Carbamates*: isolan, dimetilan
1950s	*OPs*: malathion, azinphosmethyl, phorate, vinyl phosphates *Carbamates*: carbaryl
1960s	*OPs*: fonofos, trichloronate *Carbamates*: carbofuran, aldicarb, methomyl *Pyrethroids*: resmethrin *Formamidines*: chlordimeform
1970s	*Pyrethroids*: permethrin, cypermethrin, deltamethrin, fenvalerate *New OPs*: terbufos, methamidophos, acephate *New Carbamates*: bendiocarb, thiofanox *IGRs*: methoprene, diflubenzuron *AChE receptor blockers*: cartap
1980s	*New Pyrethroids*: flucythrinate *Procarbamates*: carbosulfan, thiodicarb *New IGRs*: phenoxycarb *Microbials*: BT, BTI, *Bacillus sphaericus* *AChE receptor blockers*: bensultap *GABA agonists*: milbemycin, avermectin *Miscellaneous*: AMDRO, cyromazine

SOURCE: Adapted in part from Menn (1980).

Some of these chemicals are the result of optimization of structures within the existing classes of insecticides, such as new pyrethroids, procarbamates, and insect growth regulators. Others are totally novel, having had their genesis in the progress that is being made in our understanding of basic biology, biochemistry, and physiology, at both the organismal and molecular levels. Representatives of this effort are the acetylcholinesterase receptor blockers, the GABA agonists, and a number of other compounds such as AMDRO and cyromazine (Table 10).

Evidence of rekindled interest is seen in the small but perceptible increase in the number of new insecticides submitted to the World Health Organization for testing against mosquito and other vector species, after a strong decline in such submissions during the 1970s (Figure 14). Likewise, we now see an increased interest in research on insecticide resistance, as evidenced by the percentage of resistance papers published in the *Journal of Economic Entomology* (Figure 15).

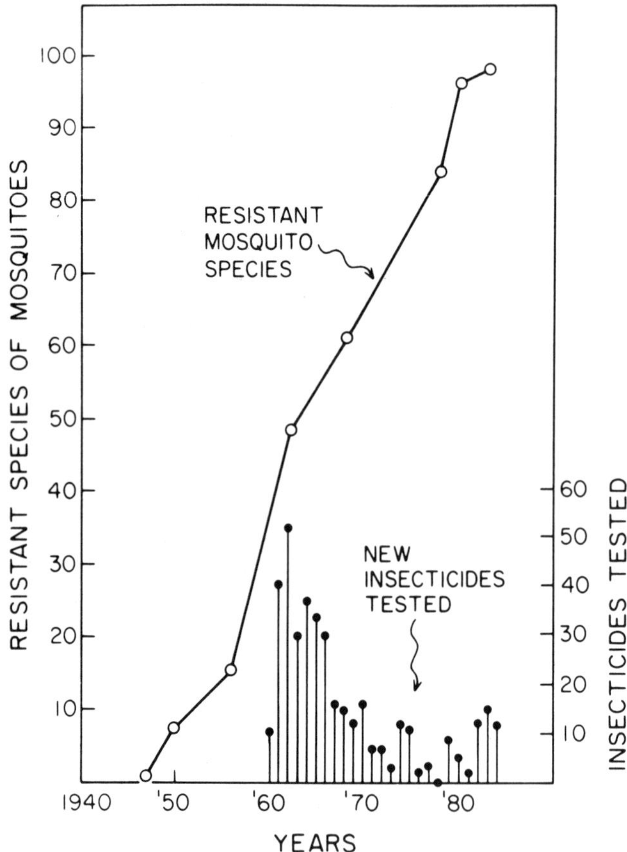

FIGURE 14 Numbers of new insecticides submitted for testing to the World Health Organization, 1960–1984, compared with the appearance of resistance in mosquito species. Source: Georghiou, unpublished.

The problem is evident, the need for action is compelling, and the opportunities for breakthroughs are substantial. It has always been axiomatic that one must intimately know one's enemy to be able to defeat him. I hope that this conference, through its exploration of the nature of pesticide resistance from all known perspectives, will enable us to develop the means and strategies for countering the adverse impact of this phenomenon on our well-being.

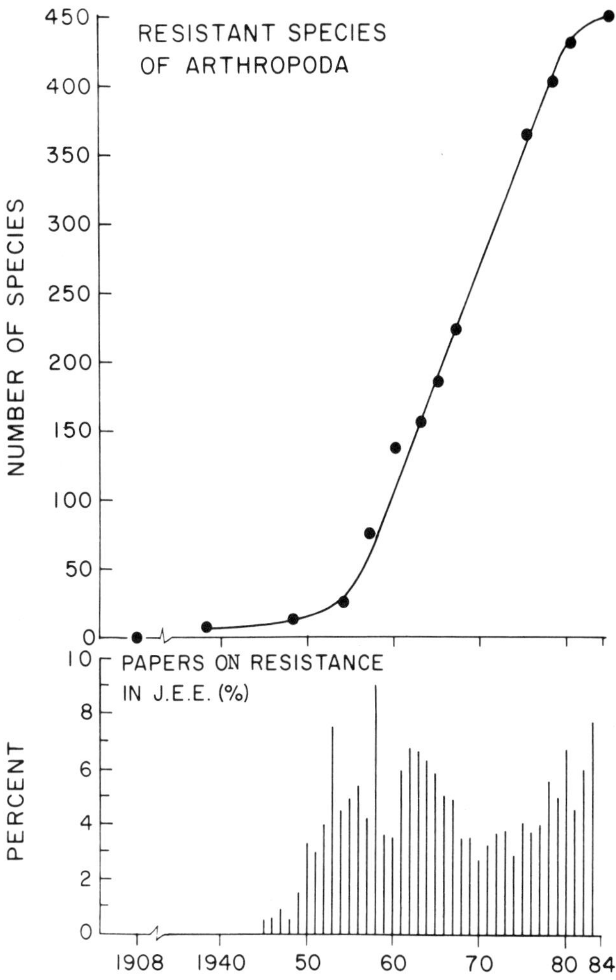

FIGURE 15 Percentage of papers concerned with insecticide resistance published in the *Journal of Economic Entomology*, 1945–1983, compared with the evolution of resistance in species of Arthropoda. Source: Georghiou, unpublished.

REFERENCES

Attia, F. I., and J. T. Hamilton. 1978. Insecticide resistance in *Myzus persicae* in Australia. J. Econ. Entomol. 71:851–853.

Braunholtz, J. T. 1981. Crop protection: The role of the chemical industry in an uncertain future. Philos. Trans. R. Soc. London, Ser. B 295:19–34.

Brown, A. W. A. 1971. Pest resistance to pesticides. Pp. 457–552 *in* Pesticides in the Environment, Vol. 1, Part II, R. White-Stevens, ed. New York: Marcel Dekker.

Campbell, C. J., and D. G. Finlayson. 1976. Comparative efficacy of insecticides against tuber flea beetle and aphids in potatoes in British Columbia. Can. J. Plant Sci. 56:869–875.

Champ, B. R., and M. J. Campbell-Brown. 1970. Insecticide resistance in Australian *Tribolium castaneum* (Herbst) (Coleoptera: Tenebrionidae). II. Malathion resistance in eastern Australia. J. Stored Prod. Res. 6:111–131.

Craig, I. A., G. P. Conway, and G. A. Norton. 1982. The consequences of resistance. Pp. 43–60 *in* Pesticide Resistance and World Food Production, G. Conway, ed. London: Imperial College, Mineral Resources Engineering Department.

Davidson, G. 1980. Insecticide resistance in Old World anopheline mosquitoes. World Health Organization Unpubl. Doc. VBC/EC/80.4.

Dekker, J. 1972. Resistance. Pp. 156–174 *in* Systemic Fungicides, E. Marsh, ed. New York: John Wiley and Sons.

Dekker, J., and S. G. Georgopoulos, eds. 1982. Fungicide Resistance in Crop Protection. Wageningen, Netherlands: Centre for Agricultural Publishing and Documentation.

Delp, C. J. 1979. Resistance to plant disease control agents: How to cope with it. Pp. 253–261 *in* Proc. 9th Int. Congr. Plant Prot., Vol. 1, T. Kommédahl, ed. Minneapolis, Minn.: Burgess.

El-Guindy, M. A., S. M. Madi, M. E. Keddis, Y. H. Issan, and M. M. Abdel-Satto. 1982. Development of resistance to pyrethroids in field populations of the Egyptian cotton leafworm, *Spodoptera littoralis* Boisd. Int. Pest Control 24(1):6,8,10–11,16–17.

Farkas, A., and J. Aman. 1940. The action of diphenyl on *Penicillium* and *Diplodia* moulds. Palest. J. Bot. Jerusalem Ser. 2:38–45.

Fontaine, R. E., J. H. Pull, D. Payne, G. D. Pradhan, G. P. Joshi, J. A. Pearson, M. K. Thymakis, and M. E. Ramos Camacho. 1978. Evaluation of fenitrothion for the control of malaria. Bull. W.H.O. 56:445–452.

Forgash, A. J. 1981. Insecticide resistance of the Colorado potato beetle, *Leptinotarsa decemlineata* (Say). Pp. 34–46 *in* Advances in Potato Pest Management, J. W. Lashomb and R. Casagrande, eds. Stroudsburg, Pa.: Hutchinson Ross.

Forgash, A. J. 1984a. History, evolution, and consequences of insecticide resistance. Pestic. Biochem. Physiol. 22:178–186.

Forgash, A. J. 1984b. Insecticide resistance of the Colorado potato beetle, *Leptinotarsa decemlineata* (Say). Paper presented at 17th Int. Congr. Entomol., Hamburg, Federal Republic of Germany, August 1984.

Gauthier, N. L., R. N. Hofmaster, and M. Semel. 1981. History of Colorado potato beetle control. Pp. 13–33 *in* Advances in Potato Pest Management, J. H. Lashomb and R. Casagrande, eds. Stroudsburg, Pa.: Hutchinson Ross.

Georghiou, G. P. 1980. Insecticide resistance and prospects for its management. Residue Rev. 76:131–145.

Georghiou, G. P. 1981. The occurrence of resistance to pesticides in arthropods: An index of cases reported through 1980. Rome: Food and Agriculture Organization of the United Nations.

Georghiou, G. P. 1982. The implication of agricultural insecticides in the development of resistance by mosquitoes with emphasis on Central America. Pp. 95–121 *in* Resistance to Insecticides Used in Public Health and Agriculture. Proc. Int. Workshop, 22–28 February, 1982. Colombo, Sri Lanka: Nat. Sci. Council Sri Lanka.

Georghiou, G. P., and R. Mellon. 1983. Pesticide resistance in time and space. Pp. 1–46 *in* Pest Resistance to Pesticides, G. P. Georghiou and T. Saito, eds. New York: Plenum.

Georghiou, G. P., and T. Saito, eds. 1983. Pest Resistance to Pesticides. New York: Plenum.

Georghiou, G. P., and C. E. Taylor. 1976. Pesticide resistance as an evolutionary phenomenon. Pp. 759–785 *in* Proc. 15th Int. Cong. Entomol., Washington, D.C. College Park, Md.: Entomological Society of America.

Georghiou, G. P., S. G. Breeland, and V. Ariaratnam. 1973. Seasonal escalation of organophosphorus and carbamate resistance in *Anopheles albimanus* by agricultural sprays. Environ. Entomol. 2:369–374.

Georgopoulos, S. G. 1976. Mutational resistance to site-specific fungicides. Pp. 1057–1061 *in* Proc. 3rd Int. Biodegrad. Symp., J. M. Sharpley and A. M. Kaplan, eds. London: Applied Science Publishers, Ltd.

Georgopoulos, S. G., and C. Zaracovitis. 1967. Tolerance of fungi to organic fungicides. Annu. Rev. Phytopathol. 5:109–130.

Gunning, R. V., C. S. Easton, L. R. Greenup, and V. E. Edge. 1984. Pyrethroid resistance in *Heliothis armiger* (Hübner) (Lepidoptera: Noctuidae) in Australia. J. Econ. Entomol. 77:1283–1287.

Ho, S. H., B. H. Lee, and D. Lee. 1983. Toxicity of deltamethrin and cypermethrin to the larvae of the diamondback moth, *Plutella xylostella*. Toxicol. Lett. 19:127–131.

Hobbs, J. H. 1973. Effect of agricultural spraying on *Anopheles albimanus* densities in a coastal area of El Salvador. Mosq. News 33:420–423.

Jackson, W. B., and A. D. Ashton. 1980. Vitamin K-metabolism and vitamin K-dependent proteins. 8th Steenbock Symp., J. W. Suttie, ed. Baltimore, Md.: University Park Press.

Jackson, W. B., P. J. Spear, and C. G. Wright. 1971. Resistance of Norway rats to anticoagulant rodenticides confirmed in the United States. Pest Control 39:13–14.

Kimura, Y., and K. Nakazawa. 1973. Local variations of susceptibility to organophosphorus insecticides in the green rice leafhopper in Hiroshima prefecture. Chugoku Agric. Res. 47:100–102.

LeBaron, H. M. 1984. Principles, problems, and potentials of plant resistance. Pp. 351–356 *in* Biosynthesis of the Photosynthetic Apparatus, J. P. Thornberg, L. A. Staehelin, and R. B. Hallick, eds. New York: Alan R. Liss.

LeBaron, H. M., and J. Gressel, eds. 1982. Herbicide Resistance in Plants. New York: John Wiley and Sons.

Liu, M-Y., Y-J. Tzeng, and C-N. Sun. 1981. Diamondback moth resistance to several synthetic pyrethroids. J. Econ. Entomol. 74:393–396.

Magee, P. S., G. K. Kohn, and J. J. Menn, eds. 1984. Pesticide Synthesis through Rational Approaches. Washington, D.C.: American Chemical Society.

Malcolm, C. A. 1983. The genetic basis of pyrethroid and DDT resistance interrelationships in *Aedes aegypti*. II. Allelism of R_{DDT2} and R_{py}. Genetica 60:221–229.

Martin, H., and C. R. Worthing, eds. 1977. Pesticide Manual. Basic Information on the Chemicals Used as Active Components of Pesticides. 5th ed. Croydon, England: British Crop Protection Council.

Martinez-Carrillo, J. L., and H. T. Reynolds. 1983. Dosage-mortality studies with pyrethroids and other insecticides on the tobacco budworm from the Imperial Valley, California. J. Econ. Entomol. 76:983–986.

Menn, J. J. 1980. Contemporary frontiers in chemical pesticide research. J. Agric. Food Chem. 28:2–8.

Metcalf, R. L. 1983. Implications and prognosis of resistance to insecticides. Pp. 703–733 *in* Pest Resistance to Pesticides, G. P. Georghiou and T. Saito, eds. New York: Plenum.

Miller, T. A., V. L. Salgado, and S. N. Irving. 1983. The *kdr* factor in pyrethroid resistance. Pp. 353–366 *in* Pest Resistance to Pesticides, G. P. Georghiou and T. Saito, eds. New York: Plenum.

Muir, R. C. 1979. Insecticide resistance in damson-hop aphid, *Phorodon humuli* in commercial hop gardens in Kent. Ann. Appl. Biol. 92:1–9.

Mullison, W. 1976. The cost of developing pesticides. Down Earth 32(2):34–36.

National Research Council. 1951. Conference on Insecticide Resistance and Insect Physiology. Washington, D.C.: National Academy of Sciences.

Nolan, J. 1981. Current developments in resistance to amidine and pyrethroid tickicides in Australia. Pp. 109–114 *in* Tick Biology and Control, Tick Research Unit, G. B. Whitehead and J. D. Gibson, eds. Grahamstown, S. Africa: Tick Research Unit.

Nolan, J., W. R. Roulston, and R. H. Wharton. 1977. Resistance to synthetic pyrethroids in a DDT-resistant strain of *Boophilus microplus*. Pestic. Sci. 8:484–486.

Ogawa, J. M., J. D. Gilpatrick, and L. Chiarappa. 1977. Review of plant pathogens resistant to fungicides and bactericides. FAO Plant Prot. Bull. 25:97–111.

Ogawa, J. M., B. T. Manji, C. R. Heaton, J. Petrie, and R. M. Sonada. 1983. Methods for detecting and monitoring the resistance of plant pathogens to chemicals. Pp. 117–162 *in* Pest Resistance to Pesticides, G. P. Georghiou and T. Saito, eds. New York: Plenum.

Omer, S. M., G. P. Georghiou, and S. N. Irving. 1980. DDT/pyrethroid resistance interrelationships in *Anopheles stephensi*. Mosq. News 40:200–209.

Parrella, M. P. 1983. Evaluation of selected insecticides for control of permethrin-resistant *Liriomyza trifolii* on chrysanthemum. J. Econ. Entomol. 76:1460–1464.

Patton, S., I. A. Craig, and G. R. Conway. 1982. The pesticide industry. Pp. 61–76 *in* Pesticide Resistance and World Food Production, G. Conway, ed. London: Imperial College, Mineral Resources Engineering Department.

Pimentel, D., D. Andow, R. Dyson-Hudson, D. Gallahan, S. Jacobson, M. Irish, S. Kroop, A. Moss, I. Schreiner, M. Shepard, T. Thompson, and B. Vinzant. 1980. Environmental and social costs of pesticides: A preliminary assessment. Oikos 34:126–140.

Pimentel, D., D. Andow, D. Gallahan, I. Schreiner, T. Thompson, R. Dyson-Hudson, S. Jacobson, M. Irish, S. Kroop, A. Moss, M. Shepard, and B. Vinzant. 1979. Pesticides: Environmental and social costs. Pp. 99–158 *in* Pest Control: Cultural and Environmental Aspects, D. Pimentel and J. H. Perkind, eds. Boulder, Colo.: Westview.

Priester, T. M., and G. P. Georghiou. 1980. Cross-resistance spectrum in pyrethroid-resistant *Culex quinquefasciatus*. Pestic. Sci. 11:617–664.

Quisenberry, S. S., L. A. Lockwood, R. L. Byford, H. K. Wilson, and T. C. Sparks. 1984. Pyrethroid resistance in the horn fly, *Haematobia irritans* (L.) (Diptera: Muscidae). J. Econ. Entomol. 77:1095–1098.

Radosevich, S. R. 1983. Herbicide resistance in higher plants. Pp. 453–479 *in* Pest Resistance to Pesticides, G. P. Georghiou and T. Saito, eds. New York: Plenum.

Roulston, W. J., R. H. Wharton, J. Nolan, J. D. Kerr, J. T. Wilson, P. G. Thompson, and M. Schotz. 1981. A survey for resistance in cattle ticks to acaricides. Aust. Vet. J. 57:362–371.

Ryan, G. F. 1970. Resistance of common groundsel to simazine and atrazine. Weed Sci. 18:614–616.

Sawicki, R. M., and A. D. Rice. 1978. Response of susceptible and resistant aphids *Myzus persicae* (Sulz.) to insecticides in leaf-dip bioassays. Pestic. Sci. 9:513–516.

Sawicki, R. M., A. L. Devonshire, A. D. Rice, G. D. Moores, S. M. Petzing, and A. Cameron. 1978. The detection and distribution of organophosphorus and carbamate insecticide-resistant *Myzus persicae* (Sulz.) in Britain in 1976. Pestic. Sci. 9:189–201.

Sawicki, R. M., A. W. Farnham, I. Denholm, and K. O'Dell. 1981. Housefly resistance to pyrethroids in the vicinity of Harpenden. Pp. 609–616 *in* Proc. Br. Crop Prot. Conf.: Pests and Diseases, Vol. 2. Croydon, England: British Crop Protection Council.

Schmidt, D. C., S. E. Kunz, H. D. Petersen, and J. L. Robertson. 1985. Resistance of horn flies (Diptera: Muscidae) to permethrin and fenvalerate. J. Econ. Entomol. 78:402–406.

Schnitzerling, H. J., P. J. Noble, A. Macqueen, and R. J. Dunham. 1982. Resistance of the buffalo fly, *Haematobia irritans exigua* (DeMeijere), to two synthetic pyrethroids and DDT. J. Aust. Entomol. Soc. 21:77–80.

Sinègre, G. 1984. La résistance des diptères culicides en France. Pp. 47–58 *in* Colloque sur la Réduction d'Efficacité des Traitements Insecticides et Acaricides et Problèmes de Résistance. Paris: Société Française de Phytiatrie et de Phytopharmacie.

Smirnova, A. S., M. I. Levi, M. V. Niyazova, E. I. Kapanadze, A. I. Bromberg, V. I. Zagroba, A. A. Budylina, R. M. Kuznetsova, A. M. Lautsin, and I. A. Kurkina. 1979. Resistance of some common cockroach *Blatella germanica* subpopulations to neopynamin and other insecticides. Med. Parazitol. Parazit. Bolezni. 48:60–66.

Steiner, H. 1973. Cost-benefit analyses in orchards where integrated control is practiced. Eur. and Mediterr. Plant Prot. Org. Bull. 3:27–36.

Sudderuddin, K. I., and P-F. Kok. 1978. Insecticide resistance in *Plutella xylostella* collected from the Cameron Highlands of Malaysia. FAO Plant Prot. Bull. 26:53–57.

Wardlow, L. R., G. A. Lewis, and A. W. Jackson. In press. Pesticide resistance in glasshouse whitefly, *Trialeurodes vaporariorum*. Res. and Dev. Agric. 2.

Weismann, R. 1947. Differences in susceptibility to DDT of flies from Sweden and Switzerland. Mitt. Schweiz. Entomol. Ges. 20:484–504.

Wood, K. A., B. H. Wilson, and J. B. Graves. 1981. Influence of host plant on the susceptibility of the fall armyworm to insecticides. J. Econ. Entomol. 74:96–98.

World Health Organization. 1976. Resistance of vectors and reservoirs of disease to pesticides. 22nd Rep. WHO Exp. Comm. Insectic. W.H.O. Tech. Rep. No. 585.

World Health Organization. 1980. Resistance of vectors of disease to pesticides. 5th Rep. WHO Exp. Comm. Vector Biol. Control. W.H.O. Tech. Rep. No. 655.

Worthing, C. R., ed. 1979. Pesticide Manual. Croydon, England: British Crop Protection Council.

Wright, J. W., R. F. Fritz, and J. Haworth. 1972. Changing concepts of vector control in malaria eradication. Annu. Rev. Entomol. 17:75–102.

2

Genetic, Biochemical, and Physiological Mechanisms of Resistance to Pesticides

SIMILAR MECHANISMS FOR RESISTANCE to pesticides have been observed in insects, fungi, bacteria, plants, and vertebrates. These include changes at target sites, increased rates of detoxification, decreased rates of uptake, and more effective storage (compartmentalization) mechanisms. The relative importance of these mechanisms varies among the classes of pests. Most resistance to pesticides is inherited in a typical Mendelian fashion, but in some cases, resistance can be attributed to, or influenced by, relatively unique genetic and biochemical characteristics, e.g., extranuclear genetic elements in bacteria and higher plants. A thorough understanding of the genetic, biochemical, and physiological mechanisms of pesticide resistance is essential to the development of solutions to the pesticide-resistance problem.

GENETIC BACKGROUND

Insects, vertebrates, most higher plants, and fungi of the class Oomycetes are diploid, and some fungi are dikaryotic. Therefore, the gene or genes responsible for resistance may exist in duplicate. Multiple allelic forms are known for many resistance genes. These alleles often produce an effect that is greater than additive. In some cases a resistance gene may exist in multiple copies, a condition called gene amplification. This is known to occur, for example, in the insects *Myzus* and *Culex*. Several genes at different loci also can be involved in resistance.

Most fungi are haploid in their vegetative state, as are bacteria generally, although multiple genomes are found in actively growing cultures. In a haploid state, the expression of each gene involved in resistance is not modified by another allele as in the case of the diploid organism. Many fungal cells, however, are multinucleate and heterokaryotic with respect to resistance genes, and these genes can produce a modification of resistance expression analogous to that found in diploid organisms. Furthermore, the resistance level of the organism is frequently the result of the interaction of alleles of several genes at different loci. This interaction is known as polygenic resistance. An additional complication in bacteria is the existence of extrachromosomal genes, which can act alone, or in concert with chromosomal genes, to confer resistance. In plants, herbicide resistance can be inherited in the plastid genome.

Genes that can mutate to confer resistance to a pesticide may be either structural or regulatory (Plapp, this volume). Some structural genes are translated into products (enzymes, receptors, and other cell components, such as ribosomes and tubulin) that are targets for pesticides. The mutation of structural genes can result in a critical modification of the gene products, such as decreases in target site sensitivity or increased ability to metabolize pesticides. Regulatory gene products may control rates of structural gene transcription. They may also recognize and bind pesticides and thereby control induction of appropriate detoxifying enzymes.

A clear and detailed understanding of the molecular genetic apparatus of the resistant organism can provide essential information for devising tools and strategies for avoidance and management of practical pesticide resistance problems. Specific examples of the utilization of genetic information for these purposes have been discussed elsewhere in this volume (Gressel, Hardy, Plapp). Some examples include: (1) the construction of genetically defined organisms for investigation of the biochemical mechanism of pesticide action and for studies on population dynamics of biotypes that are heterozygous or polygenic for pesticide resistance; (2) the rational design of synthetic antagonists to combine with regulatory proteins and block the induction of detoxifying enzymes; (3) genetic engineering of herbicide-resistant plants, insecticide-resistant beneficial insects, and microbial antagonists; and (4) preparation of monoclonal antibodies for rapid and specific detection of resistance in a pest population. Ideally, this research should lead to the isolation, cloning, and sequencing of alleles conferring resistance and elucidation of their structure relative to their susceptible alleles.

BIOCHEMICAL MECHANISMS

In insects and plants the principal biochemical mechanisms of resistance are (Plapp, Gressel, this volume): (1) reduction in the sensitivity of target sites; (2) metabolic detoxication of the pesticide by enzymes such as esterases, monooxygenases, and glutathione-sulfotransferases; and (3) decreased penetration and/or translocation of the pesticide to the target site in the insect. Alleles involving alteration of target sites include altered acetylcholinesterase resistance to organophosphates and carbamates, alterations in the gene for the receptor protein target of DDT and pyrethroids, and changes in the receptor protein target for cyclodiene insecticides. Metabolic resistance in the house fly seems to be under the control of a single gene whose product is a receptor protein. This protein binds insecticides, and the protein: insecticide complex induces synthesis of multiple detoxifying enzymes. Whether or not similar metabolic receptor proteins exist in other insects is not known. Decreased penetration has a minor or modifying effect on the level of resistance. A minor change in penetration, however, may have a profound effect upon the pharmacokinetics of a toxicant.

In plant pathogenic fungi, resistance has been attributed mainly to single gene mutations that (1) reduce the affinity of fungicides for target sites (e.g., ribosomes, tubulin, enzymes); (2) change the absorption or excretion of the fungicides; (3) increase detoxication, for example, reducing the toxicity of Hg^{++} and captan by an increase in the thiol pool of the cell (see Georgopoulos, this volume, for details). Most cases of practical fungicide resistance can be attributed to the first mechanism, which often results in a striking increase in resistance level brought about by mutation of a single gene. For this reason, fungicides that act at a single target site are at great risk with respect to the possibility of resistance development (Dekker, this volume).

Resistance to other fungicides, such as ergosterol biosynthesis inhibitors and polyene antibiotics, occurs through a polygenic process. Each gene mutation produces a relatively small, but additive, increase in resistance. When many mutations are required to achieve a significant level of resistance, there is an increased likelihood for a substantial loss of fitness in the pathogen. There have been no major outbreaks of resistance to these fungicides in the field, but this situation is changing rapidly and problems are beginning to occur with the ergosterol biosynthesis inhibitors (Butters et al., 1984; Gullino and DeWaard, 1984).

Three bactericides are used to control plant diseases in the United States: copper complexes, streptomycin, and oxytetracycline. Resistance to streptomycin in *Erwinia amylovora*, the pathogen of fireblight disease of pear and apple trees, has been a widespread problem. Resistance appears to be controlled by alteration (or mutation) of a structural chromosomal gene that reduces the affinity of the bacterial ribosome for streptomycin, an inhibitor of protein synthesis (Georgopoulos, this volume). In contrast, the most common mechanism of streptomycin resistance in human bacterial pathogens is mediated by an extrachromosomal (plasmid) gene that regulates the production of an enzyme (phosphorylase) that detoxifies streptomycin. The application of oxytetracycline to control streptomycin-resistant strains of *Erwinia amylovora* on pear trees is a relatively new practice, and reports of tetracycline resistance have not yet appeared. Oxytetracycline has been injected into palm trees and stone fruit trees for several years to control mycoplasmalike organisms, apparently without the development of resistance. In *Xanthomonas campestris* pv. *vesicatoria* (which causes bacterial leaf spot of tomatoes and peppers), resistance to copper is conferred by a plasmid gene that appears to regulate the absorption of copper ion by the bacterial cell.

Plants utilize the same general resistance mechanisms as insects. The efficacious use of herbicides on crops is made possible because many crop plants are capable of rapid metabolic inactivation of the chemicals, thereby avoiding their toxic action. Target weeds are notably deficient in this capacity. It is apparent, though, that the capability to metabolize herbicides to innocuous compounds constitutes a potentially important basis of evolved resistance to herbicides in weeds. Documented cases of resistance have been due to other mechanisms, however, such as alteration of the herbicide's target site. For example, newly appearing *s*-triazine-resistant weeds have plastid-mediated resistance that involves a reduced affinity of the thylakoids for triazine herbicides (Gressel, this volume).

The herbicide paraquat disrupts photosynthesis in target weeds by intercepting electrons from photosystem I, part of the metabolic cycle that fixes energy from sunlight into plant constituents via a complicated flow of electrons. Transfer of electrons from paraquat to oxygen gives rise to highly reactive oxygen radicals that damage plant membranes. Paraquat-resistant plants have higher levels of the enzyme superoxide dismutase, which quenches the reactive oxygen radicals.

The mechanisms of weed resistance to the dinitroaniline herbicides and to diclofop-methyl have not yet been identified.

A number of herbicides act on the photosynthetic mechanism in the chloroplasts. Although the frequency of resistant plants arising from plastid mutations would normally be very low, a plastome mutator gene has been recognized that increases the rate of plastome mutation in weeds. This factor could be largely responsible for the plastid-level resistance to herbicides that has emerged in some weeds (Gressel, this volume).

Resistance to anticoagulants is the most widespread and thoroughly investigated heritable resistance in vertebrates. Warfarin resistance in rats has been observed in several European countries, and in 1980 more than 10 percent of rats were resistant to warfarin in 45 out of 98 cities surveyed in the United States (Jackson and Ashton, this volume).

Warfarin interferes with the synthesis of vitamin K-dependent blood-clotting factors in vertebrates. Resistance in rats (*Rattus norvegicus*) appears to involve a reduced affinity of a vitamin K-metabolizing enzyme or enzymes for warfarin. The affinity of the target site is controlled by one (of four) allelic forms of a gene in linkage group I. In the mouse, there are indications that increased resistance to warfarin is due to metabolic detoxication and that the detoxication system (mixed function oxidase) is controlled by a gene cluster on chromosome 7 (MacNicoll, this volume). Our knowledge of resistance mechanisms in rodents and other vertebrate pests is fragmentary.

PROMISING RESEARCH DIRECTIONS AND THEIR IMPLEMENTATION

Synthetic chemicals will probably continue for some time as the major weapon against most pests because of their general reliability and rapid action, and their ability to maintain the high quality of agricultural products that is demanded by urban consumers today. Although new chemicals offer a short-term solution, this approach to pest control alone will rarely provide a viable, long-term strategy. Moreover, a few years of commercial exploitation may not justify the investment required to develop a new pesticide today, except where there are reasonable prospects that a pesticide's mode of action may be beyond the capability of the pest for genetic adaptation.

Despite the continual threat of resistance, we may still be able to exploit our expanding knowledge of the genetic and biochemical makeup of pests by designing pesticides that can circumvent existing resistance mechanisms, at least long enough to provide chemical manufacturers a reasonable rate of financial return on the investment needed to develop a new pesticide. Realistically, though, it is difficult to be optimistic on this point in practical situations where a synthetic pesticide is applied repeatedly to the same crop or environment to control a well-adapted pest. History promises no encouragement, at least for most pests, for the discovery of a "silver bullet." On the other hand, it is indeed encouraging that there are examples of pesticides, both selective and nonselective (e.g., the polyene fungistat pimaricin, the widely used herbicide 2,4-D, and the insecticides azinphosmethyl and carbofuran), that have been used for years in certain situations without setting off rapid, extensive resistance. The phenoxy herbicides (e.g., 2,4-D) and the broad-spectrum fungicides (captan, dithiocarbamates, and fixed coppers) have been used successfully for decades without serious resistance problems. Still, the wisest course for future research appears to be the integration of a diversity of approaches to pest control—chemical, biological, and cultural (or ecological)—because an integrated application of multiple methods will produce minimum selection pressure for development of resistance to pesticides. Evolution of resistance to minimally selective or multitarget synthetic chemicals might be delayed indefinitely if the selection pressure were kept within "reasonable" limits. The pressure might be reduced with crop rotations and careful management, but may be virtually impossible in agricultural areas typified by repeated monocultures.

The development of resistance is encouraged by pesticides that act upon single biochemical targets. Unfortunately, the modes of action of many systemic plant fungicides, and most modern synthetic insecticides and herbicides, are biochemically site-specific. Many of these fungicides and insecticides have produced a rapid, major buildup of resistance genes in pest populations after just a few seasons of use. Undoubtedly, the potential for resistance development to such compounds will continue to be a limiting factor in the widespread use of these compounds, although compounds differ in the degree of risk for rapid development of resistance. In addition, some compounds lend themselves to relatively effective resistance management strategy. Others do not. The genetic and biological reasons that some compounds rapidly select for resistance, whereas others do not, are presently obscure in nearly all cases. Further research in this area will greatly facilitate the development of efficacious strategies to manage resistance.

RECOMMENDATIONS

RECOMMENDATION 1. **A major increase in research on the genetics, biochemistry, and physiology of resistance is recommended for all pest classes—insects, fungi, bacteria, weeds, and vertebrates.**

Research support should not be restricted to or allocated primarily on the basis of the economic importance of crops. Research should include studies of genetic mechanisms in wild and resistant populations, with emphasis on common gene pools, gene flow between related species, gene sequencing, and population dynamics. Biochemical and physiological studies should be encouraged on pesticidal mode of action, characterization of target site enzymology, pharmacokinetics, and the transport, metabolism, and excretion of xenobiotics in pest species. The compilation and dissemination of data in these areas is essential to the identification of unique target sites less apt to develop resistance. Such data are essential in designing novel pesticides that exploit genetic weaknesses and bypass genetic capabilities to develop resistance. It is reasonable to anticipate that agents could be developed, for example, that are superior to existing cholinesterase inhibitors for insect pests, or to chemicals that inhibit macromolecular synthesis integral to the function of microorganisms.

The research agenda is formidable. For most plant pathogens, virtually nothing is known that would be useful in the rational design of new fungicides and bactericides. To a lesser extent, this also appears to be the case for insects, weeds, and rodent pests. Significant efforts are in progress for the design of herbicides, however.

RECOMMENDATION 2. **Use molecular biology and recombinant DNA technology to isolate, identify, and characterize the genes and gene products (enzymes and receptors) conferring resistance to pesticides and to compare these products with their alleles in susceptible pests. Use of microbial models, as appropriate, may facilitate progress in this area.**

Molecular biology has much to offer as a tool for elucidating the nature of pesticide target sites, particularly in proteins. These techniques can define resistance due to changes in structural genes, amplification of a structural gene, and alteration in regulation. Using bacteria to clone structural genes (or DNA fragments) coding for pesticide-metabolizing enzymes can provide a means for determining how these genes are regulated. These techniques can help determine the mechanism of operation of genes that appear to carry out common regulatory functions in insects, such as controlling the coordinated expression of structural genes that code for different enzymes involved in pesticide degradation.

Other applications of molecular biology techniques could involve the insertion of genes for toxin production into insect-inhabiting bacteria, fungi, or viruses. Genes for resistance to insects or plant pathogens based on the production of allelochemicals might also be transferred from nonhost species to crop plant hosts.

RECOMMENDATION 3. **Conduct research on pesticide target biochemistry to identify unique sites in pests that can serve as models for the design of new pesticides.**

The development of fungicides that inhibit ergosterol biosynthesis is a good example of the successes that can evolve from such a research program. It may also be possible to design pesticides that attack more than one target site, at least for most pests. "Target site" research should reveal opportunities for the systematic combination of compounds that possess negatively correlated cross-resistance traits that exploit structural differences in the "target site" in resistant biotypes. Several clear-cut examples of compounds that are negatively correlated with respect to cross-resistance can be found in some carbamate pesticides (Georgopoulos, Plapp, this volume).

To further the development of new rodenticides, research is required to establish the selective affinities of anticoagulants and substrates for the target site. Such understanding would greatly facilitate the rational design of chemical agents to potentiate the action of anticoagulants and/or minimize detoxication. A major focus of target biochemistry should be the identification of novel systems for exploitation, rather than exclusively studying and characterizing the targets of existing compounds.

In the future, greater understanding of target site biochemistry may make it possible to design pesticides that are themselves resistant to pests' detoxication mechanisms, as is already being done for some of the semisynthetic penicillins that inhibit bacterial β-lactamase (see Hardy, this volume). Also, possibilities for the development of new synergistic relationships would be greatly expanded by detailed information on receptor/inhibitor interactions and the metabolism of pesticides in resistant mutants.

RECOMMENDATION 4. **Conduct research on the enzymology and pharmacokinetics of pesticides in both target and nontarget species.**

Classical enzyme kinetics does not accurately describe the behavior of potential xenobiotics that are reactive at extremely low concentrations. A slight reduction in the rate of penetration of the xenobiotic into the pest may result in a drastic reduction in the reaction with the enzyme. In addition to inhibitors of detoxifying enzymes, other potentially fruitful areas for synergist research include compounds that interfere with the induction of detoxifying enzymes, agents that block active secretion (e.g., the fungicide fenarimol), and compounds that inhibit binding of anticoagulants by serum albumin in rats.

RECOMMENDATION 5. **Initiate research on new pesticides and on new ways to use existing pesticides that emphasizes compounds and procedures that result in minimum selection pressure on the pest population.**

Pesticides with one or more of the following properties would be useful in resistance management: (1) compounds that suppress target pest populations while allowing predators and parasites to multiply; (2) compounds (such as insect growth regulators) that are not lethal, but which effectively prohibit normal reproduction; (3) microbial pesticides, including bacteria, fungi, and viruses; (4) compounds related to the broad-spectrum fungicides (e.g., multisite electrophiles) that have been used for many years under high selection pressure with few problems with resistance; and (5) agents that control fungus diseases of plants by intensifying the natural defense reactions of the plant, such as the localized death of plant cells when infection by the pathogen is attempted (e.g., probendazole). Furthermore, broad-spectrum fungicides give satisfactory control in many disease situations; selective systemic compounds should be restricted to use in situations where systemic activity or postinfection activity is essential to disease control.

REFERENCES

Butters, J., J. Clark, and D. W. Hollomon. 1984. Resistance to inhibitors of sterol biosynthesis in barley powdery mildew. Meded. Fac. Landbouwwet. Rijksuniv. Gent. 49/2a:143–151.

Gullino, M. L., and M. A. DeWaard. 1984. Laboratory resistance to dicarboximides and ergosterol biosynthesis inhibitors in *Penicillium expansum*. Neth. J. Plant Pathol. 90:177–179.

WORKSHOP PARTICIPANTS

Genetic, Biochemical, and Physiological Mechanisms of Resistance to Pesticides

JOSEPH W. ECKERT (*Leader*), University of California, Riverside
HUGH D. SISLER (*Leader*), University of Maryland
S. G. GEORGOPOULOS, Athens College of Agricultural Sciences, Greece
JONATHAN GRESSEL, The Weizmann Institute of Science

BRUCE D. HAMMOCK, University of California, Davis
JOHN M. HOUGHTON, Monsanto Agricultural Products Company
DALE KAUKEINEN, ICI Americas, Inc.
ALAN MACNICHOLL, Ministry of Agriculture, Fisheries and Food, Great Britain
R. L. METCALF, University of Illinois
TOM O'BRIEN, Brigham and Women's Hospital, Boston
FREDERICK W. PLAPP, JR., Texas A&M University
NANCY RAGSDALE, U.S. Department of Agriculture
JAMES E. TAVARES, National Research Council

Pesticide Resistance: Strategies and Tactics for Management.
1986. National Academy Press, Washington, D.C.

Modes and Genetics of Herbicide Resistance in Plants

JONATHAN GRESSEL

Herbicide resistance is becoming an increasing problem throughout the world, but one that can be managed with the right tools and by understanding how plants develop resistance to the various herbicides. Population genetics models can help scientists to discern, broadly, why resistance occurs or will occur in some situations and not in others (i.e., why resistance has not developed in monoculture, monoherbicide wheat, but why it has developed in corn). Genetics and molecular biology allow scientists to understand the details of resistance development and the types of inherited resistance: nuclear with dominance, recessiveness, monogenic, polygenic, organelle, and gene duplication. Herbicides act on plants in different ways. By understanding all the processes, better methods and strategies of delaying or managing resistance to herbicides can be devised.

INTRODUCTION

The idea of weeds becoming resistant to herbicides is not new. Warnings about the possibility of weeds evolving resistance were issued soon after the phenoxy herbicides were introduced (Abel, 1954); however, as no confirmed cases of resistance to phenoxy herbicides occurred, the warnings were ignored—even after the first triazine-resistant weeds appeared. In Europe and the United States triazine resistance has become a serious problem: at least 42 species have resistant biotypes. Six weed species are resistant to paraquat; one weed species each is resistant to diclofop-methyl and trifluralin. All evolved from sensitive biotypes in agricultural situations (Figure 1). For example, more than 75 percent of Hungary's (the Eastern block's major

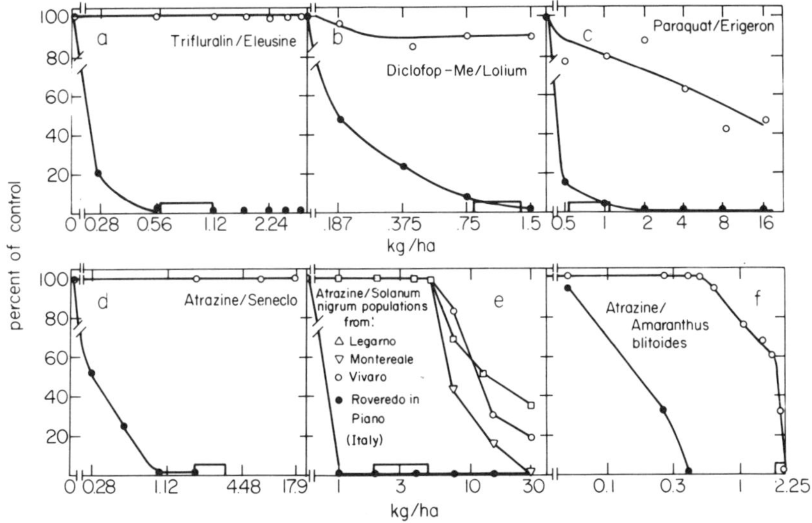

FIGURE 1 Dose response curves of the wild type and the herbicide-resistant weeds that have evolved. (Box near base of each graph denotes the recommended agricultural rates for each herbicide; concentrations are on log scales.) **A:** Resistance of *Eleusine indica*, the fifth worst weed in the world (Holm et al., 1977), evolved in South Carolina, after about 10 years of trifluralin use as the sole herbicide in monoculture cotton. Dose response curves vary among separately evolved resistant biotypes. (Figure plotted from tabular data in Mudge et al., 1984.) **B:** Resistance of *Lolium rigidum* to diclofop-methyl in legume fields receiving six applications in four years. (Redrawn from data of Heap and Knight, 1982.) **C:** Tolerance of *Erigeron philadelphicus* (=*Conyza philadelphicus*) to paraquat. Multiple yearly applications of paraquat were used as the sole herbicide in mulberry plantations. (Redrawn from data of Watanabe et al., 1982.) **D:** Resistance of *Senecio vulgaris* to atrazine appeared in a nursery where atrazine and simazine were used once or twice annually for 10 years. Data measured as survival after preemergence treatment. (Plotted from tabular data in Ryan, 1970.) **E:** Variable response of *s*-triazine-resistant accessions of *Solanum nigrum*. Seeds of the resistant biotypes were gathered from the four isolated places listed (in Northern Italy) and were assayed in pot tests. (Plotted from tabular data in Zanin et al., 1981.) **F:** The appearance of atrazine-resistant *Amaranthus blitoides*. Monoculture maize fields were treated for 17 years with atrazine. A 1 m^2 area was found with this accession in Hungary. (Plotted from unpublished data supplied by Dr. P. Solymosi, Plant Protection Inst., Budapest, 1982.)

maize-growing area) agricultural land is infested with triazine-resistant pigweeds (Hartmann, 1979; Solymosi, 1981); *s*-triazine can be used only in mixtures. Tolerance[1] to herbicides continues to increase (LeBaron and Gres-

[1]Tolerance is defined as any decrease in susceptibility, compared with the wild type. Resistance is complete tolerance to agriculturally used levels of a herbicide (LeBaron and Gressel, 1982).

sel, 1982), and resistance to other herbicides may soon appear in the field (Gressel, 1985).

The problem has been compounded because of the low price of atrazine and its superior season-long control of weeds in corn when compared with the phenoxy herbicides. It is 20 percent cheaper to treat corn with atrazine than with 2,4-D (Ammon and Irla, 1984). Thus farmers use more atrazine per year, and many stop rotating crops and herbicides. Resistance to the triazines and other herbicides has appeared in the agricultural areas of monoculture, monoherbicide use. The potential economic risk is great: while it now costs ca. $12/ha to treat sensitive weeds with atrazine, if all major corn weeds become resistant the alternative treatments would cost ca. $125/ha (Ammon and Irla, 1984).

A second problem involving resistant weeds is "problem soils." Repeated applications of herbicides can create problem soils when soil-applied herbicides can no longer control susceptible weeds. In such soils herbicides are degraded more quickly than in nonproblem soils (Kaufman et al., in press). For example, the rate of EPTC degradation more than doubles in soils that receive multiple treatments of EPTC, and there is a 50-fold increase in degradation in soils with a 12-year history of repeated diphenamid applications (Kaufman et al., in press). The problem becomes greater because the microbial enzymes degrading these pesticides often have a broad specificity that leads to cross-resistances within herbicides and between groups of herbicides and some other pesticides (Kaufman et al., in press). It is possible to conceive of the use of herbicide "extenders" that would act by inhibiting the specific soil microorganisms or the degradative enzymes' systems. By analogy it is possible to conceive the scientific feasibility of doing this from the effective specific inhibition of ammonia-oxidizing bacteria by nitrapyrin.

In this chapter we will look at the basic genetics, biochemistry, and physiology of resistance so that we can make recommendations that will delay the appearance and spread of resistance to our most cost-effective herbicides.

POPULATION GENETICS

Simple population genetics models suggest why there has been no resistance to phenoxy herbicides in monoculture, monoherbicide wheat and why resistance to triazines, especially in corn, has become so widespread (Gressel and Segel, 1978, 1982). These models, along with common sense and a closer look at crop and weed ecologies and agronomy, can help us develop strategies to delay resistance.

The appearance of resistance depends on characteristics of the different weeds and herbicides, which can be mathematically integrated into models. If a gene or genes for resistance do not exist at some low frequency in the population, resistance will never appear in that species unless introduced by genetic engineering. When resistant biotypes are grown in competition with susceptible (wild-type) biotypes of the same species without herbicides, their seed yield is often about one-half that of the wild type (Radosevich and Holt, 1982; Gressel, 1985; Gressel and Ben-Sinai, in press). Fitness will decrease the rate of enrichment for resistance when nonpersistent herbicides such as 2,4-D are used, since much of the season will be available for the remaining susceptible individuals to exert their superiority. This competition between fit and unfit biotypes is especially fierce when seedlings are established.

Persistence of herbicides interrelates not only with fitness but also with dormancy characteristics that separate weeds from crops and from other pests—the spaced germination of weed seeds. Weeds germinate not only throughout the season, but also over many seasons. Susceptible weed seeds can germinate after a rapidly degraded herbicide has disappeared; they then produce more seeds before the season is over, considerably lowering the effective selection pressure. Selection pressure is a result of "effective kill," which is not the same as the "knock down" after herbicide treatment. Effective kill is a measure of the number of surviving seeds or propagules at the end of a season, not after treatment.

Every time we enrich for resistant individuals by using a herbicide the resistant seeds are diluted by a seed bank of susceptible seeds from previous years. These seeds exert a buffering effect and delay the appearance of resistance. The first weed reported to evolve triazine resistance, *Senecio vulgaris* (Ryan, 1970), does not have an appreciable seed bank. The interaction of selection pressure, herbicide persistence, and seed bank on the rates of enrichment for resistance can be modeled to visualize how each parameter affects the rate at which resistance should appear. Similar modeling has been done for the evolution of insecticide resistance (Georghiou and Taylor, 1977), for fungicide resistance (Delp, 1981), and for resistance of cancer cells to antitumor drugs (Goldie and Coldman, 1979).

In our model (Gressel and Segel, 1978, 1982) the factors governing the rates of evolution of herbicide-resistant weeds, including the effects of the seed bank, are expressed in the equation:

$$N_n = N_0 (1 + f\alpha/\bar{n})^n,$$

where N_n is the proportion of resistants in a population in the nth year of continued treatment of a herbicide, and N_0 is the initial frequency of resistant individuals in the field before herbicide treatment. N_0 is a steady state achieved by natural mutation to resistance, lowered by the fitness of a biotype. The factor in parentheses governs the rate of increase of resistance. The overall fitness f (measured without the presence of herbicide) is that of the resistant compared with the susceptible biotype. With triazine resistance f is usually between 0.3 and 0.5 (Gressel, 1985; Gressel and Ben-Sinai, in press). Selection pressure (α) is defined as the proportion of the resistant propagules

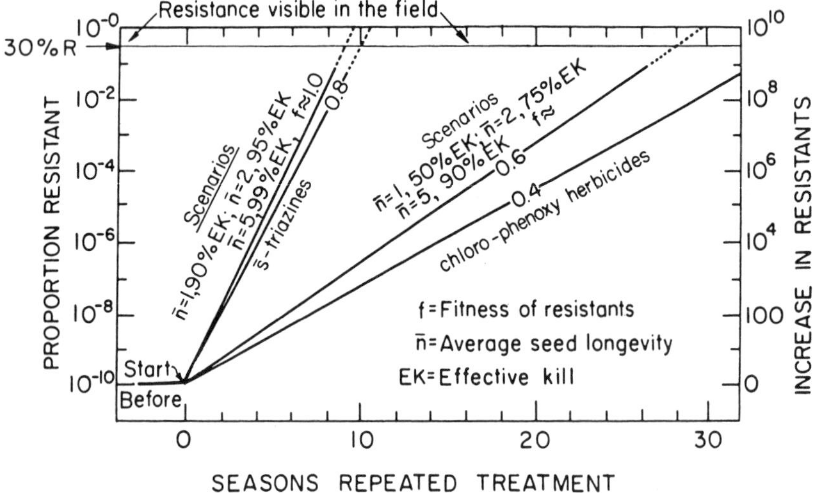

FIGURE 2 Effects of various combinations of selection pressure α (measured as effective kill, EK) fitness and of soil seed-bank longevity ($\bar{n}$) on the rates of enrichment of herbicide-resistant individuals over many seasons of repeated treatment. The values are plotted for fitnesses that would develop after the herbicide degrades. With the persistent triazine herbicide the fitness (f = 1.0; f = 0.8) would be high, as the fitness differential has no time to become apparent. With the phenoxy type, fitness differentials (f = 0.6; f = 0.4) will have time to be influential. Resistance (R) would become apparent in the field only when more than 30 percent of the plants are resistant. The scale on the right indicates the increase in resistance from any unknown initial frequency of resistant weeds in the population; whereas the scale on the left starts from a theoretically expected frequency of a recessive monogene character in a diploid organism. (Plotted from equations in Gressel and Segel, 1978.)

divided by the proportion of susceptibles at the end of the season. For example, if all resistants remain and all but 5 percent of the susceptibles are lost, $\alpha = 1/0.05 = 20$. Selection pressure and fitness are divided by $\bar{n}$, approximately the half-life of seed in soil. In weeds that germinate immediately, such as *Senecio*, $\bar{n} = 1$. With most weed species, $\bar{n}$ is between two and five years. An increase in $\bar{n}$ depresses the rate at which resistance will increase.

The interrelationships are clearer when we use the equation to generate hypothetical lines from different scenarios (Figure 2). In Figure 2 we arbitrarily started in year zero from a frequency of 10^{-10}, but the frequency scale can be moved to fit any initial field frequency. More important are the slopes showing the ratio at which enrichment occurs. The slopes show that we always enrich herbicide-resistant individuals when we treat with herbicides.

It takes many years for the frequency of resistant weeds to become noticeable (i.e., more than the 1 to 10 percent that remain after a herbicide treatment). Thus, we do not realize we are enriching for herbicide resistance until it is upon us.

Note in Figure 2 the rates of enrichment for the triazine versus the phenoxy herbicides. The selection pressures are estimated, since the proper ecological studies have not yet been done. The phenoxy herbicides have a much lower selection pressure because their shorter soil persistence allows late-season weed germination. Even without this late-season germination the actual effective selection pressure of the phenoxies is lower than the triazines. A midseason survey of North Dakota wheat fields showed that the best control of *Amaranthus retroflexus*, *Chenopodium album*, and *Brassica campestris* with a phenoxy herbicide gave 0.2 plants (probably all "escaped" susceptibles) of each weed per m^2 (Dexter et al., 1981). Considering the plasticity of these weeds, there would be hundreds of seeds per m^2 for a good stand of susceptible weeds the following year.

The population genetics models (Figure 2) can also predict what happens when a monoherbicide culture is not used. In the model the number of weed generations that are treated affects enrichment. If it takes 10 years with no herbicide rotation to obtain resistance, it would take 20 and 30 years in 1 in 2 or 1 in 3 herbicide rotations, respectively. Indeed, all *s*-triazine resistance has come from monoherbicide cultures.

If these theories are true, triazine resistance should have developed in the U.S. corn belt, where corn with atrazine has been grown in a one- in two-year rotation. This has not happened, but it may still be too soon to expect resistance to appear, or it may be that rotation is a more potent tool to decrease the rate of resistance than previously thought. Herbicide mixtures (atrazine plus an acetamide) are also used widely in the U.S. corn belt. No triazine resistance has appeared where such mixtures are used. Gressel and Segel (1982) and Gressel (in press [a]) provide theoretical analyses of the effects of such mixtures.

BIOCHEMICAL AND PHYSIOLOGICAL MODES OF RESISTANCE

s-*Triazines*

The *s*-triazine herbicides, as well as many phenyl-urea and uracil herbicides, inhibit photosynthetic electron transport on the reducing side of photosystem II in leaf plastids. These herbicides loosely bind to the thylakoids. Death occurs from release of free radicals, or chlorophyll photo-bleaching, or starvation for photosynthate. The first sign of damage is an immediate rise in chlorophyll fluorescence (Figure 3**A**).

These herbicides also inhibit photosynthesis in the crops where they are

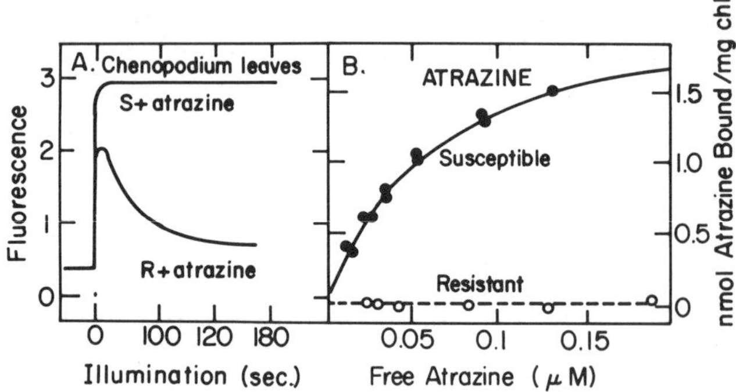

FIGURE 3 Special properties of most evolved atrazine-resistant weeds. **A:** Lack of increased chlorophyll fluorescence due to treatment with atrazine. Whole leaves of *Chenopodium album* R (resistant) and S (susceptible) biotypes were treated with 15 μM atrazine 2h before scanning. (Redrawn from Ducruet and Gasquez, 1978.) The scan of the R biotype with atrazine is similar to the scans of R and S biotypes without atrazine. **B:** Specific loss of triazine binding site in thylakoids from triazine resistant weeds. Binding of ^{14}C-atrazine to susceptible and resistant chloroplast membranes was measured. (Redrawn from Pfister and Arntzen, 1979.)

used: corn and orchards. Corn, however, is unique; it has high levels of a glutathione-*S*-transferase (GST) that conjugates glutathione to atrazine and simazine and detoxifies them before they can do lasting damage. In orchards simazine binds to the upper layer of soil; it does not reach the roots of trees, but lethally partitions into weed seedlings growing through this layer.

The herbicide-resistant weeds that appear in orchards and corn fields, however, do not have the enhanced rate of atrazine degradation as appears in corn. Instead, the plastids of these weeds are resistant because the triazines did not bind to thylakoids (Figure 3**B**) (Arntzen et al., 1982; Gressel, 1985). The simplest field test for this type of resistance is to use a field fluorometer modified from the designs of Ducruet and Gasquez (1978). One takes a fluorescence reading on a leaf, applies atrazine, and later takes another reading (Ahrens et al., 1981; Ali and Souza-Machado, 1981). Fluorescence in the resistant biotypes will not change, but it will increase in the susceptible types (Figure 3**A**).

The levels of resistance in weeds with the plastid-type triazine resistance are quite variable. Most evolved biotypes have the type of resistance shown in Figure 1**D**; saturating doses of atrazine, many times the levels used in agriculture, have no effect on the weed. Some resistant biotypes are inhibited differently by such rates (Figure 1**E**); marginally resistant biotypes have similar reactions at normal concentrations (Figure 1**F**). Weed germination in the last probably occurs after some of the atrazine has been biodegraded.

Triazine tolerance and resistance evolve differently even in the same spe-

TABLE 1 Differences in Inherent Tolerance to Atrazine

Rate for 90–100% Necrosis	Species
0.03 kg/ha	*Chenopodium album, Amaranthus retroflexus*
0.1 kg/ha	*Poa pratensis*[a], *Digitaria* sp., *Stellaria media*
0.3 kg/ha	*Echinochloa crus-galli, Avena fatua*[a], *Bromus inermis*[a], *Agropyron repens*[a], *Alopecuris pratensis*[a], *Sinapis arvenis, Datura stramonium,* soybean, *Chrysanthemum segetum*

NOTE: The rate used by farmers in corn varies between 2.2 and 4.4 kg/ha.

[a]Members of subfamily Poaceae.

SOURCE: Data from a commercial screen provided by P. F. Bocion, Dr. Maag Ltd., Dielsdorf, Switzerland (1984).

cies. For example, one population of *Senecio vulgaris* slowly increased in tolerance to sublethal triazine doses (Holliday and Putwain, 1980), yet another population "suddenly" evolved plastid resistance to high levels of atrazine (Scott and Putwain, 1981). The biochemical reasons for the increases in tolerance could not be discerned (Gressel et al., 1983b). Similarly, some populations of *Echinochloa crus-galli* have slowly increased in tolerance (Grignac, 1978), while other *Echinochloa* biotypes evolved plastid and nonplastid resistance (Gressel et al., 1982b).

Tolerance to triazines also varies among species (Table 1). The first species to evolve resistance were those with the greatest inherent susceptibility to atrazine; selection pressure was higher with fewer nonresistant escapees. Higher levels of atrazine are needed to control some species, especially the Poaceous grasses, which possess higher levels of the GST that conjugates atrazine to glutathione.

An interesting development for managing resistance is the use of a tridiphane, an herbicide "extender" that inhibits GST in the Poaceae; thus, much lower levels of atrazine need to be used (Lamoureux and Rusness, 1984). Lowering the triazine levels should decrease the rate at which dicots evolve triazine resistance (i.e., the slopes in Figure 2 would be less acute) but should not affect the rate that resistance evolves in the Poaceae. If, however, the triazine rates applied are not reduced when tridiphane is used, triazine-resistant grasses should evolve more rapidly.

Paraquat

The mode of tolerance to paraquat has been studied in two systems: *Lolium perenne* and *Conyza bonariensis* (= *C. linefolia*). Paraquat, at the levels used in agriculture, seems to be a specific acceptor of electrons from photosystem I of photosynthesis. The electrons are transferred from paraquat to oxygen, giving rise to highly reactive oxygen radicals that rapidly cause membrane damage due to lipoxidation. Paraquat reacts with photosystem I in the paraquat-resistant weeds, but damage is minimal. The tolerance has been correlated with a 50 percent higher level of superoxide dismutase in *Lolium* and a three-fold higher level in *Conyza* (Harvey and Harper, 1982). Superoxide dismutase forms hydrogen peroxide from the oxygen radicals. As peroxide is also toxic the resistant species must have sufficient levels of other enzymes to further detoxify the peroxide. These enzymes probably are in the plastids where the peroxide is formed and the first membrane lipoxidation occurs.

Diclofop-methyl

Wheat detoxifies diclofop-methyl and is thus resistant (Shimabukuro et al., 1979). It is not yet known if the resistant biotype of *Lolium rigidum* has evolved this system or some other mode of resistance.

Trifluralin

There is no information thus far on the mode of dinitro-aniline resistance that has evolved in *Eleusine*, nor is there adequate information on modes of selectivity in the species on which they act.

CROSS-RESISTANCE

The appearance of cross-resistances to totally unrelated groups of insecticides is even more disturbing because of the unpredictability of such resistances to compounds with totally different modes of action. Fortunately, with herbicides cross-resistance has been more logical and thus more predictable.

Triazines

The weeds that evolved plastid-level resistance to atrazine and simazine are resistant to all *s*-triazine herbicides and to some, but not all, asymmetric triazines (triazinones) such as metribuzin. Initially all the plastid-level triazine-resistant weeds were thought to be susceptible to diuron, a phenyl-urea herbicide with a similar mode of action as the triazines. Until triazine resistance occurred the phenyl-ureas were believed to have a totally identical binding site with the triazines (Pfister and Arntzen, 1979; Arntzen et al., 1982). Triazine-resistant biotypes, however, were found to have different cross-tolerances to the various phenyl-urea and uracil herbicides (Table 2).

TABLE 2 Cross-Resistances to Photosystem II Herbicides of Resistant Biotypes

Herbicide[a]	*Amaranthus retroflexus*[b]	*A. hybridus*[c]	*Chenopodium album*[d,e]	*Brassica campestris*[e]	*Chlamydomonas reinhardii*[f]
	(inhibitory dose for R versus inhibitory dose for S)				
Atrazine	251	1,000.0	201.0	501.0	50.0
Metribuzin	1,500	260.0	100.0	45.0	1,740.0
Diuron	4	2.6	1.2	0.7	14.5
Chloroxuron	794	1.4	6.2	—	—
Bromacil	2,000	2.0	24.0	—	106.0

[a]Atrazine is a symmetric triazine; metribuzin is an asymmetric triazine (triazinone); diuron and chloroxuron are phenyl-ureas; bromacil is a uracil derivative. All act on photosystem II and bind to wild-type thylakoids when they are competitive with each other for overlapping or allosteric binding site(s).

SOURCE: [b]Oettmeier et al. (1982); [c]Pfister and Arntzen (1979) (incorrectly designated therein as *A. retroflexus*); [d]Arntzen et al. (1982); [e]Thiel and Boger (1984); [f]Janatkova and Wildner (1982). A more complete comparison table may be found in Gressel (1985).

This, along with the data depicted in Figure **1D**–**1F**, suggests that the mutations can be at different loci in each of the biotypes, which was further borne out by the molecular biology. So far all triazine-resistant weed biotypes are susceptible to diuron, even if not to other phenyl-urea herbicides, but this need not continue (Table 2). There is probably a spectrum or continuum of binding sites that can be mutated in organisms that gives varying cross-specificities of herbicides affecting photosystem II.

Plants seem to have more substrate specificity of GSTs than found in mammalian (liver) systems. Three different GST systems in corn are substrate-specific for three herbicide groups: chloro-*s*-triazines (atrazine and simazine), acetamides (e.g., alachlor), and thiocarbamates (e.g., EPTC) (Mozer et al., 1983). The GST for atrazine is usually at a high constitutive level, but it probably can be induced to higher levels (Jachetta and Radosevich, 1981). The GST for alachlor can vary, but can be increased greatly by the protectant flurazole (Mozer et al., 1983). The GST of EPTC can be induced to higher levels, which has been correlated with resistance, by a dichloracetamide-type protectant (Lay and Casida, 1976). No cross-protection has been found in corn systems; induction of protection to one herbicide group does not grant protection to the others. Cross-protection has not been checked in the Poaceous weeds.

Paraquat

The biochemical nature of tolerance suggests that there should be ways to chemically induce tolerance (Lewinsohn and Gressel, 1984) and that there should be cross-tolerance of paraquat-resistant species with other herbicides and xenobiotics (Gressel et al., 1982a). There is no perfect cross-tolerance within the bipyridillium group: the paraquat-resistant conyzas are partially tolerant to diquat (M. Parham, I.C.I. Bracknell, United Kingdom, personal communication, 1981; Watanabe et al., 1982).

Cross-resistance has positive effects, as seen in the following examples. A *Lolium perenne* biotype, which evolved sulfur dioxide tolerance downwind from a coal-fired power plant (Horsman et al., 1978), had a modicum of tolerance to paraquat. The paraquat-resistant *Conyza bonariensis* is tolerant to sulfite (which releases SO_2) and to oxyfluorfen (a diphenyl-ether herbicide causing photoenergized membrane lipoxidation). Some ozone-tolerant tobacco varieties are also paraquat-tolerant. It might be possible, therefore, to design protectants that will guard against herbicides from more than one group as well as protect against environmental pollutants, such as sulfur dioxide and possibly ozone. It is also apparent that if a farmer were to rotate the use of two herbicides such as paraquat and oxyfluorfen, the final effect on enrichment for resistance would be the same as using a single herbicide.

Trifluralin

The *Eleusine* biotype, selected for by repeated trifluralin treatments, is resistant to all other dinitroaniline-type herbicides but not to herbicides in six other chemical types (Mudge et al., 1984).

Diclofop-methyl

The *Lolium* biotype that is tolerant to diclofop-methyl is not cross-tolerant to oxyfluorfen, as might be expected from its different mode of action. The diclofop-methyl-tolerant material, however, was tolerant to fluazifop-butyl and chlorazifop-propynil, diphenyl-ether herbicides that probably possess similar modes of action as diclofop-methyl (I. Heap and R. Knight, Waite Institute, Adelaide, Australia, personal communication, 1984). As diphenyl-ether herbicides are being developed with selectivity to different crops, they may be considered for use without herbicide rotation. This *Lolium* biotype can be used to further study cross-tolerances to ascertain which diphenyl-ether rotations are not really rotations (i.e., whether cross-tolerance occurs).

GENETICS AND MOLECULAR BIOLOGY OF RESISTANCE

During the few years in which herbicide resistance has appeared and has been studied, we have reports of possibilities of all types of inheritance: nuclear with dominance, recessiveness, monogenic and polygenic, and organelle inherited. There are even cases, studied only in tissue culture, of possible gene duplications (Gressel, in press [a]). The discussions that follow are concerned with only those cases where genetics has been studied in weeds or where the data obtained bear on what is expected to happen in weeds.

s-*Triazines and Other Photosystem II Herbicides*

Triazine tolerance and resistance can be inherited in many ways. The GST that degrades atrazine is inherited as a single dominant gene in corn (Shimabukuro et al., 1971). Thus, only one parent of each inbred line used in hybrid seed production needs to bear the trait. The increased levels of tolerance to triazines that evolved in *Senecio vulgaris* are inherited polygenically with a low heritability (Holliday and Putwain, 1980). The plastid-level resistance to triazines is maternally inherited, most probably on the plastome (chloroplast genome) (Souza-Machado, 1982). This has many implications for the appearance and spread of triazine resistance. Once resistance has appeared in weeds it cannot spread by pollen, only by seed. This should considerably slow the spread of resistance.

Each plastid has more than one DNA molecule, and each cell has more than one plastid; however, a mutation in a single plastome DNA molecule can create resistance. Most known plastome-mutant plants are a result of using mutagens. From the mode of action, triazine-resistant mutations should be the equivalent of recessive; all thylakoids must not bind triazines, otherwise lethal products would be produced. The natural rate of recessive mutations resulting in mutant plants is very low; most plastome-DNA specialists refuse to guess their actual natural frequency.

Two factors seem to converge to quicken the natural evolution of populations of triazine-resistant weeds. The first is population genetics. The second may be a nuclear gene, a plastome mutator, that increases the frequency of plastome mutations. This gene has been found in only four species (Arntzen and Duesing, 1983). Original triazine-resistant plants from which populations evolved probably were in a subpopulation that had a plastome mutator. Therefore, a given mutant is more likely to appear in a population of mutagen-treated plants than plants without mutagen. The selection pressure of triazine treatments enriches for triazine-resistant plants (which are almost always less fit than the wild type) and stabilized resistance in the population. The plastome mutator, which causes other plastome mutations, drains the population and is slowly bred out by actual hybrid selection.

It is easier to use unicellular algae with one chloroplast for basic studies on the selection, inheritance, and molecular biology of resistance than to use weeds. For example, resistance to phenyl-urea and uracil-type herbicides is maternally inherited in the green alga *Chlamydomonas* (Galloway and Mets, 1984); therefore, we can get mutants to other photosystem II-inhibiting herbicides. This has implications to proposed uses of the other herbicides as mixtures or in sequence with triazines. Population genetics theory states that whenever we treat with atrazine, we enrich for resistant alleles. If triazine-resistant alleles are found in plants with plastome mutators, enrichment for triazine-resistant individuals also enriches for individuals carrying the plastome mutator. The plastome mutator should increase mutation frequency for all plastome mutants including resistance to phenyl-ureas and uracil herbicides. Thus, enrichment for triazine resistance should also carry enrichment for resistance to other photosystem II-inhibiting herbicides, which may not be beneficial.

For example, if it took 10 years to get triazine-resistant weed populations in a given orchard, a diuron-resistant population would appear in even less time, if diuron is the replacement for atrazine. If atrazine and diuron are used together, as has been proposed for roadside weed control, or used in rotation, each could help enrich for resistance to the other by coenriching for the plastome mutator. If diuron is used for roadside weed control where atrazine resistance has occurred, the rapid appearance of diuron resistance is expected even though there is no cross-resistance between diuron and atrazine. Cross-resistance between high levels of atrazine and diuron may be precluded on molecular grounds (Table 3). Large spans of railroad rights-of-way in the United States and Europe and roadsides in Europe and Israel (Gressel et al., 1983a) are covered with recently evolved triazine-resistant weeds. Adequate long-term recommendations are needed for weed control along these roadways and the new areas where resistant biotypes continually appear.

The involvement of a peculiar protein in membranes of the plastids (thylakoids) may be responsible for susceptibility or resistance to photosystem II herbicides (Arntzen et al., 1982; Arntzen and Duesing, 1893; Gressel, 1985). Unlike most membrane proteins this protein (often called "the 32 kD" protein) has a very high turnover rate, which is under positive photocontrol, suggesting important plastid functions. This protein also is one of the most highly conserved proteins in biology. It should be very important in plastid functions—and mutations in structure should negatively affect photosynthesis and thus growth potential (Radosevich and Holt, 1982; Gressel, 1985). Mutations in the plastid-coded gene for this protein confer resistance to photosystem II-inhibiting herbicides (Table 3). Transversions at different places in the sequence lead to different resistances and cross-resistances.

Unfortunately, only two triazine-resistant weeds have been sequenced; they do not differ in amino acid transversion. The Italian *Solanum nigrum* biotypes (Figure 1**E**), however, might have different transversions than the French biotype sequenced because of the different dose-response curves. *Amaranthus blitoides* with its marginal resistance (Figure 1**F**) and *A. retroflexus* with its different cross-tolerance to chloroxuron (Table 2) should have different transversions from the *A. hybridus* sequence (Table 3). These algal mutations

TABLE 3 Amino Acid Transversions in the 32 kD Thylakoid Protein Conferring Resistance to Photosystem II-Inhibiting Herbicides

Species	Transversion				Reference
	Resistance to[a]	at Position[b]	from WT[c]	to Resistance	
Amaranthus hybridus	atrazine	264	serine	glycine	Hirschberg and McIntosh (1983)
Solanum nigrum	atrazine	264	serine	glycine	Goloubinoff et al. (1984) and Hirschberg et al. (1984)
Euglena gracilis	diuron	264	serine	alanine	U. Johanningmeier and R. B. Hallick, Univ. of Colo., Boulder, pers. comm. (1984)
Chlamydomonas reinhardtii	diuron (atrazine)	264	serine	alanine	Erickson et al. (1984)
	atrazine	255	phenylalanine	tyrosine	J. M. Erickson and J. D. Rochaix, Univ. of Geneva, pers. comm. (1984)
Chlamdomonas reinhardtii	diuron	219	valine	isoleucine	J. M. Erickson and J. D. Rochaix, Univ. of Geneva, pers. comm. (1984)

NOTE: The amino acid sequence was deduced from DNA base sequences. The triplet at position 264 in the wild-type weeds cannot, with a single base change, mutate to the triplet for alanine, and the triplet at 264 in *Euglena* and *Chlamydomonas* cannot mutate to one coding for glycine.

[a]Secondary (partial) cross-resistance given in parentheses.
[b]According to the numbering system of Zurawski et al. (1982).
[c]This amino acid is constant in the wild type (WT) of all 10 species in which it has been checked.

TABLE 4 Single Nuclear-Gene Herbicide Resistance

Herbicide[a]	Site of Action	Mode of Resistance	Inheritance	Reference
picloram	auxin type	nondegradation	dominant	Chaleff (1980)
phenmedipham	PSII	unknown	recessive	Radin and Carlson (1978)
bentazon	PSII	unknown	recessive	Radin and Carlson (1978)
chlorsulfuron	acetolactate synthase	modified enzyme	dominant	Chaleff and Ray (1984)
atrazine	PSII	degradation	dominant	Shimabukuro et al. (1971)
glyphosate	EPSP synthase	modified enzyme	—	Comani et al. (1983)

[a]Resistant plants (except atrazine and glyphosate) were selected in the laboratory using tissue culture techniques with tobacco. Atrazine was in corn, and glyphosate in bacteria.

also have different degrees of fitness loss, which has implications on the biotechnological uses of mutants in this gene for conferring atrazine resistance in crops (Gressel, in press [a]).

Other Herbicides

The genetics of other herbicide-resistant weeds have not been reported to date. Tolerance of *Lolium perenne* to paraquat and of *Senecio* to atrazine have polygenic inheritance (Faulkner, 1982). Faulkner (1982) and Gressel (1985) have reviewed the inheritance of herbicide resistances in crops.

LESSONS FROM BIOTECHNOLOGY

There are compelling commercial reasons for biotechnologically conferring cost-effective herbicide resistance to crop species (Gressel, in press [a]). The ease with which nuclear monogenic mutants resistant to many herbicides have been obtained in the laboratory (Table 4) is cause to pause and consider the implications for weed control practices.

Resistances can be dominant or recessive, with the genetics clearly related to mode of action. When resistance is due to degradation of the herbicide or to overcoming a herbicide-caused metabolic blockage of a vital pathway, resistance is dominant (Table 4).

Phenmedipham is thought to act on photosystem II similarly to atrazine and diuron, and it competes with them (Tischer and Strottmann, 1977). Presumably resistance is due to a nonbinding of the herbicide, since the mutation is recessive (Radin and Carlson, 1978). If the mutation was dominant, part of the thylakoids in a heterozygote would suicidally bind the herbicide. Bentazon is also a photosystem II, electron-transport inhibitor; thus, the reasons for resistance are similar to those for phenmedipham.

It has been easy to obtain bacterial mutants with a modified enol-pyruvate-shikimate-phosphate-synthase (EPSP synthase), the enzyme thought to be the sole target of glyphosate. Should one then expect to obtain glyphosate resistance in the field as its use increases? It depends: glyphosate is an "ephemeral" herbicide; it affects only those plants on which it is sprayed. This lack of persistence should give the herbicide the selection pressure needed for long field-life. With paraquat, a similar ephemeral herbicide, lack of soil persistence can be compensated for by the persistence of the farmers. Most paraquat resistance happened when the farmers sprayed about 10 times a year. If farmers do the same with glyphosate, they can expect resistant weeds.

Chlorsulfuron and other sulfonyl-urea herbicide-resistant mutants are easily obtained and regenerated to resistant plants (Chaleff and Ray, 1984). Resistance in tobacco is from a single dominant gene that modifies aceto-lactate synthase, the sole enzyme target of this group. A new imidazole-type herbicide affects the same enzyme site, but no data are available on cross-resistance. The specific sites affected on the enzyme may be different, as with atrazine and diuron, although neither are reversed by pyruvate, one of the substrates. Chlorsulfuron, at the rates used for weed control in wheat, has long soil persistence, rivalling that of the triazines. The models (Figure 2) predict that if sulfonyl-ureas are used without rotation or are not mixed with other herbicides, resistance will rapidly appear. The initial gene frequency of sulfonyl-urea resistant mutants in weed populations should be many orders of magnitude higher than triazine-resistant mutants; therefore, resistance should appear in a few years of widespread monoculture. There also may be enrichment for soil organisms that degrade chlorsulfuron, as in the problem soils.

Once we know whether resistance is dominant or recessive we can estimate the initial frequency in the population and plug this information into Figure 2. The frequency for dominant mutations in diploid species should be 10^{-5} to 10^{-7}. The frequency for recessive mutations should be 10^{-10} to 10^{-14} according to theory, but classical theory may be wrong because of somatic recombinations, and the frequencies may be 10^{-7} to 10^{-9} (Williams, 1976). Even these orders of magnitude differences between dominant and recessive will affect the time until resistance appears (Figure 2).

CONCLUSION

If good, cost-effective herbicides are judiciously used (only where and when needed, and in rotations and in mixtures), costly resistances can be considerably delayed. To make educated recommendations one must know the modes of action, cross-actions and cross-resistances, genetics, and molecular biology of the weeds and herbicides. The basic sciences have helped us understand the nature of excesses in agronomic practices and resistance and have given us information on how to slow down the process. We must learn from this short history. Since each herbicide and resistance may have very different properties, we must have this basic information, otherwise knowledgeable extrapolations are hard to make. ''Spray and pray'' must become a concept of the past if we wish to keep the most effective herbicides in our arsenal to fight the continual battle against loss of yields caused by weeds.

ACKNOWLEDGMENTS

I thank the many scientists around the world who have supplied yet unpublished data for use in this review. The author's own work on oxidant resistance is supported in part by the Israel Academy of Sciences and Humanities program in basic research. The author is the Gilbert de Botton professor of plant science.

REFERENCES

Abel, A. L. 1954. The rotation of weedkilling. P. 249 *in* Proc. Br. Weed Cont. Conf., Cliftonville, Margate, England, 1953.

Ahrens, W. H., C. J. Arntzen, and E. W. Stoller. 1981. Chlorophyll fluorescence assay for the determination of triazine resistance. Weed Sci. 29:316.

Ali, A., and V. Souza-Machado. 1981. Rapid detection of triazine resistant weeds using chlorophyll fluorescence. Weed Res. 21:191.

Ammon, H. U., and E. Irla. 1984. Bekampfung resistenter Unkrauter in Mais-Erfahrungen mit mechanischen und chemischen Verfahren. Die Grüne 112:12.

Arntzen, C. J., and J. H. Duesing. 1983. Chloroplast encoded herbicide resistance. Pp. 273–294 *in* Advances in Gene Technology: Molecular Genetics of Plants and Animals, K. Downey, R. W. Voellmy, F. Ahmad, and J. Schultz, eds. New York: Academic Press.

Arntzen, C. J., K. Pfister, and K. E. Steinback. 1982. The mechanism of chloroplast triazine resistance: Alterations in the site of herbicide action. Pp. 185–213 *in* Herbicide Resistance in Plants, H. LeBaron and J. Gressel, eds. New York: John Wiley and Sons.

Chaleff, R. S. 1980. Further characterization of picloram tolerant mutants of *Nicotiana tabacum*. Theor. Appl. Genet. 58:91.

Chaleff, R. S., and T. B. Ray. 1984. Herbicide resistant mutants from tobacco cell cultures. Science 223:1148.

Comai, L., L. C. Sen, and D. M. Stalker. 1983. An altered aroA gene product confers resistance to the herbicide glyphosate. Science 221:370.

Delp, C. J. 1979. Resistance to plant disease control agents: How to cope with it. Pp. 253–261 *in* Proc. Symp. 9th Int. Cong. Plant Prot., Vol. 1, T. Kommédahl, ed. Minneapolis, Minn.: Burgess.

Dexter, A. G., J. D. Nalewaja, D. D. Rasmusson, and J. Buchli. 1981. Survey of wild oats and other weeds in North Dakota: 1978 and 1979. N. D. Res. Rep. No. 79. Fargo: North Dakota State Extension Service.

Ducruet, J. M., and J. Gasquez. 1978. Observation of whole leaf fluorescence and demonstration of chloroplastic resistance to atrazine in *Chenopodium album* L. and *Poa annua* L. Chemosphere 8:695.

Erickson, J. M., M. Rahire, P. Bennoun, P. Delepelaire, B. Diner, and J. D. Rochaix. 1984. Herbicide resistance in *Chlamydomonas reinhardtii* results from a mutation in the chloroplast gene for the 32 kD protein of photosystem II. Proc. Natl. Acad. Sci. 81:3617.

Faulkner, J. S. 1982. Breeding herbicide-tolerant crop cultivars by conventional methods. Pp. 235–256 *in* Herbicide Resistance in Plants, H. M. LeBaron and J. Gressel, eds. New York: John Wiley and Sons.

Galloway, R. E., and L. J. Mets. 1984. Atrazine, bromacil and diuron resistance in *Chlamydomonas*. Plant Physiol. 74:469.

Georghiou, G. P., and C. E. Taylor. 1977. Operational influences in the evolution of insecticide resistance. J. Econ. Entomol. 70:653–658.

Goldie, J. H., and A. J. Coldman. 1979. A mathematical model for relating drug sensitivity of tumors to their spontaneous mutation rate. Cancer Treat. Rep. 63:1727.

Goloubinoff, P., M. Edelman, and R. B. Hallick. 1984. Chloroplast-coded atrazine resistance in *Solanum nigrum*: psbA loci from susceptible and resistant biotypes are isogenic except for a single codon change. Nucleic Acids Res. 12:9489–9496.

Gressel, J. 1985. Herbicide tolerance and resistance: Alteration of site of activity. Pp. 159–189 *in* Weed Physiology, Vol. 2, S. O. Duke, ed. Boca Raton, Fla.: CRC Press.

Gressel, J. In press(a). Biotechnologically conferring herbicide resistance in crops: The present realities. *In* Molecular Form and Function of the Plant Genome, L. Van Vloten-Doting, G. S. P. Groot, and T. C. Hall, eds. New York: Plenum.

Gressel, J. In press(b). Strategies for prevention of herbicide resistance in weeds. *In* Rational Pesticide Use, K. J. Brent and R. Atkin, eds. Cambridge: Cambridge University Press.

Gressel, J., and G. Ben-Sinai. In press. Low intra-specific competitive fitness in a triazine resistant, nearly nuclear-isogenic line of *Brassica napus*. Plant Sci. Lett. 38.

Gressel, J., and L. A. Segel. 1978. The paucity of plants evolving genetic resistance to herbicides: Possible reasons and implications. J. Theor. Biol. 75:349–371.

Gressel, J., and L. A. Segel. 1982. Interrelating factors controlling the rate of appearance of resistance: The outlook for the future. Pp. 325–348 *in* Herbicide Resistance in Plants, H. M. LeBaron and J. Gressel, eds. New York: John Wiley and Sons.

Gressel, J., G. Ezra, and S. M. Jain. 1982a. Genetic and chemical manipulation of crops to confer tolerance to chemicals. Pp. 79–91 *in* Chemical Manipulation of Crop Growth and Development, J. S. McLaren, ed. London: Butterworth.

Gressel, J., H. U. Ammon, H. Fogelfors, J. Gasquez, Q. O. N. Kay, and H. Kees. 1982b. Discovery and distribution of herbicide-resistant weeds outside North America. Pp. 32–55 *in* Herbicide Resistance in Plants, H. M. LeBaron and J. Gressel, eds. New York: John Wiley and Sons.

Gressel, J., Y. Regev, S. Malkin, and Y. Kleifeld. 1983a. Characterization of an *s*-triazine resistant biotype of *Brachypodium distachyon*. Weed Sci. 31:450.

Gressel, J., R. H. Shimabukuro, and M. E. Duysen. 1983b. *N*-dealkylation of atrazine and simazine in *Senecio vulgaris* biotypes, a major degradation pathway. Pestic. Biochem. Physiol. 19:361.

Grignac, P. 1978. The evolution of resistance to herbicides in weedy species. Agro-Ecosystems 4:377.

Hartmann, F. 1979. The atrazine resistance of *Amaranthus retroflexus* L. and the expansion of resistant biotype in Hungary. Novenyvedelem 15:491.

Harvey, B. M. R., and D. B. Harper. 1982. Tolerance to bipyridylium herbicides. Pp. 215–234 *in* Herbicide Resistance in Plants, H. M. LeBaron and J. Gressel, eds. New York: John Wiley and Sons.

Heap, J., and R. Knight. 1982. A population of ryegrass tolerant to the herbicide diclofop-methyl. J. Aust. Inst. Agri. Sci. 48:156.

Hirschberg, J., and L. McIntosh. 1983. Molecular basis of herbicide resistance in *Amaranthus hybridus*. Science 222:1346.

Hirschberg, J., A. Bleecker, D. J. Kyle, L. McIntosh, and C. J. Arntzen. 1984. The molecular basis of triazine-herbicide resistance in higher plant chloroplasts. Z. Naturforsch. 39c:412.

Holliday, R. J., and P. D. Putwain. 1980. Evolution of herbicide resistance in *Senecio vulgaris*: Variation in susceptibility to simazine between and within populations. J. Appl. Ecol. 17:779.

Holm, L. G., D. L. Plucknett, J. V. Pancho, and J. P. Herberger. 1977. World's Worst Weeds. Honolulu: University Press of Hawaii.

Horsman, D. A., T. M. Roberts, and A. D. Bradshaw. 1978. Evolution of sulphur dioxide tolerance in perennial ryegrass. Nature (London) 276:493–494.

Jachetta, J. J., and S. R. Radosevich. 1981. Enhanced degradation of atrazine by corn. Weed Sci. 29:37.

Janatkova, H., and G. F. Wildner. 1982. Isolation and characterisation of metribuzin-resistant *Chlamydomonas reinhardtii* cell. Biochim. Biophys. Acta 682–227.

Kaufman, D. D., Y. Katan, D. F. Edwards, and E. G. Jordan. In press. Microbial adaptation and metabolism of pesticides. *In* Agricultural Chemicals of the Future. BARC symposium, No. 8., J. L. Hilton, ed. Totowa, N.J.: Rowman and Allanheld.

Lamoureux, G. L., and D. L. Rusness. 1984. Glutathione-*S*-transferase as the basis of Dowco 356 (tridiphane) synergism of atrazine. Am. Chem. Soc. Meet. Abstr. 181.

Lay, M. M., and J. E. Casida. 1976. Dichloracetamide antidotes enhance thiocarbamate sulfoxide detoxification by elevating corn root glutathione content and glutathione-*S*-transferase activity. Pestic. Biochem. Physiol. 6:442.

LeBaron, H. M., and J. Gressel, eds. 1982. Herbicide Resistance in Plants. New York: John Wiley and Sons.

Lewinsohn, E., and J. Gressel. 1984. Benzyl viologen mediated counteraction of diquat and paraquat phytotoxicities. Plant Physiol. 76:125.

Mozer, T. J., D. C. Tiemeier, and E. G. Jaworski. 1983. Purification and characterization of corn glutathione-*S*-transferase. Biochemistry 22:1068.

Mudge, L. C., B. J. Gossett, and T. R. Murphy. 1984. Resistance of goosegrass (*Eleusine indica*) to dinitro-aniline herbicides. Weed Sci. 32:591.

Oettmeier, W., K. Masson, C. Fedtke, J. Konze, and R. R. Schmidt. 1982. Effect of different photosystem II inhibitors on chloroplasts isolated from species either susceptible or resistant toward *s*-triazine herbicides. Pestic. Biochem. Physiol. 18:357.

Pfister, K., and C. J. Arntzen. 1979. The mode of action of photosystem II-specific inhibitors in herbicide resistant weed biotypes. Z. Naturforsch. 34c:996.

Radin, D. N., and P. S. Carlson. 1978. Herbicide resistant tobacco mutants selected in situ recovered via regeneration from cell culture. Genet. Res. 32:85.

Radosevich, S. R., and J. S. Holt. 1982. Physiological responses and fitness of susceptible and resistant weed biotypes to triazine herbicides. Pp. 163–184 *in* Herbicide Resistance in Plants, H. M. LeBaron and J. Gressel, eds. New York: John Wiley and Sons.

Ryan, G. F. 1970. Resistance of common groundsel to simazine and atrazine. Weed Sci. 18:614.

Scott, K. R., and P. D. Putwain. 1981. Maternal inheritance of simazine resistance in a population of *Senecio vulgaris*. Weed Res. 21:137.

Shimabukuro, R. J., D. S. Frear, R. Swanson, and W. C. Walsh. 1971. Glutathione conjugation: An enzymatic basis for atrazine resistance in corn. Plant Physiol. 47:10.

Shimabukuro, R. J., W. C. Walsh, and R. A. Hoerauf. 1979. Metabolism and selectivity of diclofop-methyl in wild oat and wheat. J. Agric. Food Chem. 27:615.

Solymosi, P. 1981. Az *Amaranthus retroflexus* triazine resistenciujanak oroklodese. Novenytermeles 30:57.

Souza-Machado, V. 1982. Inheritance and breeding potential of triazine tolerance and resistance in plants. Pp. 257–274 *in* Herbicide Resistance in Plants, H. M. LeBaron and J. Gressel, eds. New York: John Wiley and Sons.

Thiel, A., and P. Boger. 1984. Comparative herbicide binding by photosynthetic membranes from resistant mutants. Pestic. Biochem. Physiol. 22:232.

Tischer, W., and H. Strotmann. 1977. Relationship between inhibitor binding by chloroplasts and inhibition of photosynthetic electron transport. Biochim. Biophys. Acta 460:113.

Watanabe, Y., T. Honma, K. Ito, and M. Miyahara. 1982. Paraquat resistance in *Erigeron philadelphicus*. Weed Res. (Japan) 7:49.

Williams, K. L. 1976. Mutation frequency at a recessive locus in haploid and diploid strains of a slime mould. Nature (London) 260:785.

Zanin, G., B. Vecchio, and J. Gasquez. 1981. Indagini sperimentali su popolazioni di dicotiledoni resistenti alli atrazine. Riv. Agron. 5:196.

Zurawski, G., H. J. Bohnert, P. R. Whitfeld, and W. Bottomley. 1982. Nucleotide sequence of the gene for the M,32,000 thylakoid membrane protein from *Spinacia oleracea* and *Nicotiana debneyi* predicts a totally conserved primary translation product of M,38,950. Proc. Natl. Acad. Sci. 79:7699.

Pesticide Resistance: Strategies and Tactics for Management.
1986. National Academy Press, Washington, D.C.

Genetics and Biochemistry of Insecticide Resistance in Arthropods: Prospects for the Future

FREDERICK W. PLAPP, JR.

Insecticide resistance in the house fly has a fairly simple genetic basis. There is one gene for decreased uptake of insecticides, one gene for target-site resistance to each insecticide type, and one major gene for metabolic resistance to all insecticides. The last interacts with minor genes located elsewhere in the genome. Based on limited data, resistance patterns are similar in other species.

Evidence is presented that target-site resistance to pyrethroids/DDT and to cyclodienes is controlled by changes in regulatory genes determining the number of receptor protein molecules synthesized. Resistance in both is recessive to susceptibility.

The product of the major gene for metabolic resistance appears to be a receptor protein that recognizes and binds insecticides and then induces synthesis of appropriate detoxifying enzymes. Different types of enzymes, for example, oxidases, esterases, and glutathione transferases, are coordinately induced. The effect of the gene is qualitative, that is, it determines the specific form of detoxifying enzyme synthesized. Inheritance is codominant.

Possible solutions to resistance include using synergists such as chlordimeform, which appear to act by increasing the binding of pyrethroid insecticides to their target-site proteins; using agonists, which successfully compete with insecticides for recognition by the receptor protein; and using either mixtures of insecticides or insecticides composed of multiple isomers.

INTRODUCTION

Resistance to insecticides in arthropods is widespread (Georghiou and Mellon, 1983), with at least 400 species resistant to one or more insecticides.

In some species, populations are resistant to nearly every insecticide ever used to control them and, often, to related chemicals to which the population has never been exposed. Resistance, at least in the house fly, has a fairly simple and straightforward genetic basis. Extensive genetic studies in other species, most notably *Lucilia cuprina* and *Drosophila melanogaster*, have indicated a similar situation. The biochemistry of resistance is also comprehensible, particularly when there is an adequate understanding of the genetics of resistant populations.

GENETIC MECHANISMS CONFERRING RESISTANCE

A very important question is, How many genes for resistance are there? Are there multiple genes for resistance, each conferring resistance to a narrow range of insecticides, or are there only a few genes, each conferring resistance to a wide array of insecticides? If there are numerous genes then cross-resistance associated with each gene should be limited, and new insecticides would solve the problem. Conversely, if a limited number of genetic mechanisms is involved, then resistant populations should show resistance to insecticides to which they have never been exposed. The second hypothesis is more frequently true. Thus, developing new insecticides that are closely related to existing insecticides in either mode of action or pathways of metabolism will not solve the problem.

If only a few major genes confer resistance to insecticides, it should be possible to characterize the mechanisms controlled by each gene. Once this is done, it may be possible to devise solutions and regain our ability to deal with populations recalcitrant to chemical control.

Standard neo-Darwinian models (Moore, 1984) suggest that change occurs as a result of accumulation of multiple mutations, each mutation contributing a minute amount to the total; that is, insecticide resistance should be polygenic, but it is not (Whitten and McKenzie, 1982). In field populations resistance is almost invariably due to a single major gene. Therefore, standard evolutionary theory does not seem to apply to the development of resistance.

A regulatory gene hypothesis is a more likely model to account for change, particularly at the population or subspecific level. Such genes, which control time and nature of expression of structural genes, are more likely to provide the genetic basis of adaptive variation such as the development of resistance (Levin, 1984). In my opinion, available data on resistance offer considerable support for Levin's hypothesis. In this paper I shall summarize both genetic and biochemical evidence that changes in regulatory genes are of major importance in insecticide resistance.

Two types of regulatory genes seem to be present, and both differ in inheritance and biochemistry. One type exhibits all-or-none inheritance (fully dominant or recessive) and appears to involve changes in the amount of protein synthesized. The second shows codominant (intermediate) inheritance and involves changes in the nature of proteins synthesized.

Quantitative resistance (that type involving differences in amount of proteins synthesized) is similar in nature to certain bacterial operons. Resistance of this type apparently involves regulatory elements located adjacent to the structural genes in question. Change does not occur in the structural gene, but in an adjacent, distinct, genetic element. If it were in the structural gene, inheritance would be additive. Since it is not, the evidence is that a separate protein (i.e., the product of a distinct gene) must be the site of variation. Regulators of this type have been defined as "near" regulators (Paigen, 1979).

The second type, qualitative resistance, appears to represent a mechanism allowing for production of altered forms of particular detoxifying enzymes in resistant as compared to susceptible insects. Genetic studies with the house fly (Plapp, 1984) show that change at a single genetic locus appears to control resistance associated with multiple detoxification enzymes. A similar mechanism can be inferred from earlier studies with *D. melanogaster* (Kikkawa, 1964a,b). Since one locus appears to act on a variety of enzymes, the gene probably is not adjacent to the enzymes whose activity it regulates. Such regulators have been defined as "distant" regulators (Paigen, 1979), and such systems can be considered "regulons" (Plapp, 1984). According to Paigen, these systems are characterized by their codominant inheritance rather than the all-or-none type of similar bacterial systems.

GENETICS OF RESISTANCE

The number of major genes conferring resistance to insecticides in the house fly (and presumably other species) is limited. The list of known resistance genes includes:

- *pen*—for decreased uptake of insecticides. This chromosome III gene is inherited as a simple recessive. By itself, *pen* confers little resistance to any insecticide, seldom more than two- to three-fold. It appears to be more important as a modifier of other resistance genes. In such cases *pen* may double resistance levels, for example, from 50- to 100-fold.
- *kdr*—for knockdown resistance to DDT and pyrethroids. This gene is a chromosome III recessive at a locus distinct from *pen*. It confers resistance to DDT and all analogs and to pyrethrins and all synthetic analogs. Low-level (*kdr*) and high-level (super *kdr*) alleles have been reported. The gene probably involves modifications at the target site of the insecticides.
- *dld-r*—for resistance to dieldrin and all other cyclodienes. This is a chromosome IV gene whose inheritance is incompletely recessive. Resistance appears to involve change at the target site of these insecticides.
- *AChE-R*—for altered acetylcholinesterase, the target site for organo-

phosphate (OP) and carbamate insecticides. The gene is located on chromosome II and is inherited as a codominant. Different alleles appear to confer different levels of resistance to multiple organophosphate and carbamate insecticides (Oppenoorth, 1982).

The house fly's metabolic resistance to many types of insecticides, including OPs, carbamates, pyrethroids, DDT, and juvenile hormone analogs, is associated with a gene or genes on chromosome II. This type of resistance was long thought to be due primarily to mutations in structural genes for the specific enzymes. Earlier work had shown that resistance genes were located at a variety of loci on chromosome II (Hiroyoshi, 1977; Tsukamoto, 1983). More recent work (Wang and Plapp, 1980; Plapp and Wang, 1983) suggests that inversions or other rearrangements of the chromosome are present in many resistant strains and are of sufficient extent to explain the apparent differences in gene location on the chromosome, that is, only one gene seems to be present, but it is not always located at the same place relative to other genes on chromosome II. Based on these results the idea of multiple structural genes for metabolic resistance on chromosome II becomes more tenuous, and the idea of a common resistance gene becomes more logical.

Close linkage (and, therefore, possible allelism) exists among genes for metabolic resistance to insecticides in other insect species as well. Examples include the gene RI (for resistance to insecticides) located at 64.5–66 on chromosome II of *Drosophila melanogaster*, a locus conferring resistance to organophosphates, carbamates, and DDT (Kikkawa, 1964a,b), and major genes for metabolic resistance to diazinon and malathion in numerous populations of *Lucilia cuprina* (Hughes et al., 1984). Other evidence for allelism has been reported for malathion resistance in different populations of *Tribolium castaneum* (R. W. Beeman, U.S. Department of Agriculture, Manhattan, Kansas, personal communication, 1983). In fact, our knowledge of the genetics of resistance in insects other than dipterans is so inadequate that we can only guess as to the precise nature of the genetic mechanisms involved.

Research has shown that resistance to different classes of insecticides is associated with a particular linkage group, but the number of genes involved is unknown. Genetically, the most feasible approach to this problem is to perform allelism tests. This method has demonstrated allelism of genes for reduced uptake of insecticides (*pen*) in American and European house fly populations (Sawicki, 1970) and for organophosphate resistance in spider mites (Ballantyne and Harrison, 1967). I have recently been doing such tests on several house fly strains with metabolic resistance to various organophosphates associated with chromosome II and with chromosome II resistance to DDT and organophosphates within a strain. All data indicate allelism of the genes.

Although chromosome II has been shown to make a major contribution to metabolic resistance in the house fly, minor genes on other chromosomes make additional contributions. An assay of total levels of resistance is made by crossing resistant strains with susceptible strains containing mutant markers on multiple chromosomes. Recent work in my laboratory has shown that the contribution to metabolic resistance of chromosomes other than II is not expressed in the absence of chromosome II and is inherited as incomplete recessives. Such resistance is similar in inheritance to that described previously for *pen*, *kdr*, and *dld-r*.

Position also affects the expression of resistance associated with chromosome II. Strains showing a major (20 to 30 percent) reduction in recombination values between the resistance gene and the mutation carnation eye (*car*) have increased levels of resistance, compared with strains showing smaller reductions in recombination values (Plapp and Wang, 1983). Thus, the location of the gene on chromosome II is important in determining the level of resistance present.

In summary four types of resistance, *pen*, *kdr*, *dld-r*, and metabolic, associated with chromosomes other than II, are inherited as incompletely or fully recessive characters. In contrast, altered acetycholinesterase resistance and metabolic resistance on chromosome II are inherited as codominants. The level of resistance associated with the major chromosome II gene for metabolic resistance varies with the location of the gene on the chromosome.

BIOCHEMISTRY OF RESISTANCE

This area has been intensively studied for the last 30 years. Earlier work concentrated on mechanisms associated with metabolic resistance and identified a number of enzyme systems concerned with resistance (Tsukamoto, 1969; Oppenoorth, 1984). Recent studies have dealt with mechanisms involved in nonmetabolic (target site) resistance. The availability of genetic stocks purified to contain individual mechanisms proved invaluable to these studies.

High-affinity receptors for DDT and pyrethroids are present in insects (Chang and Plapp, 1983a,c). House flies possessing the gene *kdr* for target-site resistance bound less insecticide than susceptible flies. Resistant flies had fewer target-site receptors than susceptible flies (Chang and Plapp, 1983b). Further, binding affinity between preparations from R and S strains did not differ. Therefore, the major difference between strains was strictly quantitative, that is, in receptor numbers, and not qualitative, that is, in receptor affinity.

Similar studies on cyclodiene mode of action/mechanism of resistance have been reported by Matsumura and coworkers. Kadous et al. (1983) reported that cyclodiene-resistant cockroaches were cross-resistant to the plant-derived neurotoxicant picrotoxinin and, further, that nerve components from resistant cockroaches bound significantly less [^{3}H] α-dihydropicrotoxinin than similar preparations from susceptible insects. The receptor was sensitive to all cyclodiene insecticides (Tanaka et al., 1984). Similar studies with susceptible and cyclodiene-resistant house flies have shown reduced binding in resistant insects (K. Tanaka and F. Matsumura, Michigan State University, East Lansing, Michigan, personal communication, 1984), suggesting that the number of receptor binding sites is decreasing.

Thus, quantitative decreases in numbers of target sites may be involved in target-site resistance to both DDT/pyrethroids and cyclodienes. At first glance it may appear contradictory for resistant insects to have fewer target-sites than susceptible insects. Decreased receptor numbers probably confer resistance by a needle-in-the-haystack approach (Lund and Narahashi, 1981a,b); the decrease in number may make it less likely for a toxicant to reach target-sites.

Decreases in target-site numbers are consistent with the genetics of resistance to these insecticides. If the change were in the target-sites themselves, inheritance would be additive; R/S heterozygotes would be intermediate between the parents in resistance. Inheritance being all-or-none agrees with the idea of quantitative change. The specific mutations conferring resistance are probably in genes coding for proteins that determine the number of target-site proteins synthesized. Here, heterozygotes would have the normal number of receptors since the diffusible protein product of the wild-type regulatory gene would act on both structural genes. Only the resistant homozygotes, those with two mutant genes, would produce fewer target-site receptor proteins than normal. This activity is an example of *trans* dominance; the protein product of a regulatory gene influences the expression of a specific structural gene on both members of a chromosome pair.

The precise biochemical mechanism of the major gene for metabolic resistance to insecticides is not yet known with certainty, although a single gene locus is probably involved. Since all structural genes coding for detoxification enzymes are probably not at the same site, a common controlling mechanism might be responsible.

The key to metabolic resistance is induction. Induction of different detoxifying enzymes is coordinate (Plapp, 1984); that is, exposure to chemicals that induce one detoxifying enzyme induces several. Mixed-function oxidases, glutathione transferases, and DDT dehydrochlorinase are coordinately induced in the house fly (Plapp, 1984), as are oxidases and glutathione transferases in *Spodoptera* (Yu, 1984). When the products of several structural genes (enzymes) respond to the same stimulus, they must be responding to the protein product of a separate gene, a genetic element that is distinct from the elements that define the enzymes themselves.

The finding is not original. It comes from the research of Monod and Jacob on induction in *E. coli*. As reviewed by Judson the critical idea in their work, which led to the discovery of regulatory genes, was the realization that the only way two enzymes, β-galactosidase and galactose permease, could be induced together was through the action of a third gene (Judson, 1979). The product of the third gene was a regulatory protein.

The Monod-Jacob work showed that the product of the regulatory gene, the repressor protein, functioned by recognizing inducers of the *lac* operon. The same was true of metabolic resistance. The resistance gene product must be a protein that recognizes and binds insecticides with high affinity. The next step is activating structural genes for the detoxifying enzymes conferring resistance. Since the structural genes are probably not located close to each other, the product of the regulatory protein is the so-called distant regulator. Overall the mechanism is similar to that by which steroid hormones act.

Such xenobiotic-recognizing receptor proteins appear to occur in house flies and *Heliothis* (Plapp, 1984) and probably exist in *D. melanogaster* (Hallstrom, 1984). Hallstrom pointed out the similarity of resistance to the Jacob-Monod model and also noted the basic agreement with the Britten-Davidson (1969) model of eukaryotic gene regulation. In this model there are three levels of genes in eukaryotes: structural, integrator, and sensor. The distant regulator proposed for insecticides acts like sensor genes, which, it is believed, act by recognizing external signals such as insecticides.

A similar system for xenobiotic recognition and induction resulted from research with mice. The so-called *Ah* (for aromatic hydroxylation) locus in mice (Nebert et al., 1982) responds to many environmental chemicals, similar to that proposed for the response of insects to insecticides. The system conferring metabolic resistance to insecticides, however, differs from the *lac* operon in two distinct ways. First, inheritance is codominant as opposed to the all-or-none inheritance of inducibility in the *lac* operon. Second, the biochemistry is different. Resistant populations in insects make different enzymes than susceptibles. Further, exposure to inducers results in the production of changed forms of detoxifying enzymes, not just more of the form already present. Susceptible house flies exposed to phenobarbital produced a different cytochrome P_{450} from that present in uninduced flies (Moldenke and Terriere, 1981). It was similar to the P_{450} present in resistant flies. Similarly, Ottea and Plapp (1981; 1984) demonstrated that the glutathione transferases of resistant flies always differed from those of susceptible flies in K_m and only sometimes in V_{max}. Susceptible flies induced with phenobarbital produced a different glutathione transferase, not more enzyme.

Similar work with mice (Phillips et al., 1983) has shown that exposure to phenobarbital produced a specific mRNA at a 40-fold higher concentration than in controls but only a 3-fold increase in total P_{450}, a finding again suggesting the presence of a qualitative response in eukaryotes.

Insects with metabolic resistance may also differ from susceptible insects in enzyme amount as well as specificity. Earlier genetic studies on mixed-

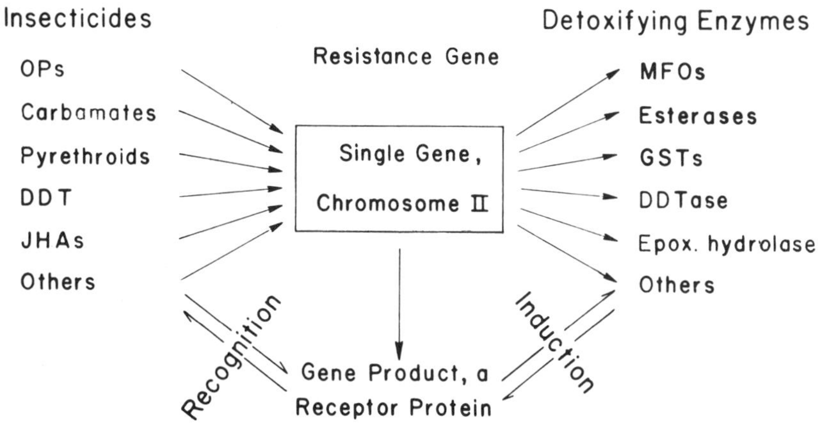

FIGURE 1 Proposed model for metabolic resistance to insecticides.

function oxidase inheritance in house flies established that the higher specific activity of cytochrome P_{450} of a resistant strain was associated with chromosome II, while the amount of P_{450} was associated with a gene or genes on chromosomes III or V. The quantitative contribution of III or V may be due to mutation at a regulatory site controlling enzyme amount, not enzyme nature. In this respect it would be similar to the control of *kdr* and *dld-r*, both of which are inherited as recessives to the normal condition.

The overall model for metabolic resistance to insecticides proposed here shows in Figure 1 that the protein product of a single gene recognizes and then presumably binds many insecticides. In turn the protein-xenobiotic combination acts to induce synthesis of appropriate forms of multiple detoxifying enzymes.

POSSIBLE SOLUTIONS TO RESISTANCE

Perhaps the best understood resistance mechanism is that involving altered acetycholinesterase. Mixtures of *N*-propyl and *N*-methyl carbamates suppress this type of resistance in the green rice leafhopper *Nephotettix cincticeps* (Yamamoto et al., 1983). The *N*-propyl carbamates are potent inhibitors of the altered enzyme of resistant insects, while the *N*-methyl carbamates inhibit the enzyme of susceptible insects. Thus, the use of combinations of the two carbamate types is more effective than the use of either type alone.

Target-site resistance to DDT/pyrethroids and cyclodienes has been the most difficult type of resistance to deal with. Typical synergists that block metabolism usually do not work well to increase toxicity since the resistance does not depend on increased metabolism, the mechanism most synergists are designed to counter. Target-site synergism may exist, however, and in at least one case the use of such a synergist has blocked the development of resistance.

Several years ago we reported that the miticide-ovicide chlordimeform was found to be strongly synergistic with several hard-to-metabolize insecticides, including toxaphene and DDT, to which resistance was present in the tobacco budworm (Plapp, 1976; Plapp et al., 1976). Since then chlordimeform synergism has been reported in the new, metabolically stable synthetic pyrethroids (Plapp, 1979; Rajakulendren and Plapp, 1982). Many formamidines are synergistic with pyrethroids and other insecticides against several arthropod species (El-Sayed and Knowles, 1984a,b). The mechanism for this synergism may be that chlordimeform is acting as a target-site synergist (Chang and Plapp, 1983c).

Chlordimeform may block pyrethroid resistance in *Heliothis* (Crowder, et al., 1984). Selection of *H. virescens* with permethrin resulted in 37-fold resistance within a few generations. Parallel selection with permethrin-chlordimeform combinations prevented resistance development.

Therefore, limited data are available suggesting that chlordimeform may synergize insecticides against insects in cases of target-site resistance and block development of such resistance. Since the new synthetic pyrethroids will probably be subject to *kdr*-type resistance, the use of such combinations offers a possible way to manage the problem.

Metabolic resistance has been attacked by a variety of approaches, primarily the use of synergists designed to poison the enzymes involved in detoxification. Since the work described in this paper indicates that a single gene is of primary importance in this resistance, different approaches may be possible. Rather than poisoning the detoxifying enzymes, it may be possible to affect the receptor protein by using agonists that compete with insecticides for recognition sites on xenobiotic receptor proteins.

This idea may already have been demonstrated. Ranasinghe and Georghiou (1979) selected an organophosphate-resistant mosquito population with three regimens. These were temephos only, temephos plus the antioxidant synergist piperonyl butoxide, and temephos plus DEF. DEF, *S,S,S*-tributyl phosphorotrithioate, is a plant defoliant that inhibits oxidases and esterases. I suggest that it is a receptor agonist. Selection with temephos resulted in the rapid development of a high level of resistance. The same thing occurred, but slightly slower, with temephos plus piperonyl butoxide. Selection with temephos plus DEF quickly restored a near-normal level of susceptibility to the test population.

The authors were unable to offer an explanation for the results of the temephos/DEF selection. I believe that DEF has a high affinity for the receptor protein, which recognizes temephos as a xenobiotic. With the temephos/DEF selection the receptor protein increased its ability to recognize and bind DEF and simultaneously lost its ability to recognize, bind, and, thus, respond to temephos.

Other work with DEF as a synergist has been done with *Lucilia* (Hughes, 1982). Preexposure to DEF significantly synergized diazinon, while simultaneous exposure to DEF and diazinon was much less effective. Again the results agree with a receptor-level effect for DEF.

Another approach to overcoming metabolic resistance involves using insecticides composed of two isomers. The major example of this effect involves phenylphosphonates of the EPN series. These insecticides have four different substituents attached to the central phosphorus atom. They exist as plus and minus isomers. Insects with metabolic resistance to the more typical dialkyl phenyl phosphorothioates show little or no cross-resistance to the phenylphosphonates. The single gene hypothesis for metabolic resistance offers an explanation. If only one receptor gene is of primary importance in metabolic resistance, its protein product can recognize either the plus or the minus isomer, but not both at once. If this is so, then synthesis of enzymes of high specific activity toward only one isomer will be induced. An example of the use of two isomer organophosphates to circumvent resistance involves profenofos to control multiresistant populations of *Spodoptera littoralis* in Egypt (Dittrich et al., 1979). I have confirmed these findings of lack of resistance to the two isomer OPs in fly strains with metabolic resistance to single isomer OPs. It may be a general phenomenon. This idea may not be practical, however, because of the delayed neurotoxicity syndrome associated with at least some of these organophosphates (Metcalf and Metcalf, 1984).

A final approach involves using multiple isomers of an insecticide. The idea is that the two will compete for the receptor protein just as the plus and minus isomers of the phenylphosphonates compete. I tested this idea by comparing the toxicity of dimethyl and diisopropyl isomers of parathion, alone and in combination, to susceptible and resistant house flies. Toxicities of the mixture were additive to susceptible flies, but synergistic with resistant flies. These results suggest that using mixed alkyl isomers of dialkyl phenylphosphates and phosphorothioates might prove quite effective for overcoming resistance. Again the mechanism responsible may be the lack of ability of a single resistance gene to handle multiple chemicals simultaneously.

CONCLUSION

Resistance genetics in the house fly is comparatively simple. The studies described here would not have been possible without the availability of mutant stocks to identify different chromosomes and to map resistance gene locations on specific chromosomes. Such studies are currently not feasible with most resistant species, due to lack of mutant markers. Nevertheless, what is true for house flies and other higher Diptera in the way of resistance genetics is probably true for other insects; that is, the genetic mechanisms involved are probably ubiquitous rather than specific.

Based on the genetics, it is possible to develop a comprehensive theory of resistance. Resistance is best understood as being due to changes in regulatory genes controlling the amount or nature of target proteins or enzymes synthesized. From this understanding, approaches to solving the problem become feasible, at least for metabolic resistance. Solutions involve using mixtures of insecticides or using insecticides composed of several isomers. The mixture approach will work because change at only a single locus is involved. Not all components of an insecticide need to be toxic; some may work primarily as receptor agonists rather than enzyme inhibitors.

Nothing in the foregoing should be interpreted, however, as an opinion that resistance is subject to perfect and/or complete suppression via chemical means. I have no doubt that, in the long term, life will always overcome chemistry and find ways to persevere. The best that can be said is that if we are lucky, we should be able to suppress resistance to such an extent that we can live with it.

REFERENCES

Ballantyne, G. H., and R. A. Harrison. 1967. Genetic and biochemical comparisons of organophosphate resistance between strains of spider mites (*Tetranychus* species). Entomol. Exp. Appl. 10:231–239.

Britten, R. J., and E. H. Davidson. 1969. Gene regulation for higher cells: A theory. Science 169:349–357.

Chang, C. P., and F. W. Plapp, Jr. 1983a. DDT and pyrethroids: Receptor binding and mode of action in the house fly. Pestic. Biochem. Physiol. 20:76–85.

Chang, C. P., and F. W. Plapp, Jr. 1983b. DDT and pyrethroids: Receptor binding and mechanism of knockdown resistance (kdr) in the house fly. Pestic. Biochem. Physiol. 20:86–91.

Chang, C. P., and F. W. Plapp, Jr. 1983c. DDT and synthetic pyrethroids: Mode of action, selectivity, and mechanism of synergism in the tobacco budworm, *Heliothis virescens* (F.), and a predator *Chrysopa carnea* Stephens. J. Econ. Entomol. 76:1206–1210.

Crowder, L. A., M. P. Jensen, and T. F. Watson. 1984. Permethrin resistance in the tobacco budworm, *Heliothis virescens*. Pp. 223–224 *in* Proc. Beltwide Cotton Conf., Atlanta, Ga., January 9–12, 1984.

Dittrich, V., N. Luetkemeier, and G. Voss. 1979. Monocrotophos and profenofos: Two organophosphates with a different mechanism of action in resistant races of the Egyptian cotton leafworm *Spodoptera littoralis*. J. Econ. Entomol. 72:380–384.

El-Sayed, G. N., and C. O. Knowles. 1984a. Formamidine synergism of pyrethroid toxicity to twospotted spider mites (Acari: Tetranychidae). J. Econ. Entomol. 77:23–30.

El-Sayed, G. N., and C. O. Knowles. 1984b. Synergism of insecticide activity to *Heliothis zea* (Boddie) by formanilides and formamidines. J. Econ. Entomol. 77:872–875.

Georghiou, G. P., and R. B. Mellon. 1983. Pesticide resistance in time and space. Pp. 1–46 *in* Pest Resistance to Pesticides, G. P. Georghiou and T. Saito, eds. New York: Plenum.

Hallstrom, I. P. 1984. Cytochrome P_{450} in *Drosophila melanogaster*: Activity, Genetic Variation and Regulation. Ph.D. dissertation. University of Stockholm, Sweden.

Hiroyoshi, T. 1977. Some new mutants and revised linkage maps of the house fly, *Musca domestica* L. Jpn. J. Genet. 52:275–288.

Hughes, P. B. 1982. Organophosphorus resistance in the sheep blowfly, *Lucilia cuprina* (Wiedemann) (Diptera: Calliphoridae): A genetic study incorporating synergists. Bull. Entomol. Res. 72:573–582.

Hughes, P. B., P. E. Green, and K. G. Reichmann. 1984. A specific resistance to malathion in laboratory and field populations of the Australian sheep blowfly, *Lucilia cuprina*. J. Econ. Entomol. 77:1400–1404.

Judson, H. F. 1979. The Eighth Day of Creation. New York: Simon and Schuster.

Kadous, A. A., F. Matsumura, J. G. Scott, and K. Tanaka. 1983. Difference in the picrotoxinin receptor between the cyclodiene-resistant and susceptible strains of the German cockroach. Pestic. Biochem. Physiol. 19:157–166.

Kikkawa, H. 1964a. Genetical analysis on the resistance to parathion in *Drosophila melanogaster*. II. Induction of a resistance gene from its susceptible allele. Botyu-Kagaku 2:37–41.

Kikkawa, H. 1964b. Genetical studies on the resistance to Sevin in *Drosophila melanogaster*. Botyu-Kagaku 29:42–46.

Levin, B. R. 1984. Science as a way of knowing—Molecular evolution. Am. Zool. 24:451–464.

Lund, A. E., and T. Narahashi. 1981a. Modification of sodium channel kinetics by the insecticide tetramethrin in crayfish giant axons. Neurotoxicology 2:213–229.

Lund, A. E., and T. Narahashi. 1981b. Kinetics of sodium channel modification by the insecticide tetramethrin in squid axon membranes. Pharmacol. Exp. Ther. 219:464–473.

Metcalf, R. L., and R. A. Metcalf. 1984. Steric, electronic, and polar parameters that affect the toxic actions of *O*-alkyl, *O*-phenyl phosphorothionate insecticides. Pestic. Biochem. Physiol. 22:169–177.

Moldenke, A. F., and L. C. Terriere. 1981. Cytochrome P_{450} in insects. 3. Increase in substrate binding by microsomes from phenobarbital-induced houseflies. Pestic. Biochem. Physiol. 16:222–230.

Moore, J. A. 1984. Science as a way of knowing—Evolutionary biology. Am. Zool. 24:467–534.

Nebert, D. W., M. Negishi, M. A. Lang, L. M. Hjelmeland, and J. J. Eisen. 1982. The Ah locus, a multigene family necessary for survival in a chemically adverse environment: Comparison with the immune system. Adv. Genet. 21:1–52.

Oppenoorth, F. J. 1982. Two different paraoxon-resistant acetylcholinesterase mutants in the house fly. Pestic. Biochem. Physiol. 18:26–27.

Oppenoorth, F. J. 1984. Biochemistry of insecticide resistance. Pestic. Biochem. Physiol. 22:187–193.

Ottea, J. A., and F. W. Plapp, Jr. 1981. Induction of glutathione S-aryl transferase by phenobarbital in the house fly. Pestic. Biochem. Physiol. 15:10–13.

Ottea, J. A., and F. W. Plapp, Jr. 1984. Glutathione S-transferase in the house fly: Biochemical and genetic changes associated with induction and insecticide resistance. Pestic. Biochem. Physiol. 22:203–208.

Paigen, K. 1979. Acid hydrolases as models of genetic control. Annu. Rev. Genet. 13:417–466.

Phillips, I. R., E. A. Shephard, B. R. Rabin, R. M. Bayney, S. F. Pike, A. Ashworth, and M. R. Estall. 1983. Factors controlling the expression of genes coding for drug-metabolizing enzymes. Biochem. Soc. Trans. 11:460–463.

Plapp, F. W., Jr. 1976. Chlordimeform as a synergist for insecticides against the tobacco budworm. J. Econ. Entomol. 69:91–92.

Plapp, F. W., Jr. 1979. Synergism of pyrethroid insecticides by formamidines. J. Econ. Entomol. 72:667–670.

Plapp, F. W., Jr. 1984. The genetic basis of insecticide resistance in the house fly: Evidence that a single locus plays a major role in metabolic resistance to insecticides. Pestic. Biochem. Physiol. 22:194–201.

Plapp, F. W., Jr., and T. C. Wang. 1983. Genetic origins of insecticide resistance. Pp. 47–70 *in* Pest Resistance to Pesticides, G. P. Georghiou and T. Saito, eds. New York: Plenum.

Plapp, F. W., Jr., L. G. Tate, and E. Hodgson. 1976. Biochemical genetics of oxidative resistance to diazinon in the house fly. Pestic. Biochem. Physiol. 6:175–182.

Rajakulendran, S. V., and F. W. Plapp, Jr. 1982. Synergism of five synthetic pyrethroids by chlordimeform against the tobacco budworm and a predator, *Chrysopa carnea*. J. Econ. Entomol. 75:1089–1092.

Ranasinghe, L. E., and G. P. Georghiou. 1979. Comparative modification of insecticide-resistance spectrum of *Culex pipiens fatigans* Wied. by selection with temephos and temephos/synergist combinations. Pestic. Sci. 10:502–508.

Sawicki, R. M. 1970. Interaction between the factor delaying penetration of insecticides and the desethylation mechanism of resistance in organophosphorus-resistant house flies. Pestic. Sci. 1:84–87.

Tanaka, K., J. G. Scott, and F. Matsumura. 1984. Picrotoxinin receptor in the central nervous system of the American cockroach: Its role in the action of cyclodiene-type insecticides. Pestic. Biochem. Physiol. 22:117–127.

Tsukamoto, M. 1969. Biochemical genetics of insecticide resistance in the house fly. Residue Rev. 25:289–314.

Tsukamoto, M. 1983. Methods of genetic analysis of insecticide resistance. Pp. 71–98 *in* Pest Resistance to Pesticides, G. P. Georghiou and T. Saito, eds. New York: Plenum.

Wang, T. C., and F. W. Plapp, Jr. 1980. Genetic studies on the location of a chromosome II gene conferring resistance to parathion in the house fly. J. Econ. Entomol. 73:200–203.

Whitten, M. J., and J. A. McKenzie. 1982. The genetic basis for pesticide resistance. Pp. 1016 *in* Proc. 3rd Australas. Conf. Grassl. Invert. Ecol., K. E. Lee, ed. Adelaide, Australia: S.A. Government Printer.

Yamamoto, I., Y. Takahashi, and N. Kyomura. 1983. Suppression of altered acetylcholinesterase of the green rice leafhopper by *N*-propyl and *N*-methyl carbamate combinations. Pp. 579–594 *in* Pest Resistance to Pesticides, G. P. Georghiou and T. Saito, eds. New York: Plenum.

Yu, S. J. 1984. Interactions of allelochemicals with detoxification enzymes of insecticide-susceptible and resistant fall armyworm. Pestic. Biochem. Physiol. 22:60–68.

Pesticide Resistance: Strategies and Tactics for Management.
1986. National Academy Press, Washington, D.C.

Resistance to 4-Hydroxycoumarin Anticoagulants in Rodents

ALAN D. MACNICOLL

There are few reported cases of development of resistance to pesticides in vertebrates. The most widespread and well-documented example is resistance to warfarin in rodents. It has been demonstrated in Rattus norvegicus *and* Mus musculus *that inheritance of warfarin resistance is monogenic and the gene is closely linked to that for coat color. The biochemistry and mechanism of resistance in the latter species has not been investigated thoroughly, but warfarin resistance may be associated with an altered metabolism of the anticoagulant. Warfarin resistance in* R. norvegicus *is probably associated with alterations in a vitamin K metabolizing enzyme or enzymes. Second-generation anticoagulants, which are more toxic than warfarin, were introduced in the 1970s and were considered effective in controlling warfarin-resistant rodent infestations. Some warfarin-resistant populations may also be cross-resistant to other 4-hydroxycoumarin anticoagulant rodenticides, and control of these infestations with more toxic compounds is less effective than using warfarin to control anticoagulant-susceptible rodents.*

INTRODUCTION

The incidence of inheritable resistance to pesticides in vertebrates is remarkably low. The mosquito fish *Gambusia affinis* (Vinson et al., 1963; Boyd and Ferguson, 1964) and other fish species (Ferguson et al., 1964; Ferguson and Bingham, 1966) have developed resistance to chlorinated hydrocarbon pesticides. Also, two frog species may have developed resistance to DDT (Boyd et al., 1963). Incidences of inheritable pesticide resistance in mammals are confined almost exclusively to rodents. Differential susceptibility to fluoroacetate, however, has been reported in some areas of Australia in populations of the grey kangaroo and tammar wallaby, as well as the bush rat *Rattus fuscipes* (Oliver et al., 1979). Inheritable tolerance in these species is thought to be a result of the abundance in some parts of Australia of leguminous plants that naturally produce fluoroacetate.

Genetically determined resistance in humans to coumarin anticoagulant drugs, some of which are also used as rodenticides, was first reported in 1964 (O'Reilly et al., 1964). Resistance to coumarin anticoagulants in rodents is the most widespread and thoroughly investigated example of inheritable pesticide resistance in vertebrates and will be discussed in detail.

A laboratory mouse strain has been developed that showed a 1.7-fold tolerance to DDT when compared to the original susceptible strain (Ozburn and Morrison, 1962). This was achieved by treating nine successive generations with DDT and breeding the survivors. Probably more significant was the discovery that pine voles, *Microtus pinetorium*, trapped in orchards with a history of endrin treatment, had a 12-fold resistance to this compound, compared with voles trapped in untreated orchards (Webb and Horsfall, 1967). These resistant animals also showed a two-fold cross-resistance to dieldrin, a stereoisomer of endrin. This example of inheritable resistance may be associated with alterations in the metabolism of endrin, as indicated by studies on the hepatic, microsomal, mixed-function oxidase system of endrin-resistant and endrin-susceptible strains (Webb et al., 1972; Hartgrove and Webb, 1973).

INCIDENCE AND GENETICS OF WARFARIN RESISTANCE IN RODENTS

Warfarin resistance in *R. norvegicus* was first noted in Scotland in 1958 (Boyle, 1960) and subsequently on the Wales-England border (Drummond and Bentley, 1967) and in Denmark (Lund, 1964), Holland (Ophof and Langveld, 1969), Germany (Telle, 1967), and the United States (Jackson and Kaukeinen, 1972). These initial observations were not isolated, and in 1979, it was reported in 36 out of 77 American cities surveyed that more than 10 percent of each *R. norvegicus* population was warfarin-resistant (Jackson and Ashton, 1980). With evolutionary pressure from the continued use of warfarin, some resistant populations can spread to cover areas of several thousand square kilometers (Greaves, 1970).

Inheritance of warfarin resistance in *R. norvegicus* is due to the inheritance of an autosomal gene, closely linked to the gene controlling coat color, which has been mapped in linkage group I (Greaves and Ayres, 1969). Further genetic studies (Greaves and Ayres, 1977, 1982) on warfarin resistance in wild rats from Wales, Scotland, and Denmark showed that there are at least three multiple alleles of the warfarin resistance gene *Rw*. Strains of *R. norvegicus* derived from wild Welsh or Danish rats have an increased requirement for vitamin K (Pool et al., 1968; Hermodson et al., 1969; Greaves and Ayres, 1973, 1977; Martin, 1973), but only the Welsh resistance gene is described as dominant (Greaves and Ayres, 1969, 1982).

Inheritable warfarin resistance in *Rattus rattus* has been observed in the United Kingdom (Greaves et al., 1973a, 1976), Australia (Saunders, 1978), and the United States (Jackson and Ashton, 1980). Warfarin resistance in this species was a significant problem in 4 of 12 American cities where populations had been sampled.

Warfarin resistance in the house mouse *Mus musculus* has followed a similar pattern to that of *R. norvegicus*. Problems in controlling house mice (Dodsworth, 1961) were initially thought to be due to inheritance of more than one gene (Rowe and Redfern, 1965; Roll, 1966). Subsequent investigations (Wallace and MacSwiney, 1976) demonstrated a major warfarin resistance gene, *War*, that was closely linked to coat color and located on chromosome 7 in the mouse, which is analogous to linkage group I in the rat. Monitoring of warfarin resistance in the house mouse is not routine, but resistance seems to be widespread (Jackson and Ashton, 1980).

WARFARIN ACTION AND RESISTANCE MECHANISM

The naturally occurring anticoagulant dicoumarol (structure I in Figure 2) was isolated from moldy sweet clover hay in 1939 (Link, 1944). Following observations that cattle that were fed on spoiled sweet clover hay developed a fatal haemorrhagic malady, dicoumarol was subsequently clinically used as a prophylactic agent against thrombosis. Oral vitamin K_3 (menadione: structure V in Figure 2) or vitamin K_1 were antidotal in excessive hypoprothrombinaemia (Cromer and Barker, 1944; Lehmann, 1943). This naturally occurring coumarin was also considered for rodent control, but it was replaced by a more toxic synthetic analogue, warfarin (structure II in Figure 2). Warfarin was also more suitable than dicoumarol for routine clinical use and for 30 years has been widely used both as a drug and as a rodenticide (Shapiro, 1953; Clatanoff et al., 1954).

Despite this widespread dual use of warfarin and the known role of vitamin K as an antidote, little progress was made in elucidating the mode of action of warfarin until the mid-1970s. Vitamin K and warfarin are antagonistic in their effects on the synthesis of blood-clotting factors II, VII, IX, and X. In 1974 γ-carboxyglutamic acid residues (GLA) were discovered (Stenflo et al., 1974) in prothrombin (factor II), which were not present in the altered proteins in the blood of cows or humans treated with coumarin anticoagulants. Post-translational γ-carboxylation of glutamyl residues appears to require the hydroquinone (or reduced form) of vitamin K as a cofactor (Sadowski et al.,

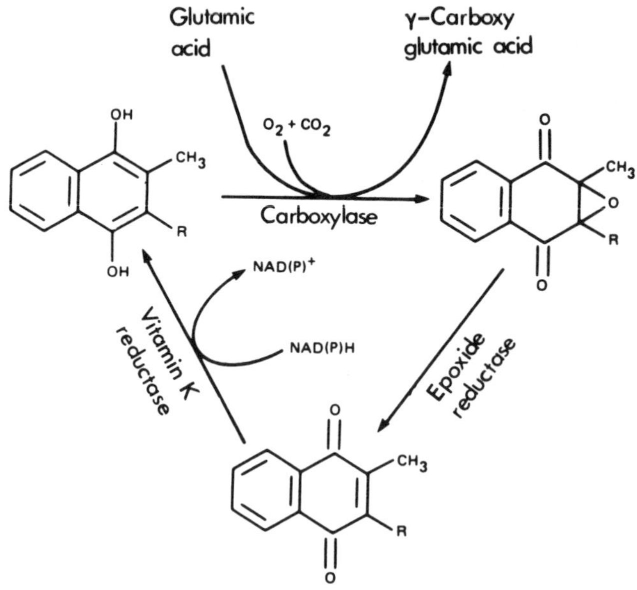

FIGURE 1 Schematic representation of the vitamin K cycle.

1980), and vitamin K 2,3-epoxide is a product of this reaction (Larson et al., 1981). An enzyme cycle (Figure 1) exists in liver microsomes to generate vitamin K hydroquinone from the epoxide, with the quinone form of the vitamin as an intermediate product (Fasco and Principe, 1980; Fasco et al., 1982). Administration of warfarin and vitamin K_1 to rats increased the ratio of vitamin K_1 2,3-epoxide to vitamin K_1 quinone in plasma and liver, when compared with animals that received vitamin K_1 alone (Bell and Caldwell, 1973). This effect was more pronounced in warfarin-susceptible than in warfarin-resistant animals. Further studies confirmed the hypothesis that 4-hydroxycoumarin anticoagulants act by inhibiting the enzyme vitamin K epoxide reductase (Ren et al., 1974, 1977; Shearer et al., 1974). In addition, S(−)-warfarin was more effective in inhibiting prothrombin synthesis and vitamin K epoxide reductase activity than the R(+)-enantiomer (Bell and Ren, 1981). An efficient method for determining the warfarin resistance genotype in *R. norvegicus* was based partly on the effect of coadministration of vitamin K_1 2,3-epoxide and warfarin on prothrombin synthesis (Martin et al., 1979). Analysis of blood-clotting time 24 hours after treatment showed that rats that were either homozygous or heterozygous for the Welsh warfarin resistance gene had normal prothrombin levels, but homozygous-susceptible animals had elongated clotting times. The implication was that warfarin-resistant animals were able to utilize vitamin K 2,3-epoxide in the presence of warfarin. Other studies showed that warfarin metabolism and excretion were not significantly altered in warfarin-resistant strains of *R. norvegicus* when compared with a related susceptible strain (Hermodson et al., 1969; Townsend et al., 1975).

This evidence, and some from other studies not described above, led to the common belief that 4-hydroxycoumarin anticoagulants inhibit the enzyme vitamin K epoxide reductase, which is altered in warfarin-resistant rats (*R. norvegicus*), and therefore indirectly inhibits the synthesis of vitamin K-dependent clotting factors. These hypotheses can be questioned on a number of points. All of the supporting evidence has been obtained from investigations of the metabolism of vitamin K_1 (phylloquinone) and its epoxide, but this form of vitamin K is present only in plant material (McKee et al., 1939). Vertebrates (Dialameh et al., 1971) as well as invertebrates (Burt et al., 1977) and bacteria (Tishler and Sampson, 1948), synthesize compounds of the vitamin K_2 (menaquinone) series. Compounds of the vitamin K_2 series have a variable-length polyisoprene (unsaturated) substituent at the 3-position of the 2-methyl 1,4-naphthoquinone nucleus, whereas the side chain of phylloquinone is 20 carbon atoms long and has only one double bond. Synthesis of vitamin $K_{2(20)}$, the equivalent of phylloquinone, by chick liver microsomes is inhibited by warfarin in vitro (Dialameh, 1978), and the effects of the S(−) and R(+)-enantiomers are proportional to the effects on prothrombin synthesis. In addition, menadione (vitamin K_3) is as effective as phylloquinone (when administered intravenously) in relieving vitamin K deficiency in chicks (Dam and Sondergaard, 1953), but it is not as effective an antidote to warfarin (Green, 1966; Griminger, 1966).

Studies on vitamin K metabolism in warfarin-resistant *R. norvegicus* until recently have only been carried out using animals derived from wild Welsh rats (Pool et al., 1968; Greaves and Ayres, 1969). These rat strains undoubtedly have an altered hepatic microsomal vitamin K epoxide reductase with reduced sensitivity to warfarin. The activity of this enzyme is, however, as sensitive to warfarin in a strain derived from wild Scottish warfarin-resistant rats as the enzyme from a closely related susceptible strain (MacNicoll, 1985). Studies of warfarin inhibition in vitro of NADH and dithiothreitol-dependent vitamin K reductase (Fasco and Principe, 1980; MacNicoll et al., 1984) have shown that this enzyme is as sensitive to warfarin as vitamin K epoxide reductase, but it probably is not the same enzyme. Similar investigations of the vitamin K-dependent γ-glutamyl carboxylase, however, have shown that this third enzyme of the vitamin K cycle is relatively insensitive to warfarin (Hildebrandt and Suttie, 1982) and is probably not inhibited directly in vivo by 4-hydroxycoumarin anticoagulants. The hypothesis that inhibition of vitamin K epoxide reductase is the only effect of warfarin on vitamin K-dependent protein synthesis and that reduced warfarin sensitivity of this enzyme is the result of expression of all of the different allelic forms of the warfarin resistance gene in *R. norvegicus* is, therefore, questionable (Bechtold et al., 1983; MacNicoll, 1985; Preusch and Sutte, 1984).

Other hypotheses on the mechanism of warfarin resistance in *R. norvegicus* have been largely discounted. For example, Ernster et al. (1972) observed that the activity of the enzyme DT-diaphorase was considerably lower in a soluble fraction prepared from the livers of warfarin-resistant rats when compared with preparations from susceptible animals. This enzyme is present (in different forms) in several liver fractions, utilizes NADH or NADPH as cofactors, and reduces quinone groups in a number of substrates including menadione (vitamin K_3) (Ernster et al., 1960). DT-diaphorase is also highly sensitive to dicoumarol, and it was concluded (Ernster et al., 1972) that altered activity of this enzyme was a result of expression of the warfarin resistance gene. A later study (Greaves et al., 1973b), however, clearly demonstrated that the different enzyme activities were more correctly assigned to differences between the Wistar stock, from which the warfarin-resistant animals were derived, and the Sprague-Dawley strain, which was used for the susceptible comparison in the earlier study. This enzyme has been implicated in the production of vitamin K hydroquinone in vivo. Highly purified rat-liver cytosolic DT-diaphorase reduced vitamin K_1 (Fasco and Principe, 1982); this reduction was dicoumarol- but not warfarin-sensitive. The results are inconsistent with the warfarin-sensitive NADH or DDT-dependent vitamin K_1 hydroquinone formation observed with crude rat-liver microsomal fractions. Recent studies (Lind et al., 1982; Talcott et al., 1983) on the action of DT-diaphorase in detoxification or activation of a wide range of quinones, including some antimalarial drugs, suggests that the capacity of this enzyme for vitamin K reduction is not associated with the ribosomal synthesis of vitamin K-dependent clotting factors.

A more recent hypothesis on the mechanism of warfarin resistance in *R. norvegicus* was based on the formation of 2- or 3-hydroxyvitamin K_1 from the epoxide by liver microsomal fractions (Fasco et al., 1983). These putative metabolites were detected in greater quantities in incubations with preparations from warfarin-resistant rats when compared with preparations from susceptible animals. This observation was associated with the reduced activity of vitamin K epoxide reductase in that resistant strain. A second report, however, showed that under certain conditions these hydroxylated compounds were formed by a chemical reaction in control incubations (Hildebrandt et al., 1984). The apparent increase in metabolism to these compounds by liver microsomes from resistant animals probably reflected the reduced rate of metabolism to the quinone form of the vitamin. The detection of hydroxyvitamin K_1 in the blood of warfarin-resistant rats that had received an intravenous injection of vitamin K 2,3-epoxide (Preusch and Suttie, 1984), therefore, is probably not associated directly with expression of the warfarin resistance gene.

Little if any work has been carried out on the mechanism of warfarin resistance in *R. rattus*, but there have been many studies conducted on *M. musculus*. Observations of the effect of warfarin on mortality and blood clotting in wild warfarin-resistant and -susceptible house mice indicated that resistant animals developed a tolerance to daily doses of warfarin (administered intravenously) up to 100 mg/kg, and susceptible animals developed a tolerance to doses of 1 mg/kg administered at 21-day intervals (Rowe and Redfern, 1968). Female mice were particularly tolerant to warfarin. Animals trapped in areas with control problems had normal clotting times when fed a diet containing 0.025 percent warfarin for 21 days.

As mentioned above, warfarin resistance in *M. musculus* is due to inheritance of the gene *War* located on chromosome 7 (Wallace and MacSwiney, 1976). This resistance may be related to a gene on the same chromosome (Wood and Conney, 1974), which is expressed as an increased rate of hydroxylation of coumarin. Subsequent investigation of 16 different strains demonstrated that warfarin resistance and rapid coumarin hydroxylation were not coinherited (Lush and Arnold, 1975). Warfarin resistance in this species may be inversely related to hexobarbitone sleeping time, but it is not stimulated by phenobarbitone (Lush, 1976). The report suggested that warfarin resistance in the house mouse may be due to an increased rate of warfarin hydroxylation. There are no reports of vitamin K deficiency in warfarin-resistant mouse strains, and it is possible that resistance in this species is related to alterations in warfarin rather than vitamin K metabolism.

SECOND-GENERATION ANTICOAGULANT RODENTICIDES

The three compounds (Figure 2) based on 4-hydroxycoumarin, commonly known as the second-generation anticoagulant rodenticides, are difenacoum (structure III: when radical = hydrogen), brodifacoum (structure III: when radical = bromine), and bromadiolone (structure IV). The mechanism of action of these compounds is assumed to be the same as for warfarin. The increased toxicity is assigned to the highly lipophilic nature of the substituents at the 3-position of the 4-hydroxycoumarin nucleus (Hadler and Shadbolt, 1975; Dubock and Kaukeinen, 1978). Initial laboratory studies and field trials indicated that these compounds could effectively control warfarin-resistant rat and mouse populations (Hadler, 1975; Hadler et al., 1975; Hadler and Shadbolt, 1975; Redfern et al., 1976; Rennison and Dubock, 1978; Redfern and Gill, 1980; Lund, 1981; Richards, 1981; Rowe et al., 1981). Studies in vitro on the mode of action of difenacoum (Whitlon et al., 1978; Hildebrandt and Suttie, 1982) and in vivo on difenacoum and brodifacoum (Breckenbridge et al., 1978; Leck and Park, 1981) indicated that these compounds inhibited the enzyme vitamin K epoxide reductase and were effective in both warfarin-susceptible and -resistant *R. norvegicus*.

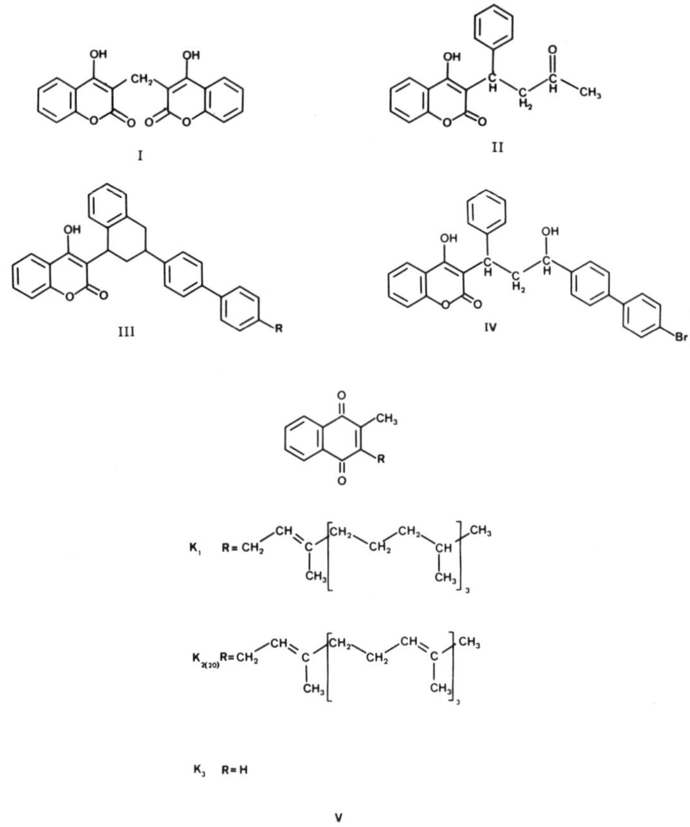

FIGURE 2 Chemical structures: I. Dicoumarol. II. Warfarin. III. When the radical (R) is hydrogen, the compound is difenacoum. When R is bromine, the compound is brodifacoum. IV. Bromadiolone. V. Vitamin K.

Some early reports on field trials of difenacoum and bromadiolone expressed concern about apparent incidences of cross-resistance observed in some warfarin-resistant populations of *R. norvegicus* and *M. musculus*. A laboratory test for difenacoum resistance in *R. norvegicus* was developed a few years after this compound was introduced as a rodenticide (Redfern and Gill, 1978). A significant widespread incidence of difenacoum resistance was detected in rat populations across an area of English farmland (Greaves et al., 1982a) where a monogenic form of resistance to warfarin had been present for several years. Resistance to difenacoum suggested that this was an example of another allele of the warfarin resistance gene, since no difficulty had been experienced previously in controlling warfarin-resistant populations of *R. norvegicus* (Rennison and Dubock, 1978). Further field trials of bromadiolone, brodifacoum, and difenacoum in this area showed that these compounds were not as effective in controlling the *R. norvegicus* populations as warfarin was for controlling warfarin-susceptible infestations (Greaves et al., 1982b). Continued use of 4-hydroxycoumarin anticoagulants in this area may apply evolutionary pressure favoring animals that may be resistant to this whole class of compounds. Since there are several forms of the warfarin resistance gene in *R. norvegicus*, and inherited resistance in *R. rattus* and *M. musculus*, it may be difficult to control rodent infestations in other areas using 4-hydroxycoumarin anticoagulants.

CONCLUSION

The development of resistance to 4-hydroxycoumarin anticoagulants in rodents may have implications for resistance to other pesticides. Studies on the biochemistry and pharmacology of warfarin resistance may have provided misleading information. Almost all such studies used rat strains derived from wild Welsh rats, and comparative studies have not always used a suitable susceptible control. At least one hypothesis of the mechanism of resistance was erroneously based on a strain difference. The current theory on altered vitamin K epoxide reductase activity may apply only to animals whose resistance is associated with an increased susceptibility to vitamin K deficiency.

When the highly toxic second-generation anticoagulants were developed, most of the evidence for the control of warfarin-resistant *R. norvegicus* was based on studies using rats of the Welsh resistant strains. Control of rat infestations in Wales and several other areas was achieved with these compounds, but in other areas resistance to the new compounds developed or was already present. It is important, therefore, that appropriate comparative studies are carried out and that when similar compounds are introduced to control pesticide-resistant populations, the potential for cross-resistance is fully investigated.

There is not a logical explanation for the apparent confinement of cross-resistance to 4-hydroxycoumarin anticoagulants to the United Kingdom. The long history of widespread use of anticoagulants for rodent control may be significant, but so could the established system for detecting and monitoring rodenticide resistance, which may not be so well developed in other countries. It is likely, therefore, that the continued use of 4-hydroxycoumarin anticoagulants in areas with known warfarin-resistant populations could result in rodent infestations that are difficult to control with any of this class of compounds.

REFERENCES

Bechtold, H., D. Trenk, T. Meinertz, M. Rowland, and E. Jahnchen. 1983. Cyclic interconversions of vitamin K_1 and vitamin K_1 2,3-epoxide in man. Br. J. Clin. Pharmacol. 16:683–689.

Bell, R. G., and P. T. Caldwell. 1973. Mechanism of warfarin-resistance. Warfarin and the metabolism of vitamin K_1. Biochemistry 12:1759–1762.

Bell, R. G., and P. Ren. 1981. Inhibition by warfarin enantiomers of prothrombin synthesis, protein carboxylation, and the regeneration of vitamin K from vitamin K epoxide. Biochem. Pharmacol. 30:1953–1958.

Boyd, C. E., and D. E. Ferguson. 1964. Susceptibility and resistance of mosquito fish to several insecticides. J. Econ. Entomol. 57:430–431.

Boyd, C. E., S. B. Vinson, and D. E. Ferguson. 1963. Possible DDT resistance in two species of frog. Copeia 2:426–429.

Boyle, C. M. 1960. Case of apparent resistance of *Rattus norvegicus* Berkenhout to anticoagulant poisons. Nature (London) 188:517.

Breckenbridge, A. M., J. B. Leck, B. K. Park, M. J. Serlin, and A. Wilson. 1978. Mechanisms of action of the anticoagulants warfarin, 2-chloro-3-phytylnapthoquinone (Cl-K), acenocoumarol, brodifacoum and difenacoum in the rabbit. Br. J. Pharmacol. 64:399.

Burt, V. T., E. Bee, and J. F. Pennock. 1977. The formation of menaquinone-4 (vitamin K) and its oxide in some marine invertebrates. Biochem. J. 162:297–302.

Clatanoff, D. V., P. O. Triggs, and O. O. Meyer. 1954. Clinical experience with coumarin anticoagulants Warfarin and Warfarin sodium. Arch. Int. Med. 94:213–220.

Cromer, H. E., Jr., and N. W. Barker. 1944. Effect of large doses of menadione bisulfite (synthetic vitamin K) on excessive hypoprothrombinaemia induced by dicoumarol. Proc. Staff Meet. Mayo Clin. 19:217–223.

Dam, H., and E. Sondergaard. 1953. Comparison of the effects of vitamin K_1, menadione, and Synkavit intravenously injected in vitamin K-deficient chicks. Experientia 9:26–27.

Dialameh, G. H. 1978. Stereobiochemical aspects of warfarin isomers for inhibition of enzymatic alkylation of menaquinone-0 to menaquinone-4 in chick liver. Int. J. Vitam. Nutr. Res. 48:131–135.

Dialameh, G. H., W. V. Taggart, J. T. Matschiner, and R. E. Olson. 1971. Isolation and characterization of menaquinone-4 as a product of menadione metabolism in chicks and rats. Int. J. Vitam. Nutr. Res. 41:391–400.

Dodsworth, E. 1961. Mice are spreading despite such poisons as warfarin. Munic. Eng. (London) 3746:1668.

Drummond, D. C., and E. W. Bentley. 1967. The resistance of rodents to warfarin in England and Wales. Pp. 56–67 *in* EPPO Report of the International Conference on Rodents and Rodenticides, Paris 1965. Paris: EPPO Publications.

Dubock, A. C., and D. E. Kaukeinen. 1978. Brodifacoum (Talon rodenticide), a novel concept. Pp. 127–137 *in* Proc. 8th Vertebr. Pestic. Conf., Sacramento, Calif.: University of California, Davis.

Ernster, L., M. Ljunggren, and L. Danielson. 1960. Purification and some properties of a highly dicoumarol-sensitive liver diaphorase. Biochem. Biophys. Res. Commun. 2:88–92.

Ernster, L., C. Lind, and B. Rase. 1972. A study of DT-diaphorase activity of warfarin-resistant rats. Eur. J. Biochem. 25:198–206.

Fasco, M. J., and L. M. Principe. 1980. Vitamin K_1 hydroquinone formation catalysed by a microsomal reductase system. Biochem. Biophys. Res. Commun. 97:1487–1492.

Fasco, M. J., and L. M. Principe. 1982. Vitamin K_1 hydroquinone formation catalysed by a DT-diaphorase. Biochem. Biophys. Res. Commun. 104:187–192.

Fasco, M. J., E. F. Hildebrandt, and J. W. Suttie. 1982. Evidence that warfarin anticoagulant action involves two distinct reductases. J. Biol. Chem. 257:11210–11212.

Fasco, M. J., P. C. Preusch, E. F. Hildebrandt, and J. W. Suttie. 1983. Formation of hydroxyvitamin K by vitamin K epoxide reductase of warfarin resistant rats. J. Biol. Chem. 258:4372–4380.

Ferguson, D. E., and C. R. Bingham. 1966. Endrin resistance in the yellow bullhead, *Ictaluris natalis*. Trans. Am. Fish Soc. 95:325–326.

Ferguson, D. E., D. D. Culley, W. D. Cotton, and R. P. Dodds. 1964. Resistance to chlorinated hydrocarbon insecticides in three species of freshwater fish. BioScience 14:43–44.

Greaves, J. H. 1970. Warfarin-resistant rat. Agriculture 77:107–110.

Greaves, J. H., and P. B. Ayres. 1969. Linkages between genes for coat color and resistance to warfarin in *Rattus norvegicus*. Nature (London) 224:284–285.

Greaves, J. H., and P. B. Ayres. 1973. Warfarin resistance and vitamin K requirement in the rat. Lab. Anim. 7:141–148.

Greaves, J. H., and P. B. Ayres. 1977. Unifactorial inheritance of warfarin resistance in *Rattus norvegicus* from Denmark. Genet. Res. 29:215–222.

Greaves, J. H., and P. B. Ayres. 1982. Multiple allelism at the locus controlling warfarin resistance in the Norway rat. Genet. Res. 40:59–64.

Greaves, J. H., C. Lind, B. Rase, and K. Enander. 1973a. Warfarin resistance and DT-diaphorase activity in the rat. F.E.B.S. Letts. 37:144.

Greaves, J. H., B. D. Rennison, and R. Redfern. 1973b. Warfarin resistance in the ship rat in Liverpool. Int. Pest. Control. 15:17.

Greaves, J. H., B. D. Rennison, and R. Redfern. 1976. Resistance of the ship rat, *Rattus rattus* L. to warfarin. J. Stored Prod. Res. 12:65–70.

Greaves, J. H., D. S. Shepherd, and J. E. Gill. 1982a. An investigation of difenacoum resistance in Norway rat populations in Hampshire. Ann. Appl. Biol. 100:581–587.

Greaves, J. H., D. S. Shepherd, and R. Quy. 1982b. Field trials of second-generation anticoagulants against difenacoum-resistant Norway rat populations. J. Hyg. 89:295–301.

Green, J. 1966. Antagonists of vitamin K. Vitam. Horm. 24:619–632.

Griminger, P. 1966. Biological activity of the various vitamin K forms. Vitam. Horm. 24:605–618.

Hadler, M. R. 1975. A weapon against the resistant rat. Pesticides 9:63–65.

Hadler, M. R., and R. S. Shadbolt. 1975. Novel 4-hydroxycoumarin anticoagulants active against resistant rats. Nature (London) 253:275–277.

Hadler, M. R., R. Redfern, and F. P. Rowe. 1975. Laboratory evaluation of difenacoum as a rodenticide. J. Hyg. 74:441–448.

Hartgrove, R. W., and R. E. Webb. 1973. The development of benzpyrene hydroxylase activity in endrin susceptible and resistant pine mice. Pestic. Biochem. Physiol. 3:61–65.

Hermodson, M. A., J. W. Suttie, and K. P. Link. 1969. Warfarin metabolism and vitamin K requirement in the warfarin resistant rat. Am. J. Physiol. 217:1316–1319.

Hildebrandt, E. F., and J. W. Suttie. 1982. Mechanism of coumarin action: Sensitivity of vitamin K metabolizing enzymes of normal and warfarin-resistant rat liver. Biochemistry 21:2406–2411.

Hildebrandt, E. F., P. C. Preusch, J. L. Patterson, and J. W. Suttie. 1984. Solubilization and characterization of vitamin K epoxide reductase from normal and warfarin-resistant rat liver microsomes. Arch. Biochem. Biophys. 228:480–492.

Jackson, W. B., and A. D. Ashton. 1980. Present distribution of anticoagulant resistance in the United States. Pp. 392–397 *in* Vitamin K Metabolism and Vitamin K-dependent Proteins, J. Suttie, ed. Baltimore, Md.: University Park Press.

Jackson, W. B., and D. Kaukeinen. 1972. Resistance of wild Norway rats in North Carolina to warfarin rodenticide. Science 176:1343–1344.

Larson, A. E., P. A. Friedman, and J. W. Suttie. 1981. Vitamin K-dependent carboxylase: stoichiometry of carboxylation and vitamin K 2,3-epoxide formation. J. Biol. Chem. 256:11032–11035.

Leck, J. B., and B. K. Park. 1981. A comparative study of the effects of warfarin and brodifacoum on the relationship between vitamin K_1 metabolism and clotting factor activity in warfarin-susceptible and warfarin-resistant rats. Biochem. Pharmacol. 30:123–128.

Lehmann, J. 1943. Thrombosis: Treatment and prevention with methylenebis-(hydroxycoumarin). Lancet 1:611–613.

Lind, C., P. Hochstein, and L. Ernster. 1982. DT-diaphorase as a quinone reductase: A cellular control device against semiquinone and superoxide radical formation. Arch. Biochem. Biophys. 216:178–185.

Link, K. P. 1944. The anticoagulant from spoiled sweet clover hay. Harvey Lect. 34:162–216.

Lund, M. 1964. Resistance to warfarin in the common rat. Nature (London) 203:778.

Lund, M. 1981. Comparative effect of the three rodenticides warfarin, difenacoum and brodifacoum on eight rodent species in short feeding periods. J. Hyg. 87:101–107.

Lush, I. E. 1976. A survey of the response of different strains of mice to substances metabolised by microsomal oxidation: hexabarbitone, zoxasolamine and warfarin. Chem. Biol. Interact. 12:363–373.

Lush, I. E., and C. J. Arnold. 1975. High coumarin 7-hydroxylase activity does not protect mice against warfarin. Heredity 35:279–281.

MacNicoll, A. D. 1985. A comparison of warfarin-resistance and liver microsomal vitamin K epoxide reductase activity in rats. Biochim. Biophys. Acta. 840:13–20.

MacNicoll, A. D., A. K. Nadian, and M. G. Townsend. 1984. Inhibition by warfarin of liver microsomal vitamin K-reductase in warfarin-resistant and susceptible rats. Biochem. Pharmacol. 33:1331–1336.

Martin, A. D. 1973. Vitamin K requirement and anticoagulant response in the warfarin resistant rat. Biochem. Soc. Trans. 1:1206–1208.

Martin, A. D., L. C. Steed, R. Redfern, J. E. Gill, and L. W. Huson. 1979. Warfarin resistance genotype determination in the Norway rat *Rattus norvegicus*. Lab. Anim. 13:209–214.

McKee, R. W., S. B. Binkley, D. W. MacCorquodale, S. A. Thayer, and E. A. Doisy. 1939. The isolation of vitamin K_2. J. Biol. Chem. 131:327–344.

Oliver, J. A., D. R. King, and R. J. Mead. 1979. Fluoroacetate tolerance, a genetic marker in some Australian mammals. Aust. J. Zool. 27:363–372.

Ophof, A. J., and D. W. Langveld. 1969. Warfarin-resistance in the Netherlands. Schriften. Ver. Wasser-, Boden-, Lufthyg. Berlin-Dahlem 32:39–47.

O'Reilly, R. A., P. M. Aggeler, M. S. Hoag, L. S. Leong, and M. Kropatkin. 1964. Hereditary resistance to coumarin anticoagulant drugs: The first reported kindred. Clin. Res. 12:218.

Ozburn, G. W., and F. O. Morrison. 1962. Development of a DDT-tolerant strain of laboratory mice. Nature (London) 196:1009–1010.

Pool, J. G., R. A. O'Reilly, L. J. Schneiderman, and M. Alexander. 1968. Warfarin resistance in the rat. Am. J. Physiol. 215:627.

Preusch, P. C., and J. W. Suttie. 1984. Formation of 3-hydroxy-2-3-dihydrovitamin K_1 in vivo: Relationship to vitamin K epoxide reductase. J. Nutr. 114:902–910.

Redfern, R., and J. E. Gill. 1978. The development and use of a test to identify resistance to the anticoagulant difenacoum in the Norway rat (*Rattus norvegicus*). J. Hyg. 81:427–431.

Redfern, R., and J. E. Gill. 1980. Laboratory evaluation of bromadiolone as a rodenticide for use against warfarin-resistant and non-resistant rats and mice. J. Hyg. 84:263–268.

Redfern, R., J. E. Gill, and M. R. Hadler. 1976. Laboratory evaluation of WBA 8119 as a rodenticide for use against warfarin-resistant and non-resistant rats and mice. J. Hyg. 77:419–426.

Ren, P., R. E. Laliberte, and R. G. Bell. 1974. Effect of warfarin, phenylindanedione and tetrachloropyridinol in normal and warfarin-resistant rats. Mol. Pharmacol. 10:373–380.

Ren, P., P. Y. Stark, R. L. Johnson, and R. G. Bell. 1977. Mechanism of action of anticoagulants: Correlation between the inhibition of prothrombin synthesis and the regeneration of vitamin K_1 from vitamin K_1 epoxide. J. Pharmacol. Exp. Ther. 201:541–546.

Rennison, B. D., and A. C. Dubock. 1978. Field trials of WBA 8119 (PA 581, brodifacoum) against warfarin-resistant infestations of *Rattus norvegicus*. J. Hyg. 80:77–82.

Richards, C. G. J. 1981. Field trials of bromadiolone against infestations of warfarin-resistant *Rattus norvegicus*. J. Hyg. 86:363–367.

Roll, R. 1966. Uber die Wirkung eines Cumarinpraparates (Warfarin) auf Hausemouse (*Mus musculus* L.). Z. Angewante Zool. 53:277–349.

Rowe, F. P., and R. Redfern. 1965. Toxicity tests on suspected warfarin-resistant house mice (*Mus musculus* L.). J. Hyg. 63:417–425.

Rowe, F. P., and R. Redfern. 1968. The effect of warfarin on plasma clotting time in wild house mice (*Mus musculus*). J. Hyg. 66:159–174.

Rowe, F. P., C. J. Plant, and A. Bradfield. 1981. Trials of the anticoagulant rodenticides bromadiolone and difenacoum against the house mouse (*Mus musculus* L.). J. Hyg. 87:171–177.

Sadowski, J. A., C. T. Esmon, and J. W. Suttie. 1980. Vitamin K-dependent carboxylase: Requirements of the rat liver microsomal enzyme system. J. Biol. Chem. 251:2770–2776.

Saunders, G. R. 1978. Resistance to warfarin in the roof rat in Sydney, N.S.W. Search 9:39–40.

Shapiro, S. 1953. Warfarin sodium derivative (coumadin sodium): Intravenous hypoprothrombinaemia-inducing agent. Angiology 4:380–390.

Shearer, M. J., A. McBurney, and P. Barkhan. 1974. Studies on the absorption and metabolism of phylloquinone (vitamin K_1) in man. Vitam. Horm. 32:513–542.

Stenflo, J., P. Fernlund, W. Egan, and P. Roepstorff. 1974. Vitamin K-dependent modifications of glutamic acid residues in prothrombin. Proc. Natl. Acad. Sci. 71:2730–2733.

Talcott, R. E., M. Rosenblum, and V. A. Levin. 1983. Possible role of DT-diaphorase in the bioactivation of antitumour quinones. Biochem. Biophys. Res. Commun. 111:346–351.

Telle, H. J. 1967. Die Auswahl von Rodentiziden für die Rattenvertilgungen und für die Beibehaltung eines rattenfreien Zustandes. Anz. Schaedlingskd. Pflanzenschutz 40:161–166.

Tishler, M., and W. L. Sampson. 1948. Isolation of vitamin K_2 from cultures of a spore-forming bacillus. Proc. Soc. Exp. Biol. Med. 68:136–137.

Townsend, M. G., E. M. Odam, and J. M. J. Page. 1975. Studies on the microsomal drug metabolism system in warfarin-resistant and susceptible rats. Biochem. Pharmacol. 24:729–735.

Vinson, S. B., C. E. Boyd, and D. E. Ferguson. 1963. Resistance to DDT in the mosquito fish, *Gambusia affinis*. Science 139:217–218.

Wallace, M. E., and F. J. MacSwiney. 1976. A major gene controlling warfarin resistance in the house mouse. J. Hyg. 76:173–181.

Webb, R. E., and F. Horsfall. 1967. Endrin resistance in the pine mouse. Science 156:1762.

Webb, R. E., W. C. Randolph, and F. Horsfall. 1972. Hepatic benzyprene hydroxylase activity in endrin susceptible and resistant pine mice. Life Sci. 11:477–483.

Whitlon, D. S., J. A. Sadowski, and J. W. Suttie. 1978. Mechanism of coumarin action: Significance of vitamin K epoxide reductase inhibition. Biochemistry 17:1371–1377.

Wood, A. W., and A. H. Conney. 1974. Genetic variation in coumarin hydroxylase activity in the mouse (*Mus musculus*). Science 185:612–613.

Pesticide Resistance: Strategies and Tactics for Management.
1986. National Academy Press, Washington, D.C.

Plant Pathogens

S. G. GEORGOPOULOS

Heritable variation for sensitivity to many of the protectant fungicides has not been demonstrated in plant pathogenic fungi, and the effectiveness of these chemicals has not changed. The remaining protectants, together with the systemics, can be classified into two groups, depending on whether resistance is controlled by a major gene or a number of interacting genes. Field populations in the former give a bimodal and in the latter a unimodal distribution for sensitivity.

Resistance to benzimidazoles, carboxamides, acylalanines, and the protein synthesis inhibitors develops by modification of the sensitive site. Changes in membrane transport systems have been shown responsible for resistance to polyoxins and the inhibitors of ergosterol biosynthesis. Finally, resistance to dihydrostreptomycin and to pyrazophos may result from a change in the ability to metabolize the chemical.

INTRODUCTION

The main causes of infectious plant diseases are fungi, bacteria, and viruses. At present, effective antiviral agents to control plant viruses in agriculture are not available. Current chemical control of plant pathogenic bacteria and other prokaryotes is based only on copper and the antibiotics streptomycin and oxytetracycline (Jones, 1982). A large variety of chemicals, however, are available against fungi. My discussion will deal mainly with resistance to fungicides, although resistance in bacteria will be mentioned. (For discussion on preventing and managing resistance, see Dekker in this volume.)

Earlier treatments of the subject include those of Georgopoulos (1977; 1982) and Dekker (1985).

Fungi are eukaryotic organisms with well-defined nuclei, each bounded by an envelope that remains intact during mitosis. The vegetative pathogenic phase of most fungi is characterized by haploid nuclei, with the exception of the members of Oomycetes, in which meiosis takes place in the oogonia and the antheridia, so that the organism is diploid throughout the asexual stages of its life cycle (Fincham et al., 1979). In haploid fungi, resistance mutations are subject to immediate selection because they may not be shielded by dominance. Complications arise, however, because many fungi can carry two or more genetically unlike nuclei in a common cytoplasm. In the Ascomycetes, this condition, known as heterokaryosis, often permits changes in the proportions of different nuclei in response to selection (Davis, 1966). By contrast the heterothallic Basidiomycetes are characterized by a stable dikaryon, with each cell containing two nuclei. The dikaryon is genetically equivalent to a diploid, but is more flexible. In heterothallic species each cell of the dikaryon contains two nuclei of different mating type. Bacteria as well as mycoplasmal- and rickettsial-like plant pathogens do not contain typical nuclei. The genetic information in a bacterium is contained in the chromosome and in a variable number of plasmids, which carry genes for their own replication in bacterial host cells and for their transmissibility from cell to cell (and often also genes conferring a new phenotype on their hosts). Most of the antibiotic resistance found in bacteria that cause disease in humans and animals is plasmid determined (Datta, 1984).

GENETIC CONTROL OF RESISTANCE

Fungi are highly variable and adaptable organisms. Plant breeders are particularly conscious of this in their attempts to achieve disease control by developing resistant varieties of crop plants. The ability of fungi to render fungicides ineffective varies greatly, however, depending mainly on the fungicide (Georgopoulos, 1984).

Appropriate Variability Apparently Unavailable

The effectiveness of most protectant agricultural fungicides has remained unchanged after decades of use. Mutational modification of fungal sensitivity to practically any of these fungicides has not been demonstrated in the laboratory. The variability required to break down the effectiveness of these chemicals apparently is unavailable to the target fungi. The multisite activity of most protectant fungicides is undoubtedly important but is not always sufficient to explain the inability for resistance to develop.

Copper fungicides, for example, have been used for 100 years against several of the major fungal pathogens of plants, with no decline in their effectiveness or isolation of resistant mutants of plant pathogenic fungi. Yet mechanisms for copper resistance do exist. In several species of higher plants, tolerance of high concentrations of copper can be achieved by mutations of chromosomal genes (Bradshaw, 1984). In the yeast *Saccharomyces cerevisiae*, copper resistance of naturally occurring resistant strains is mediated by a single gene. Sensitive strains cannot grow on media containing 0.3 mM $CuSO_4$, while resistant mutants are not inhibited at concentrations up to 1.75 mM. Enhanced resistance levels, up to 12.0 mM $CuSO_4$, reflect gene amplification (Fogel and Welch, 1982).

Unlike fungi, bacterial plant pathogens have evolved copper resistance. In *Xanthomonas campestris* pv. *vesicatoria*, copper-resistant isolates exist in nature and are not controlled by the amount of Cu^{++} available from fixed copper fungicides. The genetic determinant of this resistance is located in a conjugative plasmid. A gene for avirulence (inducing a hypersensitive response) to certain lines of pepper is located on the same plasmid (Stall et al., 1984).

Copper fungicides probably have retained the same effectiveness in controlling plant pathogenic fungi, because the genes conferring resistance to copper are not available to these fungi. Similarly, no genes substantially affect the sensitivity of fungi to sulfur, dithiocarbamates, phthalimides, quinones, chlorothalonil, or any of a few other, less important protectant fungicides. Mutants with well-defined resistance to any fungicide of this group have never been obtained. Variations in sensitivity seem to be neither heritable nor of considerable importance in practice.

One-Step Pattern

In some of the specifically acting systemic fungicides, one-step major changes in sensitivity of plant pathogenic fungi are obtained with single-gene mutations. One mutation is sufficient to achieve the highest level of resistance possible. If more loci control sensitivity a mutant allele at one locus is epistatic over wild-type alleles at other loci. All sensitive fungi appear to have the genes required for major, one-step changes in sensitivity to fungicides of this group. In nature, sensitive and resistant populations are distinct, and controlling resistant populations by increasing the dose rate of the fungicides or shortening the spray interval is not possible. Such complete loss of effectiveness has not been experienced with fungicides where development of resistance does not follow this one-step pattern.

The best known examples of this type of genetic control of sensitivity have been provided by studies on the benzimidazole fungicides, introduced in 1968. At least 50 species of fungi have developed resistance to benzimidazoles; all attempts to obtain resistance to these fungicides in any sensitive species have succeeded. In some fungi, for example, *Aspergillus nidulans*, in addition to the locus for high resistance to benzimidazoles, a few other loci may be involved in smaller decreases of sensitivity. Mutant genes at different loci, however, do not interact, and a stepwise increase of resistance does not occur (Hastie and Georgopoulos, 1971). In other species, for example, *Venturia inaequalis*, polymorphism in a single gene causes different resistance levels, and a second locus does not seem to be involved in sensitivity to benzimidazole fungicides (Katan et al., 1983).

Similar one-step development of resistance in fungi has been recognized with several other systemics and with the aromatic hydrocarbon and dicarboximide group, most of which do not show systemic activity. Major genes have been identified for carboxamides (Georgopoulos and Ziogas, 1977), kasugamycin (Taga et al., 1979), and aromatic hydrocarbons and dicarboximides (Georgopoulos and Panopoulos, 1966). Similar genes are undoubtedly involved in the development of resistance to acylalanines and to polyoxin. Although genetic studies have not demonstrated this yet, the bimodal sensitivity distribution found in field populations indicates a one-step change. As with benzimidazoles, resistance can make any of these fungicides ineffective. In practice this does not always happen, where the mutant gene adversely affects fitness (Georgopoulos, in press). Development of resistance to streptomycin, mediated either by chromosomal or plasmid-borne genes, also appears to follow the same one-step pattern (Schroth et al., 1979; Yano et al., 1979).

Multistep Pattern

The genetic control of resistance to a third category of fungicides is more complicated. Single gene mutations may have measurable effects on the phenotype, although they are generally small. High level resistance requires positive interaction between mutant genes and is acquired in a multistep fashion, for example, to dodine (Kappas and Georgopoulos, 1970) and to the ergosterol biosynthesis inhibitors (van Tuyl, 1977). The involvement of several resistance genes and of modifiers maintains a unimodal sensitivity distribution in field populations even after many exposures. Mean sensitivity may gradually decrease, but effectiveness is not completely lost and an increase in fungicide dosage improves disease control (Georgopoulos, in press). The most resistant members of field populations cannot become predominant, because the required accumulation of several resistance genes apparently affects fitness.

Similar selection of less sensitive forms and some decrease in effectiveness with time has been noticed with the 2-aminopyrimidine fungicides, fentin, and the phosphorothiolates. Differences in sensitivity to these fungicides, which are found in nature, either have not been studied genetically or cannot

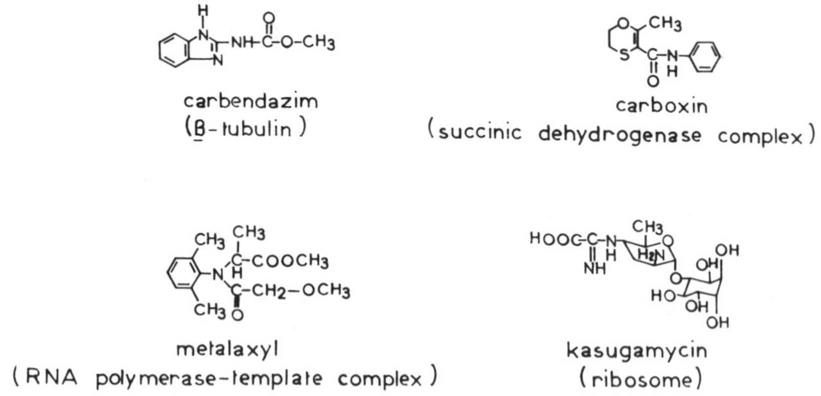

carbendazim
(β-tubulin)

carboxin
(succinic dehydrogenase complex)

metalaxyl
(RNA polymerase-template complex)

kasugamycin
(ribosome)

FIGURE 1 Structural formulas of fungicides to which resistance develops by modification of the sensitive site (indicated in parentheses).

be attributed to specific genes (Hollomon, 1981). Variation within populations however, is continuous, and no discrete classes can be distinguished for sensitivity, excluding the possibility of involvement of major genes. Resistance to the above fungicides probably develops in a stepwise manner.

RESISTANCE MECHANISMS

The few biochemical studies on fungicide resistance indicate that resistance mutations either modify the sensitive site or the membrane transport systems involved in influx and efflux of the fungicidal molecule, or they affect the ability for toxification or detoxification. Examples illustrating the operation of these mechanisms follow.

Modification of Sensitive Site

The benzimidazole fungicides, such as carbendazim (Figure 1), inhibit mitotic division by preventing tubulin polymerization. In the nonpathogen *Aspergillus nidulans*, a major gene for resistance to these fungicides codes for β-tubulin, one of the subunits of the tubulin molecule. Mutational modifications of this subunit can be recognized electrophoretically and by the tubulin's ability to bind benzimidazole fungicides (this ability is inversely correlated to resistance) (Davidse, 1982). The genes for carbendazim resistance and for carbendazim extra-sensitivity are allelic and are 16 nucleotides apart (van Tuyl, 1977). Tubulin modifications that lower affinities for benzimidazole fungicides increase affinity for N-phenyl carbamate compounds, some of which possess antimitotic activity in higher plants (Kato et al.,

1984). Other modifications, however, may cause resistance to benzimidazoles and to N-phenyl carbamates.

The carboxamide fungicides, such as carboxin (Figure 1), inhibit respiration by preventing the transport of electrons from succinate to coenzyme Q. In the corn smut pathogen, *Ustilago maydis*, two allelic mutations modify the succinic dehydrogenase complex (SDC, succinate-CoQ oxidoreductase), resulting in moderate and high resistance of mitochondrial respiration to carboxin and to most carboxamides (Georgopoulos and Ziogas, 1977). Some specific structural groups of carboxamides, however, are selectively active against one or the other type of mutated SDC (White and Thorn, 1980). Apparently the gene controlling resistance codes for a component of SDC and, when it mutates, the component's affinity for a given carboxamide increases or decreases, depending on the structure and on the mutation. The binding site of carboxin in animal mitochondria is formed by two small peptides, C_{II-3} and C_{II-4} (Ramsey et al., 1981).

The acylalanines, such as metalaxyl (Figure 1), are fungicides selectively active against Oomycete fungi. These fungicides inhibit RNA synthesis by interfering with the activity of a nuclear, α-amanitin-insensitive RNA polymerase-template complex. Nuclei isolated from a metalaxyl-sensitive strain of the pathogenic *Phytophthora megasperma* f. sp. *medicaginis* contained RNA polymerase activity that could be partially inhibited by metalaxyl. By contrast, nuclei isolated from a resistant strain did not contain metalaxyl-sensitive polymerase activity (Davidse, 1984). Resistance, therefore, results from mutational change of one of the RNA polymerases.

Many antifungal antibiotics act on protein synthesis (Siegel, 1977), but most are not used to control plant diseases. Cycloheximide binds to the 60-S ribosomal subunit and inhibits the transfer of amino acids from aminoacyl tRNA to the polypeptide chain, preventing also the movement of ribosomes along the mRNA. In the nonpathogen *Neurospora crassa*, modifications of protein components of the 60-S subunit create cycloheximide resistance. Single gene-controlled configurational changes of the ribosomes appear to not interfere with normal ribosome functioning. In double mutants, however, where positive interactions result in higher cycloheximide resistance, the presence of two mutant ribosomal components disturbs vital functions of the ribosomes (Vomvoyanni and Argyrakis, 1979).

Kasugamycin (Figure 1) is more important than cycloheximide in plant disease control, particularly against the rice blast pathogen, *Pyricularia oryzae*. This antibiotic inhibits protein synthesis in both 80-S and 70-S ribosomes. Kasugamycin interacts with the 30-S subunit of ribosomes from sensitive strains, but it does not bind to ribosomes from resistant strains of bacteria. Resistance mutations either inactivate an RNA methylase or alter a ribosomal protein (Cundliffe, 1980). In *P. oryzae*, kasugamycin inhibits protein synthesis, probably by preventing the binding of aminoacyl-tRNA to the ribosome. In a cell-free system with ribosomes from a resistant mutant, protein synthesis is not inhibited, indicating that mutations modify some component of the ribosome (Misato and Ko, 1975).

Membrane Transport Systems

Polyoxins, for example, polyoxin D (Figure 2), block the biosynthesis of chitin, acting as competitive inhibitors for uridine diphosphate-N-acetylglucosamine in the chitin synthesis reaction. The presence of polyoxin leads to a pronounced accumulation of the normal metabolite. In strains of *Alternaria kikuchiana*, a pathogen of Japanese pear, polyoxin sensitivity and chitin synthesis inhibition correlate in vivo but not in vitro, indicating that the site of action of the antibiotic remains equally sensitive. Polyoxin resistance is associated with a very ineffective system for dipeptide uptake. Sensitive strains are capable of high active uptake of polyoxin in media without dipeptides, but not in media containing glycyl-glycine. In contrast, polyoxin uptake is very low in resistant strains, whether dipeptides are present or absent (Hori et al., 1977). Thus, reduced activity of dipeptide permease appears to be responsible for polyoxin resistance.

Resistance to ergosterol biosynthesis-inhibiting fungicides such as fenarimol (Figure 2), however, is not related to fungicide influx, which is passive. In wild-type strains of the nonpathogen *Aspergillus nidulans*, passive fenarimol influx results in considerable accumulation that induces an efflux activity that is energy-dependent. In strains of the same organism carrying a mutation for fenarimol resistance, the efflux activity appears to be constitutive, preventing initial fungicide accumulation within the cells. When efflux

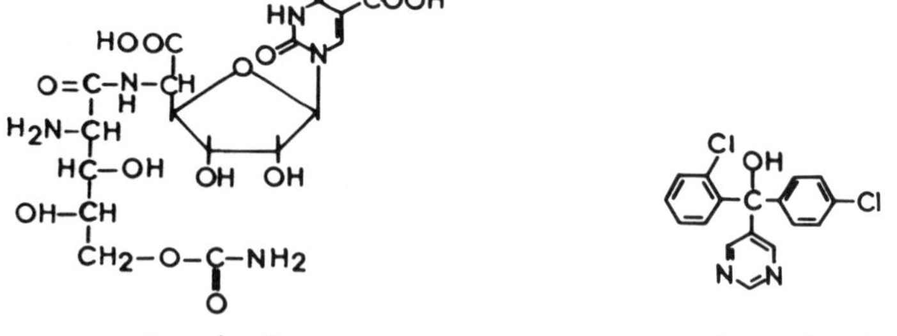

polyoxin D
(chitin biosynthesis)

fenarimol
(ergosterol biosynthesis)

FIGURE 2 Structural formulas of fungicides to which resistance develops by modification of membrane transport systems (mechanism of action indicated in parentheses).

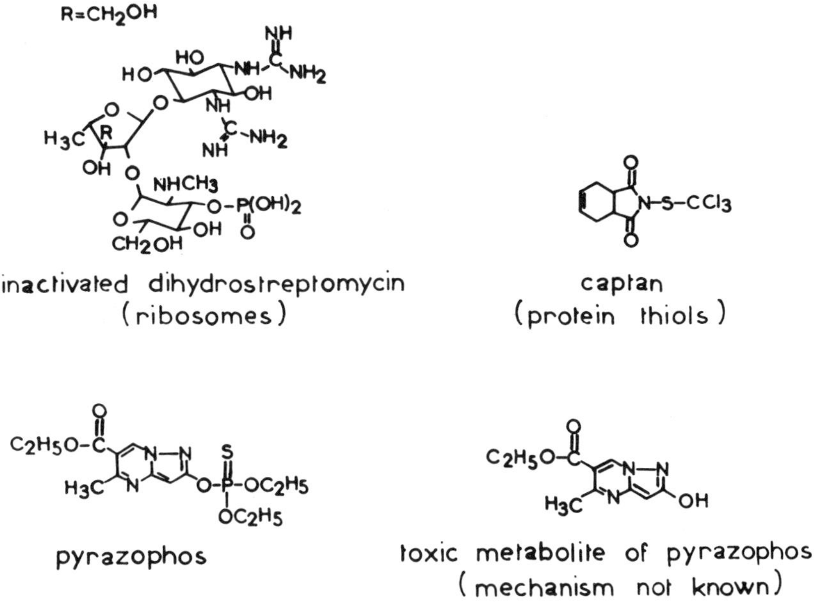

inactivated dihydrostreptomycin (ribosomes)

captan (protein thiols)

pyrazophos

toxic metabolite of pyrazophos (mechanism not known)

FIGURE 3 Structural formulas of dihydrostreptomycin 3′-phosphate, captan, pyrazophos, and the toxic metabolite 2-hydroxy-5-methyl-6-ethoxycarbonylpyrazolo (1-5-a)-pyrimidine (information on the mode of action given in parentheses).

activity is inhibited by respiration or phosphorylation inhibitors, net fungicide uptake by the mutant strains may be as high as that by the wild type. Mutant genes, therefore, affect the efficiency of fungicide excretion from the mycelium (de Waard and Fuchs, 1982).

Detoxification or Nontoxification

Streptomycin resistance in the fireblight pathogen *Erwinia amylovora* is believed to result from a chromosomal mutation modifying the ribosome (Schroth et al., 1979). In *Pseudomonas lachrymans* (the bacterium causing cucumber angular leaf spot), however, resistance to dihydrostreptomycin is plasmid mediated; the antibiotic is detoxified by phosphorylation. From resistant isolates, one can obtain a cell-free system that can inactivate the antibiotic in the presence of ATP. The product of the enzymatic inactivation is dihydrostreptomycin 3′-phosphate (Figure 3). The antibiotic can be regenerated by alkaline phosphatase treatment (Yano et al., 1978b).

A difference in captan sensitivity (Figure 3) between two isolates of *Botrytis cinerea* could be correlated with the rate of synthesis of reduced glutathione in response to the fungicide (Barak and Edgincton, 1984). Increased amounts of nonvital soluble thiolic compounds may inactivate fungicides reacting with thiol, thus preventing the damage to cellular protein thiols. Widespread occurrence of this type of resistance to multisite fungicides, however, has not been reported.

The systemic fungicide pyrazophos (Figure 3) is toxic to fungi that convert it to 2-hydroxy-5-methyl-6-ethoxycarbonylpyrazolo (1-5-a)-pyrimidine (PP) (Figure 3), which has a much broader fungitoxic spectrum than pyrazophos. In *Ustilago maydis*, a fungus incapable of this toxification, mutants with resistance to PP could not be obtained. Pyrazophos resistance in *Pyricularia oryzae* comes from mutational loss of the ability to metabolize the fungicide and to produce the toxic product (de Waard and van Nistelrooy, 1980). Apparently, resistance develops more easily by loss of ability for toxification than by modification of the site(s) of action of the toxic product.

CONCLUSION

Research is greatly needed to increase our understanding of the genetic and biochemical mechanisms of resistance to chemicals used to control plant diseases. Unfortunately methods for such research are either unavailable or time-consuming. At the same time, the study of resistant mutants has contributed considerably to our understanding of the action of several selective antifungal substances and of some basic cellular processes. Although a better knowledge of the genetics and biochemistry of plant pathogenic microorganisms will facilitate future efforts to understand fungicide resistance, scientists must not overweigh present difficulties to achieve their goals.

REFERENCES

Barak, E., and L. V. Edgincton. 1984. Glutathione synthesis in response to captan: A possible mechanism for resistance of *Botrytis cinerea* to the fungicide. Pestic. Biochem. Physiol. 21:412–416.

Bradshaw, A. D. 1984. Adaptation of plants to soils containing toxic metals—a test for conceit. Pp. 4–19 *in* Origins and Development of Adaptation. Ciba Found. Symp. 102. London: Pitman.

Cundliffe, E. 1980. Antibiotics and prokaryotic ribosomes: Action, interaction and resistance. Pp. 555–581 *in* Ribosomes: Structure, Function, and Genetics, G. Chambliss, G. R. Craven, J. Davies, K. Davis, I. Kahan, and M. Nomura, eds. Baltimore, Md.: University Park Press.

Datta, N. 1984. Bacterial resistance to antibiotics. Pp. 204–218 *in* Origins and Development of Adaptation. Ciba Found. Symp. 102. London: Pitman.

Davidse, L. C. 1982. Benzimidazole compounds: Selectivity and resistance. Pp. 60–70 *in* Fungicide Resistance in Crop Protection, J. Dekker and S. G. Georgopoulos, eds. Wageningen, Netherlands: Centre for Agricultural Publishing and Documentation.

Davidse, L. C. 1984. Antifungal activity of acylalanine fungicides and related chloroacetanilide herbicides. Pp. 239–255 *in* Mode of Action of Antifungal Agents, A. P. J. Trinci and J. F. Ryley, eds. Cambridge: British Mycological Society.

Davis, R. H. 1966. Heterokaryosis. Pp. 567–588 *in* The Fungi: An Advanced Treatise, Vol. 2, G. C. Ainsworth and A. S. Sussman, eds. New York: Academic Press.

Dekker, J. 1985. The development of resistance to fungicides. Prog. Pestic. Biochem. Toxicol. 4:165–218.

de Waard, M. A., and A. Fuchs. 1982. Resistance to ergosterol biosynthesis inhibitors II. Genetic and physiological aspects. Pp. 87–100 *in* Fungicide Resistance in Crop Protection, J. Dekker and S. G. Georgopoulos, eds. Wageningen, Netherlands: Centre for Agricultural Publishing and Documentation.

de Waard, M. A., and J. G. M. van Nistelrooy. 1980. Mechanism of resistance to pyrazophos in *Pyricularia oryzae*. Neth. J. Plant Pathol. 86:251–258.

Fincham, J. R. S., P. R. Day, and A. Radford. 1979. Fungal Genetics, 4th ed. Oxford: Blackwell.

Fogel, S., and J. W. Welch. 1982. Tandem gene amplification mediates copper resistance in yeast. Proc. Natl. Acad. Sci. 79:5342–5346.

Georgopoulos, S. G. 1977. Development of fungal resistance to fungicides. Pp. 439–495 *in* Antifungal Compounds, Vol. 2, M. R. Siegel and H. D. Sisler, eds. New York: Marcel Dekker.

Georgopoulos, S. G. 1982. Genetical and biochemical background of fungicide resistance. Pp. 46–52 *in* Fungicide Resistance in Crop Protection, J. Dekker and S. G. Georgopoulos, eds. Wageningen, Netherlands: Centre for Agricultural Publishing and Documentation.

Georgopoulos, S. G. 1984. Adaptation of fungi to fungitoxic compounds. Pp. 190–203 *in* Origins and Development of Adaptation. Ciba Found. Symp. 102. London: Pitman.

Georgopoulos, S. G. In press. The development of fungicide resistance. *in* Populations of Plant Pathogens: Their Dynamics and Genetics, M. S. Wolfe and C. E. Caten, eds. Oxford: Blackwell.

Georgopoulos, S. G., and N. J. Panopoulos. 1966. The relative mutability of the *cnb* loci in *Hypomyces*. Can. J. Genet. Cytol. 8:347–349.

Georgopoulos, S. G., and B. N. Ziogas. 1977. A new class of carboxin resistant mutants of *Ustilago maydis*. Neth. J. Plant Pathol. (Suppl. 1) 83:235–242.

Hastie, A. C., and S. G. Georgopoulos. 1971. Mutational resistance to fungitoxic benzimidazole derivatives in *Aspergillus nidulans*. J. Gen. Microbiol. 67:371–374.

Hollomon, D. W. 1981. Genetic control of ethirimol resistance in a natural population of *Erysiphe graminis* f. sp. *hordei*. Phytopathology 71:536–540.

Hori, M., K. Kakiki, and T. Misato. 1977. Antagonistic effect of dipeptides on the uptake of polyoxin A by *Alternaria kikuchiana*. J. Pestic. Sci. 2:139–149.

Jones, A. L. 1982. Chemical control of phytopathogenic prokaryotes. Pp. 399–413 *in* Phytopathogenic Prokaryotes, Vol. 2, M. S. Mount and G. H. Lacy, eds. New York: Academic Press.

Kappas, A., and S. G. Georgopoulos. 1970. Genetic analysis of dodine resistance in *Nectria haematococca*. Genetics 66:617–622.

Katan, T., E. Shabi, and J. D. Gilpatrick. 1983. Genetics of resistance to benomyl in *Venturia inaequalis* isolates from Israel and New York. Phytopathology 73:600–603.

Kato, T., K. Suzuki, J. Takahashi, and K. Kamoshita. 1984. Negatively correlated cross-resistance between benzimidazole fungicides and methyl N-(3,5-dichlorophenyl) carbamate. J. Pestic. Sci. 9:489–495.

Misato, T., and K. Ko. 1975. The development of resistance to agricultural antibiotics. Environmental Qual. Sa. Suppl. 3:437–440.

Ramsay, R. R., B. A. C. Ackrell, C. J. Coles, T. P. Singer, G. A. White, and G. D. Thorn. 1981. Reaction site of carboxanilides and of thenoyltrifluoroacetone in Complex II. Proc. Natl. Acad. Sci. 78:825–828.

Schroth, M. N., S. V. Thompson, and W. J. Moller. 1979. Streptomycin resistance in *Erwinia amylovora*. Phytopathology 69:565–568.

Siegel, M. R. 1977. Effect of fungicides on protein synthesis. Pp. 399–438 *in* Antifungal Compounds, Vol. 2, M. R. Siegel and H. D. Sisler, eds. New York: Marcel Dekker.

Stall, R. E., D. C. Loshke, and R. W. Rice. 1984. Conjugational transfer of copper resistance and avirulence to pepper within strains of *Xanthomonas campestris* pv. *vesicatoria*. Phytopathology 74:797. (Abstr.)

Taga, M., H. Nakagawa, M. Tsuda, and A. Ueyama. 1979. Identification of three different loci controlling kasugamycin resistance in *Pyricularia oryzae*. Phytopathology 69:463–466.

van Tuyl, J. M. 1977. Genetics of fungal resistance to systemic fungicides. Meded. Landbouwhogesch. Wageningen Ser. 77–2.

Vomvoyanni, V. E., and M. P. Argyrakis. 1979. Pleiotropic effects of ribosomal mutations for cycloheximide resistance in a double-resistant homocaryon of *Neurospora crassa*. J. Bacteriol. 139:620–624.

White, G. A., and G. D. Thorn. 1980. Thiophene carboxamide fungicides: Structure activity relationships with the succinate dehydrogenase complex from wild-type and carboxin-resistant mutant strains of *Ustilago maydis*. Pestic. Biochem. Physiol. 14:26–40.

Yano, H., H. Fujii, and H. Mukoo. 1978a. Drug-resistance of cucumber angular leaf spot bacterium, *Pseudomonas lachrymans* (Smith et Bryan) Carsner. Ann. Phytopathol. Soc. Jpn. 44:334–336.

Yano, H., H. Fujii, H. Mukoo, M. Shimura, T. Watanabe, and Y. Sekizawa. 1978b. On the enzymic inactivation of dihydrostreptomycin by *Pseudomonas lachrymans*, cucumber angular leaf spot bacterium: Isolation and structural resolution of the inactivated product. Ann. Phytopathol. Soc. Jpn. 44:413–419.

Pesticide Resistance: Strategies and Tactics for Management.
1986. National Academy Press, Washington, D.C.

Chemical Strategies for Resistance Management

BRUCE D. HAMMOCK and DAVID M. SODERLUND

The possible roles of chemical and biochemical research in alleviating the problems caused by pesticide resistance are explored. Pesticides play a central role in current and future crop protection strategies, and there is a need for the continued discovery of new compounds. Constraints, both real and perceived, have limited the discovery and development of new compounds by the agrochemical industry. Industry has responded to these constraints in a variety of ways. Several areas of research must be emphasized if chemical approaches are to have significant impact on the management of resistance. Administrative changes also might foster increased research activity in these areas or might increase the probability that novel approaches will be developed by the agrochemical industry or otherwise be made available for use in integrated pest-management programs.

INTRODUCTION

The Critical Role of Insecticides in Insect Control

The overuse and misuse of insecticides[1] have caused target pest resurgence, secondary pest outbreaks, and environmental contamination (Metcalf and McKelvey, 1976). Nevertheless, it is difficult to foresee how insect pests can be controlled effectively without chemical intervention. Highly produc-

[1]We use the term insecticide in its broadest meaning as any foreign ingredient introduced to control insects.

tive agricultural practices and the high density of human population have been achieved at the expense of ecological balance. To maintain this imbalance in our favor, we must continue to use ecologically disruptive tools, including insecticides. Even novel pest-control strategies such as pest-resistant plant cultivars will not eliminate the need for chemical pest control. Given the choice of a more expensive and pest-infested food supply or pesticide use, we will continue to use pesticides (Boyce, 1976; Krieger, 1982; Ruttan, 1982; Mellor and Adams, 1984). Therefore, the chemicals available for insect control must lend themselves to rational and environmentally sound use.

Integration of Chemical and Nonchemical Control Tactics

During the past two decades the concept of the judicious use of pesticides has been formalized in integrated pest management (IPM). A key strategy of IPM is to use insecticides only when damage is likely to exceed clearly defined economic thresholds. Such procedures constitute the most fundamental approach to resistance management by minimizing the selection pressure leading to resistance. Reduced pesticide use not only decreases selection pressure on pest insects but preserves natural enemies and other nontarget species, reduces environmental contamination, reduces the exposure of farm workers and consumers to potentially toxic materials, and may reduce phytotoxicity. Thus, IPM increases agricultural profitability, improves public health, and reduces environmental contamination. Most IPM programs consider pesticides as nonrenewable resources and stress their judicious use. The limited availability of compounds that are compatible with IPM may restrict the broad application of this approach.

The Need for New Insecticides

Effective insect control requires not only the continued use of existing insecticides but also the continued availability of new insecticides. Existing compounds will probably continue to vanish from the market because of problems with human or environmental safety. Compounds that survive these challenges may still be lost, owing to the development of resistance. Other compounds, although technically still available, may become obsolete as a result of changing agricultural practices or may be replaced by compounds that offer a greater profit margin to the user.

Of these new agricultural practices, the one having the greatest impact on pesticide use patterns is likely to be low-till (or conservation-till) agriculture. Adoption of this practice will be encouraged by the lower costs resulting from reductions in energy consumption, erosion, and loss of tilth (Lepkowski, 1982; Hinkle, 1983). Since tillage is a major means of pest control, this practice will change pesticide use patterns and increase pesticide usage. Without suitable compounds, low-till agriculture will probably increase environmental and resistance problems.

The potential for loss of effective compounds to resistance has provided impetus for formulating resistance management strategies. The effective management of pesticide resistance, however, involves not only the judicious use of existing compounds but also the discovery and development of new chemical control agents. No management strategy can prolong the useful life of pesticides indefinitely. New chemical tools will be needed, particularly those that exploit new biochemical targets. Thus, rather than removing us from a "pesticide treadmill," IPM and resistance management will only slow the treadmill, thereby extending the usefulness of available chemicals.

Integrated pest management also requires new insecticides. That IPM programs use existing compounds is a credit to the skills of agricultural entomologists, because few if any of these compounds were developed for IPM. At best they are marginally compatible with IPM programs.

TRENDS IN INSECTICIDE DISCOVERY AND DEVELOPMENT

The Declining Rate of Insecticide Development

Although new and better insecticides are needed, there are fewer insecticides on the market, fewer compounds being developed, and fewer companies searching for novel compounds than a decade ago. A number of reasons for this decline have been proposed (Metcalf, 1980). The following four constraints are of particular concern.

Increased Cost of Discovery The cost of discovering new insecticides has increased dramatically. First, the cost of synthesis of new compounds for evaluation has increased because most of the simple molecules have been made and multistep, expensive syntheses are now required. Second, the discovery of highly potent groups of compounds, such as the pyrethroid insecticides and sulfonylurea herbicides, has raised the standards of comparison for new compounds. Levels of insecticidal activity that seemed highly competitive a decade ago are no longer competitive, particularly if the chemistry involved is complex. Third, the abandonment of complete dependence on random screening requires a commitment to the rational discovery and optimization of insecticidal activity. Such a commitment requires more sophisticated, and hence more expensive, biological assays.

Increased Costs of Registration The costs of registration can be reduced. Long-term toxicology testing accounts for most of the registration costs. Despite their imperfections these studies are essential to ensure that insecticide-related hazards are identified and minimized. The development of short-term assays may reduce registration costs, but the Environmental Protection Agency (EPA) generally requires new short-term assays while continuing to require the major long-term toxicology studies. In the absence of regulatory requirements, insecticide manufacturers would still conduct many of these studies to protect themselves against unanticipated adverse effects. Administrative delays and apparently capricious policy shifts also increase costs and stifle the development of new compounds.

Increased Costs of Production Increased chemical complexity increases production costs. Recently introduced compounds require expensive starting materials, multistep syntheses, isomer separations, and sometimes the preparative resolution of optical isomers. These costs are also indirectly increased by the costs of energy and petroleum-based feedstocks, transportation, and more stringent regulations regarding worker safety and chemical waste disposal. Although high production costs increase the level of profitability required of a product, they are not the most serious barrier to development. When a company has a promising product, careful market evaluations provide data needed to support decisions regarding capital investment. Continued improvements in production technology alone are unlikely to have a major impact on the rate at which new compounds are made available for use.

Increased Competition The market for agrochemicals is mature and diversified, and growth in most product areas is less than 5 percent per year (Storck, 1984). Most major insecticide markets are divided among several similar products. This competition increases the requirements for developing a successful compound.

Relative Importance of Problems Limiting Development of New Compounds

The four factors interact synergistically to make the development of insecticides unattractive despite the promise of one of the highest profit margins in the chemical industry. Agricultural chemical companies often emphasize the costs of production and registration as the major roadblocks to developing new compounds. Although high, these costs are not the only barriers to development. The cost and risk involved in the discovery process are significant and often unrecognized impediments. Discovery requires a large long-term investment that is separated by years or even decades from ultimate profit. Moreover, it can be addressed most readily by changes in policy.

Current Strategies and Approaches in the Agrochemical Industry

Industry has adopted several conservative strategies to minimize risk. The most drastic has been withdrawal from the agrochemical field. As some of these companies retire from the marketplace, society loses tremendous expertise in the development of pest control agents. This also reduces the diversity of chemicals that will become available, a diversity that is essential if IPM is to be a sophisticated management strategy rather than simply an exercise in timing insecticide applications.

A second strategy is for a company to emphasize its expertise in development or marketing while leaving the high risks involved in actual discovery to other firms (i.e., licensing compounds that have been discovered and patented by other companies). Thus, fewer organizations have the responsibility for new compound discovery. A related approach is to de-emphasize insecticide development and to emphasize development of materials such as herbicides that are perceived to be less risky or less expensive to register. For example, some of the explosive growth of industrial research in agricultural biotechnology has been at the expense of research on crop chemicals.

A third strategy involves increasing a product's market life. Petitions to register tank mixtures and combinations of existing pesticides are increasing. Use of mixtures or combinations may result in less environmental contamination—a new approach in resistance management—or may lead to the development of new classes of pesticides. The toxicological and environmental effects of such combinations, however, may include phenomena not predicted from studies on the individual components; therefore, these should be closely scrutinized.

A second example of this strategy is the patenting and development of derivatives of existing compounds. Many of these derivatives are "propesticides," which degrade to give an established compound as the active ingredient. Such derivatives may improve safety or environmental behavior.

The major advantage of these approaches is that industry can capitalize on its investment in a mature product without the high risks inherent in new chemistry. Maintaining a mature product on the market has little risk. The profits from an established agricultural chemical can support a great deal of maintenance, and the profits are immediate. When they become uneconomical, they can be dropped quickly without a great loss of invested capital. The extreme measures taken by some companies to maintain cyclodiene insecticides on the market exemplify this approach. Integrated pest management systems keyed to particular chemicals can also contribute to this approach if practitioners of these systems feel that the continued availability of a certain compound is critical.

Product maintenance can also indirectly benefit the development of new compounds. The future development of new compounds becomes more attractive because recovery of development costs can be expected over a longer period.

Companies actively seeking new insecticides have attempted to minimize risk by narrowing the scope of their development efforts. Most new insecticides are developed for one of only two markets: foliar application to cotton or soil application to corn. These two markets are perceived to be sufficiently large and stable so that a company can recover development costs and make a profit during the compound's patent life. Although these compounds may be registered for other uses, they are often forced into secondary uses for which they are not well-suited. This narrow targeting severely limits the diversity of insecticides available for use in pest management.

Companies also avoid risk by emphasizing "me too" chemistry. In this approach a competitor's product is used as a lead to identify related but patentable compounds. This action results in a series of active structures and produces large families of similar pesticides. It diverts resources from the development of novel compounds and may accelerate the development of resistance. Moreover, it does not promote industrial cooperation in resistance management. There is little incentive to preserve susceptibility in pest populations because it also preserves market opportunities for competitors. In contrast, companies that are sole exploiters of a chemical family have a great incentive to preserve their market through resistance management.

CHEMICAL AND BIOCHEMICAL SOLUTIONS TO PROBLEMS CAUSED BY RESISTANCE

Understanding Resistance to Existing Insecticides

Resistance management is based on the belief that rational and informed decisions on insecticide use can be made and that these decisions will prevent, delay, or reverse the development of resistance. To make such decisions, we must know why resistant populations are resistant and know (or estimate) the frequency of resistant genotypes. Resistance management may be very difficult without a comprehensive knowledge of the mechanisms by which insects become resistant.

To date, some resistance mechanisms have been identified: reduced rates of cuticular penetration; enhanced detoxication by elevated levels of monooxygenases, esterases, or glutathione-S-transferases; and intrinsic insensitivity of target sites. Knowing these mechanisms exist, however, is not enough on which to base resistance management decisions. Simple, rapid biochemical assays to detect the presence of these mechanisms in individual insects must be developed.

With such assays resistance mechanisms in field populations can be characterized and the relative abundance of resistant and susceptible individuals in a population can be determined. This information will benefit IPM systems and programs of resistance management. Sometimes the assays will be able to distinguish between heterozygous and homozygous individuals or determine the extent of gene amplification in resistant individuals.

Assays may be developed simply on the basis of a correlation between resistance and an observed phenotype, such as the presence of a particular isozyme. Advances in immunochemical technology are such that it may be possible to identify antigens present in a resistant population, but not a susceptible population. Although they are expedient, methods of detection based on fortuitous correlation rather than the measurement of actual resistance mechanisms may be misleading and must be used with great care even when based on hybridoma technology. Techniques such as internal imaging with monoclonal antibodies may help to explain resistance phenomena.

Research resources must focus on the developing biochemical diagnostic procedures. For enhanced detoxication the challenge is simply to develop microanalytical techniques to determine the level of activity of enzymes of interest in individual insects. Simple microassays can also be developed for one major type of intrinsic insensitivity, such as the altered cholinesterase involved in organophosphate and carbamate resistance. For some mechanisms of resistance, additional fundamental research is needed before diagnostic assays can be devised. An important example is nerve insensitivity resistance to DDT and pyrethroids. Although this type of resistance is well documented in a few species and is suspected in many others, there is no way at present to detect this resistance through diagnostic assays. Behavioral mechanisms may contribute significantly to some resistance. Ultimately, behavioral resistance must have a physiological basis, but it is likely to be even more difficult to find reliable markers for such resistance mechanisms (Lockwood et al., 1984). For these areas the development of diagnostic antigens may be expedient and may even help to discover the true resistance mechanism.

Diagnostic assays such as those outlined are extremely useful in identifying and characterizing resistance that results from a single mechanism. A potentially more serious problem involves the synergistic interaction of two or more mechanisms. To evaluate the underlying causes of polygenic resistance, we must conduct more studies of the distribution and fate of insecticides in both resistant and susceptible individuals. These pharmacokinetic studies have barely been exploited in insects, yet they are essential for us to understand how specific genetic changes act and interact to modify the availability and persistence of insecticides at their sites of action in living insects.

We also must study the metabolism and mechanism of action of insecticides in insect species important in agriculture, animal health, and medicine before resistance develops. Knowledge of sites of action and critical pathways of detoxication is essential when devising strategies to impede the development of resistance to a particular compound in a particular control system. The use of insect strains that are either resistant or susceptible to related insecticides or to other widely used insecticides can enhance the predictive value of these studies. Similarly, to identify potential resistance mechanisms, these studies must use insect species that exhibit natural tolerance.

Clearly, we need to expand the research base for rational strategies of resistance management. We must support and pursue research ranging from analytical biochemistry to insecticide neuropharmacology. These approaches are a necessary adjunct to more familiar experimental approaches if the rapid detection, characterization, and management of insecticide resistance is to become an integral part of pest management.

Discovering New Insecticides

Approaches to Finding and Optimizing Biological Activity The agrochemical industry is very skilled at optimizing the biological activity of a series of chemicals (Magee, 1983; Menn, 1983). Recent technological advances, many of which have been adopted by industrial research laboratories, are certain to refine and enhance this expertise. The use of linear free-energy parameters to establish quantitative structure/activity relationships has proved very effective in optimizing activity in some series. As computer time becomes less expensive, graphics capability more sophisticated, instruments easier to use, and software more powerful, these approaches will become even more useful.

Computer-assisted design in biochemistry, analogous to procedures already used in architecture, is becoming more accessible and affordable. These techniques use X-ray crystallographic data to generate three-dimensional images of complex macromolecules. The scientist can then view the structure of a target macromolecule in three dimensions as it interacts with a ligand, inhibitor, or substrate. These tools will be of tremendous benefit in optimizing chemical structures in a rational, cost-effective manner. The creative potential of these tools is of even greater importance, because they are a powerful resource for making logical transformations, not only from one substituent to another but also from a biologically active peptide to something as dissimilar as a synthetic hydrocarbon. In the field of spectroscopy, nuclear magnetic resonance (NMR) technology is evolving rapidly, not only to support structure elucidation but as a tool to probe the active sites of biological molecules and even physiological function in vivo.

The elucidation of enzyme-substrate interactions and enzyme reaction mechanisms has provided new paradigms for the discovery of new compounds. Several laboratories are applying transition-state theory, which describes the mechanisms of enzyme-catalyst reactions, to the design of exceptionally powerful enzyme inhibitors. A related approach involves the design of compounds that interact with enzymes as suicide substrates, which trick the enzyme into self-destructing in the process of catalysis. The proliferation of these sophisticated, targeted approaches depends on the continued growth of fundamental information about enzymes, receptors, and other regulatory macromolecules.

Recent advances in genetic engineering and biotechnology are facilitating basic research on many fronts. For example, the ability to isolate and sequence small quantities of peptides and proteins, to isolate their messages and genes, and to measure them with immunochemical and other tools will provide new leads for using classical chemistry. Moreover, these biological messages may be directly useful in developing microbial pesticides or for enhancing crop resistance to pests. Microbial pesticides may bridge the gap between classical chemical and classical biological control. The current industrial effort to develop avermectins, a group of fungal toxins with high insecticidal activity, illustrates that a very complex molecule can be made by a fermentation process that is competitive with classical industrial chemistry. This concept greatly expands the variety of structural types that might be used commercially for insect control and indicates that rigorous screening of plant and microbial natural products may meet with still further success. The *Bacillus thuringiensis* toxins represent another level of complexity, in which the marketed toxins are proteins (Kirschbaum, 1985). The potential for selectivity among these toxins is very exciting. The *B. thuringiensis* gene can also be expressed in both a crop plant and a plant commensal organism and may herald a new phase in research on plant resistance, in which the insecticide chemical or biochemical is produced by the plant itself or by an associated microorganism.

Advancing biotechnology also offers the prospect of new opportunities for exploiting insect viruses (Miller et al., 1983). These highly selective agents have shown considerable promise for insect control, but their wide use has been limited by difficulties in registration and, more seriously, problems in devising in vitro production systems. Continuing improvements in insect tissue culture may improve the economic feasibility of these materials. It may also be feasible to clone messages into viruses to block a critical physiological process in insects in vivo at very low levels of infection, while still allowing the virus to propagate in vitro.

Research in these areas may drastically alter our concepts of what an insecticide is. The move toward biorational design and genetically engineered biological insecticides or insect pathogens does not mean, however, that the resulting products will be free from the hazards we associate with classical insecticides. These novel materials will still require thorough investigation for their possible toxicological and environmental effects. For pathogens, suitable registration guidelines remain to be established, and answers to the public concern over the release of genetically engineered pathogenic organisms into the environment must be formulated. Resistance to these materials could develop if they are used in ways that lead to high selection pressure.

New Targets for Insecticide Development The four major classes of synthetic organic insecticides developed since 1945 are neurotoxins. Yet, most insecticides act at only two sites in the nervous system. Thus, genetic modifications that change the sensitivity of these sites of action (altered acetylcholinesterase for carbamates and phosphates, nerve insensitivity resistance for DDT and pyrethroids) produce cross-resistance that renders entire classes of compounds ineffective against resistant populations. These resistance mechanisms cannot be overcome by synergists. Resistance management strategies based on rotating compounds that differ in their sites of action have not been tested in the field and are limited by the lack of diversity of sites of action in our current armament of insecticides.

Ample opportunities exist for discovering insecticides that act at new sites in the nervous system. The discovery that both the chlorinated cyclodienes and the avermectins apparently act at the γ-aminobutyric acid (GABA) receptor (Mellin et al., 1983; Matsumura and Tanaka, 1984) highlights the potential significance of this target. Similarly, the discovery that chlordimeform acts at the insect octopamine receptor (Hollingworth and Murdock, 1980) has stimulated renewed interest in the formamidines as a class and in novel structures acting at this site. These compounds illustrate that successful control can be achieved without kill.

Beyond these, several novel sites remain to be exploited as advances in fundamental neurobiology define their properties. Several neurotransmitter systems are promising targets: the acetylcholine receptor in the insect central nervous system, the glutamate receptor at the insect neuromuscular junction, and receptors for peptide neurotransmitters and neurohormones are just now being discovered. Both the acetylcholine and glutamate receptors have previously been targets of insecticide development in industry without great success, but their significance as targets may increase as more information about the pharmacology of these sites accumulates. Other targets also exist beyond the level of transmitter receptors. The enzymes involved in metabolizing or maintaining homeostatic levels of transmitters are potential sites of action, as are the processing enzymes involved in the release of neuropeptides from precursor proteins and the peptidases that degrade bioactive peptides. The success of the drug Captopril, which inhibits the angiotensin-converting enzyme, illustrates the potential for biological activity in compounds that interfere with normal neuropeptide processing.

Targets also exist outside the nervous system (Mullin and Croft, in press), such as compounds that act on the insect endocrine system (e.g., juvenoids) and on the biochemical processes involved in insect cuticle formation (acyl ureas). The selective action of these insect growth regulators makes them highly suitable for IPM systems. They act only at specific times in insect development, however, and the interval between application and effect can be several days rather than a few hours, as with neurotoxic compounds. (Fast-acting herbicides once were the industry standard until highly effective slow-acting compounds became available.) Many developmentally active compounds exhibit a degree of selectivity that makes them more suitable than broad-spectrum neurotoxicants for use in IPM systems. Under current economic and regulatory constraints, however, they are less effective than neuroactive compounds.

Even a cursory knowledge of insect physiology shows numerous systems that may be exploited to control insects. For instance, the regulation of oxygen toxicity and water balance are critical in an insect, and therefore are susceptible to disruption. Phytophagous insects have unique systems for using phytosteroids that may provide biochemical leverage for the design of selective compounds. Exploitation of some of these systems may lead to the fast-acting toxins we have come to expect in agriculture.

Some of these targets may yield compounds very selective for pest insects versus beneficials (Mullin and Croft, in press). The term pest has no systematic basis, however, and the bionomics of pest versus beneficial insect interaction is unknown for many cropping systems. Although there are some limited generalizations regarding the comparative biochemistry and toxicology of pest versus beneficial insects, their general applicability is unknown (Metcalf, 1975; Granett, in press). It is not necessary to develop selectivity among insects by planned exploitation of a biochemical lesion. Once high biological activity is discovered, such selectivity can be developed by synthesizing compounds to exploit differences in xenobiotic metabolism or simply by testing a series of chemicals on pest and beneficial insects as part of the evaluation process. Just as industry invested in resistance management when it became financially advantageous, many companies will eventually include selectivity as a major criterion in the future selection of compounds.

Encouraging Fundamental Research

Although there are ample opportunities to discover novel insecticides, the critical problem lies in incentives to pursue these opportunities. Historically, the agrochemical industry has succeeded by optimizing biological activity in a series of compounds. Industry has not pursued sustained in-house research to discover new leads. One reason is the expense of long-term commitments of personnel and facilities to do basic research on insect biochemistry. Moreover, scientists attempting to pursue these efforts under the cloak of industrial secrecy are isolated from the free interchange of ideas and the honing influence of peer review in publication and the pursuit of funding. Consequently, basic research in an industrial setting runs the risk of losing contact with the leading edge of knowledge, particularly in some of the more progressively fast-paced fields of academic research (Webber, 1984).

This argument may imply that such research is most appropriately pursued in academic laboratories. Yet, we found very few academic scientists actively pursuing the definition of possible new sites for insecticide action, and the funds that were spent came largely from projects funded for other reasons. More scientists must be enticed into these areas by convincing them that a career based on such research is socially responsible and professionally profitable. There are a variety of mechanisms to accomplish this end, a few of which follow. Our suggestions raise questions regarding the role of the public sector in fundamental agricultural research. Ruttan (1982) argued that incentives are not adequate to encourage private research and that social return on public investment in agricultural research may exceed private profit. He concluded that "simultaneous achievement of safety, environmental, and productivity objectives in insect pest control will require that the public sector play a larger role in research and development."

National Institutes of Health and the National Science Foundation If gold stars were to be awarded to agencies for funding work leading to the discovery of new targets for insecticide development, the National Institutes of Health (NIH) and the National Science Foundation (NSF) would receive them. Most of this work is outside the mandates of these agencies, but they have provided a base level of funding presumably because the proposed science is good and because the agencies see some social value in the research product. Our observations on pesticides appear to apply to agriculture in general (Lepkowski, 1982). Some slight changes could be made in the mandates of certain institutes at NIH to facilitate the funding of such work "up front." For instance, a great deal of work is supported on the deleterious effects of pesticides on mammalian systems. One way to improve human health would be to encourage the development of insecticides that are less risky to humans and the environment. Ironically, the National Institute of Environmental Health Statistics (NIEHS) has recently designated such research as "peripheral" and "no longer relevant."

An agency like NSF, which funds the pure pursuit of knowledge, is of tremendous value to the scientific community. Its resources must not be diluted, because much of the work on fundamental chemistry and biochemistry that it funds is of great value in the elucidation of new targets for insecticides even when insects are not the subject of investigation. Yet, NSF should not eliminate from consideration good basic research simply because a pest insect is used as a model organism to evaluate a fundamental question in biology. Among the very best models for asking basic questions in biology are those related to resistance. The excitement demonstrated in this publication from population biologists is one illustration. The availability of strains of insects either susceptible or resistant to the toxin provides an unparalleled opportunity to determine the impact of altered biochemical processes on the functioning of intact organisms. The value of insects as models when investigating fundamental biological processes has been illustrated often.

U.S. Environmental Protection Agency Research funding by the U.S. Environmental Protection Agency (EPA) is generally restricted to areas that require additional information to support a regulatory decision. Nevertheless, EPA has funded some of the most exciting and innovative work on the development of new insecticides; it has also funded research that will improve environmental quality and encourage implementation of IPM programs. Certainly, research that leads to the discovery and development of insect control agents that promise fewer environmental and nontarget problems is a logical extension of the above programs.

U.S. Department of Agriculture Responsibility to support fundamental research as a basis for pesticide development is part of the U.S. Department of Agriculture's (USDA) mandate. Unfortunately, USDA has failed to fulfill this responsibility. This failure is due partly to the negative connotations that surround the idea of promoting pesticide research or pesticide use in any way and the obvious difficulties of selling the need for such work in the present political climate. To reverse this trend USDA must take a position of informed advocacy for these research needs rather than capitulating to prevailing public opinion.

The USDA is the only federal agency with an in-house research effort capable of addressing this problem. A recent review of USDA research recommended a renewed emphasis on basic research directed toward solving agricultural problems of national importance (Lepkowski, 1982). Research to define targets for novel insecticides fits within this recommendation. Although some excellent research has been done by USDA scientists, administrative neglect of these priorities and concomitant emphasis of other programs has left USDA laboratories with little in-house expertise in this area. A renewed USDA effort in target biochemistry would require not only a policy decision but also a commitment to hire new professional staff.

Fostering an environment of creativity and free scientific interchange within the USDA is essential. There is a constant tension within the USDA between the need for directed research and the negative impact of excessive direction on innovation. Several initiatives might improve the productivity and creativity of all research programs within USDA's broad mandate. Programs to encourage collaboration between some USDA laboratories and universities have been very successful and could be expanded. Additional funds could be designated, and individuals might be encouraged to take sabbatical leave at USDA laboratories. The development of an in-house career development program could greatly increase the level of innovative work as well as research esprit de corps. Researchers could be granted salary and support funding for five years, based on past performance or a competitive proposal.

The most immediate impact of USDA support of target biochemistry would be felt in universities. Academic laboratories already possess the expertise to pursue this research. The U.S. Department of Agriculture, through its Competitive Grants Program, can provide the opportunity. Unfortunately, the current guidelines for the program virtually exclude research in this area.

Simply broadening the objectives of the Competitive Grants Program would be of little help, as the program is too small to fund even the high-quality proposals submitted under current guidelines. Instead, we suggest an increase in funding specifically to support a new program area in target biochemistry. For example, supporting 50 research projects at a level of $60,000 per year ($40,000 in direct costs and $20,000 in indirect costs) would cost $3 million per year, a modest amount compared to the nearly $20 million increase recently designated to establish funding through the Competitive Grants Program for research in agricultural biotechnology.

Despite the need for this type of funding, the future of the entire Competitive Grants Program is regularly threatened in the budget process. The most recent example is the elimination of all funding for this program in the proposed executive budget for fiscal year 1986. If competitive funding is to have a large impact on research productivity, it must be a stable, integral, and significant part of the annual USDA budget.

Another approach would be to institute a strong, competitive postdoctoral program for in-house and extramural positions. This program, patterned after the highly successful NIH program, would encourage new Ph.D.s to prepare research proposals relating to fundamental problems in agriculture. It would encourage young scientists from a variety of disciplines to enter the field and, if properly administered, would further excellence in agricultural research. A second approach would be to establish a grant program to support new assistant professors in fundamental research related to agriculture. Such a program would encourage individuals in basic science departments to exploit the exciting models offered in agriculture. Once a young scientist has established a research direction related to agriculture, long-term funding might be obtained from other agencies. A similar approach might be taken with starter grants to encourage scientists to extend their research into new areas. Ideally these grants would be limited to two or three years and would be nonrenewable for a similar period. Such a system would encourage individuals to seek other support and prevent the funding from going only to a few established laboratories. These three programs would acquire for agriculture more basic research than agriculture actually supported. Such a course may be initially defensible, but ultimately, there is also the need to establish stable, long-term support for the fundamental science that will maintain our high level of agricultural productivity and profitability while still protecting the environment.

Universities Universities can increase research on target biochemistry. Experiment station directors and land-grant institutions can immediately encourage such work. Scientists lacking experiment station appointments could be encouraged to carry out collaborative projects in these areas.

The major commitment that a university must make is to hire faculty to work in the area of target biochemistry and physiology. It takes more than a two-week short course to convert an organic chemist into a creative leader of a biorational pesticide development program. The chemist must have either extensive cross training or colleagues who speak a similar language. Who will train these individuals? Many of the pioneers of post-World War II pesticide development have retired and have not been replaced. A teaching cadre in this area is critical if work along these lines is to continue.

Although agrochemical companies have the chemical expertise to exploit a biochemical system, they lack the in-house expertise in biology and biochemistry. Acquiring such expertise by extensively retraining existing personnel or hiring new staff is an expensive, long-term commitment. Collaborating with a university laboratory having the required expertise is a more logical solution.

Collaborative arrangements benefit both parties, but they are relatively rare in this country (Webber, 1984). Therefore, universities must develop reasonable guidelines to permit and encourage interaction with industry. Collaboration means far more than just accepting money. Acceptance carries with it the obligation to conduct research that will be meaningful to the sponsoring company. In return, industry must appreciate that university laboratories do not exist solely for subcontracting proprietary research. A great deal of basic research can be accomplished on a minimal budget in a university setting, but a major professor must protect the careers of students and postgraduates. Thus, industry must be willing to make a commitment to multiyear support and must have realistic expectations of productivity for research undertaken in the context of graduate and postdoctoral training. Areas of research must be explicitly defined so that university collaborators are not barred from publishing their results, and patent agreements must respect the rights of the university as well as the research sponsor.

Private and public investment in university-based agricultural research is sound (Ruttan, 1982). Such research is complementary to graduate education in agriculture. Public investment in a university setting will draw scientists from a variety of areas into agriculture. Since industry is in need of in-house scientists capable of developing new pest-control agents by both classical and molecular procedures, industrial support of university research provides not only the data needed but a pool of well-trained potential employees as well.

Chemical Industry The pesticide chemical industry invests roughly 10 percent of its gross profits in research, making it one of the most research-intensive industries (Ruttan, 1982). Companies must establish sufficient in-house expertise in insect biochemistry and physiology and must initiate basic research programs that are relevant to the company's objectives and complementary to university research efforts. The agrochemical industry tends to hire basic scientists and then assumes that basic research is simply the screening of experimental chemicals on an elegant in vitro preparation. Such work is important, but it should be a minor portion of the duties of an industrial scientist. The scientists must be free to explore new opportunities for chemical exploitation and to define the biorational models for directed chemical synthesis programs. Another problem is that industrial scientists doing basic research are often prevented from testing the validity of their ideas through publication in peer-reviewed journals. Companies can remedy this by establishing a tradition of peer review and publication of in-house basic research after an appropriate delay to allow its proprietary use.

State IPM and Commodity Groups Funding available to state IPM programs and commodity groups varies dramatically from state to state. The funding is characteristically applied to local problems, not to fundamental research on target biochemistry. Developing selective materials is to their benefit. These groups should support legislative efforts to encourage fundamental research in agriculture even if the expected benefits extend beyond the individual state. When possible, these groups should fund long-term basic research directly, partly because they can have a more profound influence on growers to use selective materials.

ENCOURAGING REGISTRATION AND DEVELOPMENT

Industry will use any available information on target biochemistry to discover new compounds. Although broad-spectrum compounds will be developed, selective compounds are desperately needed for IPM programs, especially since regulatory law and economic constraints impede the development of diverse crop chemicals.

A variety of modifications of patent law and enforcement can encourage development. For instance, legislation to start the patent clock ticking when registration is granted has already been proposed. Patent life could be further extended for compounds considered to have exceptional value to IPM programs, especially if the compounds act by a unique mechanism. An extended patent life would give the company owning the compounds a major incentive to avoid resistance problems (Djerassi et al., 1974).

Although many regulatory costs cannot be reduced, costly delays in regulatory decisions can be eliminated. The EPA has often appeared to avoid making bad decisions by avoiding any decisions. An effort by EPA to process registration petitions as rapidly as possible would be of great benefit, particularly if extensions in patent life cannot be obtained.

Changes in the ways in which toxicological risks are evaluated would promote the development of novel, selective compounds. Current regulatory procedures may inadvertently encourage the registration of compounds that are acutely toxic to mammals over selective materials (Retnakaran, 1982; Ruttan, 1982). The evaluation of the toxicological risks of insecticides must be relevant to the expected routes and levels of exposure rather than requiring toxicological evaluation at maximum tolerated doses. To do this, we need well-trained, courageous regulators acting with legislative support. The public must understand that a blind effort to obtain zero-risk may only increase risk.

Further expanding the subsidized registration of pesticides for minor crop uses would give IPM practitioners a greater variety of compounds to work with. Eliminating some registration requirements for several closely related IPM-compatible compounds by the same company might encourage the development of highly selective compounds. Although registration cost will probably not decrease dramatically, some scientific improvements can be made. For instance, immunochemical technology can reduce the cost of residue analysis. Since efficacy and residue analyses are the major costs involved in minor crop registration, this technology could greatly expand the effectiveness of the IR-4 program with no increase in budget (Hammock and Mumma, 1980).

Another option is an orphan pesticide development program to encourage the development of compounds that cannot be developed economically by industry but are likely to be of great benefit. The recently established orphan drug program provides both a precedent for this approach and an administrative model for its operation.

CONCLUSION

Many resistance management tactics tend to focus on existing resistance problems and attempt to preserve the utility of compounds currently available. Although these efforts are valuable, we believe that the effective management of resistance to pesticides depends on the continued development of new compounds, as well as on the judicious use of existing materials. Therefore, the recent decline in the rate of development of new insecticides is a serious limitation to resistance management and the development of sophisticated pest-management strategies.

There is a great need to stimulate both basic research on the biochemistry and physiology of target species and development of selective insecticides.

We have identified many avenues of research in insect biochemistry that appear promising for the design of novel insecticides, and there are many more that we have not mentioned. Federal agencies and the agrochemical industry must recognize that research is critically needed.

The stimulation of the industrial development of new compounds is a more complex problem. Potent, broad-spectrum pesticides will continue to be developed, but economic and regulatory constraints work against the development of more selective compounds. The agrochemical industry exists to discover and sell products at a profit, not to develop ideal pesticides for pest management. They will not develop compounds that are perceived to be unprofitable or excessively risky. If, however, an increase in our knowledge of the biochemistry of target species and the impact of new technologies can decrease the cost of discovery, if the time and cost of regulatory compliance can be minimized without detriment to the public good, and if patent lives of compounds can be extended to compensate for marketing time lost in regulatory review, then the search for and development of novel insecticides will be perceived to be a sound, profitable business, and the tremendous potential that we see for the development of safe and selective pesticides by both chemical and molecular approaches will be realized.

ACKNOWLEDGMENTS

This work was supported by NIEHS Grant ES02710-05, Research Career Award 5 K04 ES500107, and a grant from the Herman Frasch Foundation to Bruce D. Hammock and by NIEHS Grant ES02160-06 to David M. Soderlund. We thank the Ciba-Geigy Corporation for supporting David Soderlund on sabbatical leave. We extend our thanks to many colleagues for critical comments on this manuscript.

REFERENCES

Boyce, A. M. 1976. Historical aspects of insecticide development. Pp. 469–488 *in* The Future for Insecticides: Needs and Prospects, R. L. Metcalf and J. J. McKelvey, Jr., eds. New York: John Wiley and Sons.

Djerassi, C., C. Shih-Coleman, and J. Diekman. 1974. Insect control of the future: Operational and policy aspects. Science 186:596–607.

Granett, J. In press. Potential of benzoylphenyl ureas in integrated pest management. *In* Chiten and Benzoylphenyl Ureas, J. E. Wright and A. Retnakaran, eds. The Hague: Junk Press.

Hammock, B. D. 1985. Regulation of juvenile hormone titer: Degradation. Pp. 431–472 *in* Comprehensive Insect Physiology, Biochemistry and Pharmacology, G. A. Kerkut and L. I. Gilbert, eds. New York: Pergamon.

Hammock, B. D., and R. O. Mumma. 1980. Potential of immunochemical technology for pesticide residue analysis. Pp. 321–352 *in* Recent Advances in Pesticide Analytical Methodology, J. Harvey, Jr. and G. Zweig, eds. Washington, D.C.: American Chemical Society.

Hinkle, M. K. 1983. Problems with conservation tillage. J. Soil Water Conserv. May-June:201–206.

Hollingworth, R. M., and L. L. Murdock. 1980. Formamidine pesticides: Octopamine-like actions in a firefly. Science 208:74–76.

Kirschbaum, J. B. 1985. Potential implication of genetic engineering and other biotechnologies to insect control. Annu. Rev. Entomol. 30:51–70.

Krieger, H. 1982. Chemistry confronts global food crisis. Chem. Eng. News. 60:9–23.

Lepkowski, W. 1982. Shakeup ahead for agricultural research. Chem. Eng. News 60:8–16.

Lockwood, J. A., T. C. Sparks, and R. N. Story. 1984. Evolution of insect resistance to insecticides: A reevaluation of the roles of physiology and behavior. Bull. Entomol. Soc. Am. 30:41–51.

Magee, P. S. 1983. Chemicals affecting insects and mites. Pp. 393–463 *in* Quantitative Structure-Activity Relationships of Drugs, J. G. Topliss, ed. New York: Academic Press.

Matsumura, F., and K. Tanaka. 1984. Molecular basis of neuroexcitatory actions of cyclodiene-type insecticides. Pp. 225–240 *in* Cellular and Molecular Neurotoxicology, T. Narahashi, ed. New York: Raven.

Mellin, T. N., R. D. Busch, and C. C. Wang. 1983. Postsynaptic inhibition of invertebrate neuromuscular transmission by avermectin BQ_{1a}. Neuropharmacology. 22:89–96.

Mellor, J. W., and R. H. Adams, Jr. 1984. Feeding the underdeveloped world. Chem. Eng. News 62:32–39.

Menn, J. J. 1983. Present insecticides and approaches to discovery of environmentally acceptable chemicals for pest management. Pp. 5–31 *in* Natural Products for Innovative Pest Management, D. L. Whitehead, ed. New York: Pergamon.

Metcalf, R. L. 1975. Insecticides in pest management. Pp. 235–274 *in* Introduction to Insect Pest Management, R. L. Metcalf and W. H. Luckmann, eds. New York: John Wiley and Sons.

Metcalf, R. L. 1980. Changing role of insecticides in crop production. Annu. Rev. Entomol. 25:219–256.

Metcalf, R. L., and J. J. McKelvey, Jr. 1976. Summary and recommendations. Pp. 509–511 *in* The Future for Insecticides: Needs and Prospects, R. L. Metcalf and J. J. McKelvey, Jr., eds. New York: John Wiley and Sons.

Miller, L. K., A. J. Lingg, and L. A. Bulla, Jr. 1983. Bacterial, viral and fungal insecticides. Science 219:715–721.

Mullin, C. A., and B. A. Croft. In press. An update on development of selective pesticides favoring arthropod natural enemies. *In* Biological Control in Agricultural Integrated Pest Management Systems, M. A. Hoy and D. C. Herzog, eds. New York: Academic Press.

Retnakaran, A. 1982. Do regulatory agencies unwittingly favor toxic pesticides? Bull. Entomol. Soc. Am. 28:146.

Ruttan, V. W. 1982. Changing role of public and private sectors in agricultural research. Science 216:23–38.

Storck, W. J. 1984. Pesticides head for recovery. Chem. Eng. News 62:35–59.

Webber, D. 1984. Chief scientist Schneiderman: Monsanto's love affair with R and D. Chem. Eng. News 62:6–13.

Pesticide Resistance: Strategies and Tactics for Management.
1986. National Academy Press, Washington, D.C.

Biotechnology in Pesticide Resistance Development

RALPH W. F. HARDY

The role and potential of biotechnology in pesticide resistance development is projected to be quite large but has been minimally used. Relevant biotechnology techniques are numerous, including cell and tissue culture and genetic and biochemical techniques.

The classic case of the role of biotechnology in resistance is antibiotic resistance. Biotechnology identified the basis of resistance and is guiding synthesis of novel antibiotics to circumvent resistance; antibiotic resistance provided a critical process for genetic engineering. In the area of pesticide resistance, the only well-developed application of biotechnology is for three different classes of herbicides. The sulfonylurea herbicides are presented as an example of the role and potential of biotechnology in any pesticide resistance case. Biotechnology has not been applied to fungicide, insecticide, or rodenticide resistances.

The opportunity for biotechnology is large, but will require a multiplicity of skills beyond those used by scientists who are working at the organismal/physiological and biochemical levels of pesticide resistance. This opportunity should be pursued aggressively, since it can provide new directions to alleviate or minimize pesticide resistance where the benefits from additional organismal, physiological, and biochemical studies may be limited.

INTRODUCTION

The new biotechnology is providing biology with a powerful array of techniques that are advancing molecular understanding of biological processes and phenomena at an unprecedented rate. Outstanding examples are antibody formation and oncogenes and cancer. From this understanding and these techniques, useful new products, processes, and services are being and will be generated. The generation will be direct in terms of biological products, processes, and services and indirect in terms of chemical products, processes, and services. Agrichemical and pharmaceutical discoveries will become increasingly driven by biotechnology or biotechnology-chemistry rather than by the current dominant process of empirical chemical synthesis coupled with biological screening. Tagamet®, an antiulcer drug that has produced the highest sales for any single pharmaceutical, is an early example of a biotechnology-chemistry-based innovation. Since the health care field has been quicker in using the new biotechnology than the field of agriculture, such a product has yet to be produced for agriculture. For example, the application of the new biotechnology has only recently begun and is limited in the area of pesticide resistance (Brown, 1971; Dekker and Georgopoulos, 1982; Georghiou and Saito, 1983; Hardy and Giaquinta, 1984).

BIOTECHNOLOGY TECHNIQUES

Biotechnology comprises cell and tissue culture techniques and genetic and biochemical-chemical techniques. Cell and tissue culture techniques range from microbial culture through higher organism cell and tissue culture to somatic cell fusion and regeneration. Somatic cell fusion has become especially useful for antibody production, where an antibody-producing cell with a limited life is fused with a transformed cell with an infinite life to produce a hybrid cell (hybridoma) that produces over an almost infinite period of time a single type of antibody called a monoclonal antibody (MAB). These MABs could become very useful in both qualitative and quantitative diagnosis of pesticide resistance, as they are becoming useful as in vitro health care diagnostics. Several start-up companies have been established for health care MAB diagnostics.

Cell culture techniques will also be useful in developing and/or selecting resistance in model systems. Resistance development may use microorganisms or cells or tissues of higher organisms. In the latter, regeneration of plants from culture often increases phenotypic variability, such as possible herbicide resistance, over that shown in the parental cells. This phenomenon is called somaclonal or gametoclonal variation, depending on the cell source.

Genetic techniques, especially molecular genetic techniques, have expanded greatly during the last decade and are propelling our understanding at the molecular level. Several of these techniques are the basis of a major biotechnology called genetic engineering, in which defined genes are introduced into foreign host cells. In theory any gene can be moved from a microbe to a plant, a plant to an animal or human, a human to a microbe, eliminating the barriers of sexual plant and animal breeding.

Production of gene fragments is the initial technique used to generate understanding and to perform genetic engineering. Restriction enzymes, of which there are about 100, cleave DNA at specific sites dictated by the DNA base sequence. The restriction enzymes cut the DNA of organisms such as fungi, insects, plants, and animals into useful fragments called gene libraries. These fragments are useful, since they are of a small size in which specific genes can be identified. A gene library is the starting point. There are few if any gene libraries available for agricultural pests, although the techniques needed are available. Separating the fragments produced by the restriction enzymes enables characterization of the genotype for polymorphisms. This technique, called restriction enzyme mapping, could be used to diagnose and characterize resistance at the genetic level so as to establish the similarities or differences of observed resistances.

Study of a gene of interest, such as a resistance gene, requires its identification, isolation, biosynthesis or chemical synthesis, and cloning, usually in a genetically well-characterized microorganism such as *E. coli*, to produce adequate quantities for characterization or further use. The gene can be isolated from the gene library, biosynthesized as a complementary DNA (cDNA) from its messenger RNA (mRNA), or chemically synthesized directly if the DNA sequence is known. If the sequence is not known, powerful DNA sequencing techniques exist for rapid sequencing. DNA sequencing will identify the similarity or difference of resistant versus susceptible genes.

Genetic engineering of organisms requires these steps so as to obtain a source of the desired gene and to generate genetic constructions with appropriate replication sites and control elements so that they can be introduced into the desired host, retained, and replicated to produce the gene product at an appropriate rate. Techniques have been developed to introduce functional foreign genes into microorganisms, embryos of mammals, and cells of at least dicotyledonous plants. Human insulin produced by microorganisms, antibiotic-resistant model plants, and super rodents with additional copies of the growth hormone gene are examples.

We are beginning to understand the molecular basis of how gene expression is regulated. Recent studies on *Drosophila* are a major example in a model sytem. As this knowledge becomes known, it should be very useful in exploring resistance on the basis of regulatory-based changes.

Overall, genetic techniques will be useful to understand, manage, circumvent, and exploit pesticide resistance. These genetic techniques, however, will need to be coupled with chemical and biochemical techniques.

The biochemical and chemical techniques of biotechnology include synthetic and analytical methods. Synthetic oligonucleotides for use as DNA probes to identify genes can be made readily with automated commercial instruments. These DNA probes will succeed MABs as even more useful diagnostic agents for pesticide resistance. Micro quantities of proteins can be sequenced with commercial instruments, and synthetic peptides up to about 20+ amino acid residues can be synthesized routinely. Biophysical techniques utilizing X ray, nuclear magnetic resonance (NMR), and other methods will provide three-dimensional structures of biological macromolecules such as proteins; thus, we will be able to correlate structure with pesticide activity or resistance.

Gene sequences and resultant protein sequences will be changed by design, using site-specific mutagenesis to change the DNA sequence. For example, β-lactamase, the antibiotic-resistant gene in bacteria, was altered to place a cysteine at the active site in place of the naturally occurring serine. The designed gene produced a novel active β-thiollactamase (Sigal et al., 1982). By combining this wealth of information (generated from chemical, biochemical, and genetic techniques) with computer graphics, we will be able to design novel pesticides and genes.

BIOTECHNOLOGY AND XENOBIOTICS

Biotechnology has been intimately involved in antibiotic resistance research and development. The techniques of biotechnology identified the basis of resistance, which provided a critical resource for genetic engineering. For example, penicillin and cephalosporins are widely used antibiotics. The β-lactam ring of these molecules is essential for their antibiotic activity. Bacteria, however, have developed resistance to these molecules. The resistance is located on small extrachromosomal circular pieces of DNA called plasmids, and the resistance is specifically due to a gene that makes an enzyme called β-lactamase, which cleaves the β-lactam ring of these antibiotics and inactivates them.

Antibiotic resistance has provided essential selectable markers for following genetic constructions introduced into cells. Cells containing the new functional genetic material are selected for their antibiotic resistance. The markers have enabled genetic engineering of microorganisms to develop rapidly. Understanding these antibiotics and antibiotic resistances facilitated the knowledge of microbial cell-wall synthesis.

The problem of antibiotic resistance has led to several ways to circumvent it. An empirical approach such as the use of clavulinic acid (a naturally occurring suicide inhibitor of β-lactamase) in combination with an antibiotic, amoxacillin, is one way to circumvent the resistance problem. Another approach is to develop commercial semisynthetic β-lactam antibiotics, which have incorporated within them the ability to also inhibit β-lactamase. Of possible greater significance, based on the understanding generated by biotechnology, are current efforts to design drugs to which resistant bacteria are susceptible.

In pesticide resistance management, biotechnology can play a key role, but much more research is necessary before we can fully exploit these benefits. For example, herbicide research and development offers opportunities and limitations. We little understand the mechanisms of action of herbicides; therefore, informed decisions on research and development, safety, and use are limited. The empirical synthesis-screening approach through which almost all herbicides are discovered is becoming increasingly inefficient; researchers must synthesize some tens of thousands of new chemical structures to find a commercial product.

Crops usually have inadequate tolerance to herbicides; thus, herbicides are selected for tolerance to specific crops. The lack of broad crop tolerance limits broad crop use of most herbicides, as do soil residues. More herbicide-resistant crops are desirable for broad use of low-cost herbicides and crop rotation. Finally, a few weeds have developed resistance to herbicides such as atrazine, and it may be desirable to manage or circumvent this resistance.

The earliest products of crop biotechnology will probably be crops with specific herbicide resistance, followed by designed herbicides. The sulfonylurea herbicides illustrate the major role that biotechnology can play in generating understanding of pesticides and, in this case, resistance. The sulfonylurea herbicides demonstrate the integrated role of a number of techniques and disciplines.

Using empirical synthesis and screening, the du Pont Company developed a novel class of herbicides, some examples of which are Ally®, Classic®, Glean®, Londax®, and Oust®. These herbicides are very potent, with unusually low application rates.

Plant physiological investigations on the active sulfonylurea compounds in Glean® and Oust® showed that these sulfonylureas rapidly inhibited cell division. Tobacco cell cultures grown on media containing the sulfonylureas yielded cell lines and regenerated plants with a chromosomally localized single resistant gene and a greater than 100-fold increase in resistance to sulfonylureas (Chaleff and Ray, 1984). Further mechanistic studies utilized less complex, more defined microorganismal systems. The sulfonylureas also inhibited the growth of several, but not all, bacteria. The biocidal target of these herbicides was an enzyme, acetolactate synthase (ALS II and III), that is involved in the synthesis of the branched-chain essential amino acids valine, isoleucine, and leucine (LaRossa and Schloss, 1984). Physiological, biochemical, and genetic analyses confirmed the target site.

Along these same lines a molecular biological characterization showed that a major form of resistance in yeast arises from an altered structural gene for ALS, in which a proline amino acid residue in the sensitive ALS is replaced by a serine in the resistant ALS. Other forms of resistance were also found. The structural ALS resistance gene may be useful as a selectable marker for genetic engineering in higher organisms, as antibiotic resistance has been in bacteria.

This rapidly generated base of information in model microorganismal systems led to the identification of ALS as the site of herbicidal activity in plants (Ray, 1984). A less-sensitive ALS was shown to be the basis of herbicidal resistance in resistant tobacco. Other plant studies showed that herbicide selectivity in crop plants arose from metabolism of the sulfonylureas to a nonherbicidal form in the tolerant crops, not to a less-sensitive ALS. Herbicidal activity can be evaluated directly on the ALS target, thus providing more rigorous structure/activity information than whole plant screens, where activity is the result of ALS activity and penetration, translocation, and detoxification of the sulfonylureas.

Biophysical studies on sulfonylureas and ALS at the kinetic and structural levels can provide information on the specific mechanism of inhibition. Opportunities for designed herbicides, designed resistance genes, and the genetic engineering of herbicide-resistant crops come from this multidisciplinary and multitechnique generation of understanding. Without microorganismal techniques and development of model resistance, the time required to generate this level of understanding on the sulfonylureas would have taken several additional years. Although sulfonylureas were used in the above study, similar examples exist for the *s*-triazine (Arntzen and Duesing, 1983) and glyphosate herbicides (Comai et al., 1983). The time required to reach an understanding of the *s*-triazines and glyphosate was much longer than for the sulfonylureas, because the newer biotechnology techniques were not available or not initially used for most of the former studies.

BIOTECHNOLOGY IN PESTICIDE RESISTANCE

Schematics of the role and potential of biotechnology in pesticide research and development, understanding, management, circumvention, and exploitation and in pesticide resistance and development are presented in Figures 1 and 2. The following sections will consider biotechnology in all phases of pesticide resistance.

Resistance Development

Xenobiotics select or generate resistance broadly in organisms (Georghiou and Mellon, 1983). Fungi, acarina, and insects have shown resistance to fungicides. Bacteria, fungi, nematodes, acarina, insects, crustacea, fish, frogs, rodents, and higher plants have shown resistance to insecticides. Bacteria, yeast, and higher plants have shown resistance to herbicides. This broad occurrence of resistance suggests that by using model systems, we can understand the molecular process of resistance. The model system should be biochemically and genetically well-characterized and as simple as possible, such as a microorganism, although some problems will require more complex

Empirical Synthesis—Pest Screening
- Biocide discovery

Model (Microbial) Systems
- Mechanism of biocide action
- Biocide target genes and gene products
- Surrogate screens

Pest-Host Systems
- Biocide target in pest
- Molecular biological characterization of target and biocide target interaction
  - In vitro biocide structure target activity relationships
  - Designed biocides
  - Host tolerance/resistance

FIGURE 1 Biotechnology—pesticide research and development.

systems. Development of model resistance in defined organisms will accelerate the understanding, management, circumvention, and exploitation of resistance.

Resistance Understanding

Most if not all resistances result from one of three genetic changes. With qualitative change a structural gene is altered so that its protein product is less affected by the pesticide, such as the sulfonylurea resistance gene with its altered ALS enzyme. The other genetic changes are quantitative: gene

Empirical Synthesis—Resistant Pest Screening
- Biocide discovery

Model (Microbial) Systems
- Resistance development
- Resistance type(s)
- Resistance gene(s) and gene products
- Surrogate screens
- MABs, DNA probes, restriction maps as diagnostics

Pest-Host Systems
- Pest resistance types
- Pest/host resistance gene(s) and gene products
- MABs, DNA probes, restriction maps as diagnostics
- In vitro biocide structure resistance relationships
- Designed biocides
- Resistant hosts by somaclonal variation
- Genetically engineered resistant hosts
- Natural biocides
- Synergists
- Agriregulators

FIGURE 2 Biotechnology—pesticide resistance research and development.

regulation and gene amplification, in which increased amounts of gene product make the organism less sensitive to the pesticide. Structural gene changes usually produce stable resistance, while gene amplification changes may be less stable.

To understand resistance we need to address the number of types; identify the general types such as target site, metabolism, penetration, reproduction, excretion, storage, and feeding; and define the genetic change responsible. Biotechnology has provided this level of understanding for at least three herbicides—glyphosates, sulfonylureas, and *s*-triazines—where altered structural genes, gene amplification of target-site genes, and altered regulation of target-site genes have been demonstrated. In some the product of the altered structural gene is as active as the unaltered (sulfonylureas and glyphosates) or highly fit; in others (*s*-triazines) the product is less active or less fit. Biotechnology should provide similar definition for other pesticide resistances where an adequate physiological, biochemical, and genetic base exists in an appropriate experimental organism. Effective programs will be highly interdisciplinary using a breadth of biotechnology techniques. Biotechnology can expand our understanding in most if not all areas of pesticide resistance. Unfortunately, biotechnology has been little used in this field. Obvious opportunities are cytochrome P_{450} in cases of some insecticide resistances and β-tubulin in the case of benomyl fungicide resistances.

Resistance Management

In the short term, biotechnology can provide the reagents and techniques for qualitative and quantitative diagnosis of pesticide-resistant organisms. MABs may be useful for measuring structurally altered gene products and an altered quantity of gene products. Restriction maps and DNA probes should be useful, but they will require an expanded base of information. These techniques should enable researchers to define the similarities or differences of observed pesticide resistances in the same or different laboratories. They would be used first as research diagnostics, but could become field diagnostics to guide pesticide use practices.

Also in the short term, biotechnology would help researchers to establish rigorous pesticide structure/resistance relationships that may differ from pesticide structure/activity relationships, especially for altered target sites. Pesticide use practice could be guided by this base of understanding.

In the midterm, increased understanding of multiplicity, type, and genetic change will result in informed, early decisions on agronomic use practices that will minimize the impact of resistance. For example, gene amplification-based resistances are probably less stable than altered structural-gene resistances, suggesting alternation of pesticide use as a desirable practice in the first case. Further, an expanded use of biotechnology will provide significant new opportunities for the more effective management of pesticide resistance.

Resistance Circumvention

Circumvention of resistance may be sought through new pesticides, natural pesticides, synergists, and agriregulants. New pesticides could be discovered by empirical synthesis or design. Empirical synthesis could be coupled with a screen using resistant organisms under continuous pesticide selection pressure to discover chemical structures that inhibit critical, nonalterable enzyme or protein sites. This approach realistically assumes the existence of critical sites that cannot be changed and still maintain an adequately fit activity for the pest. Designed synthesis would use target-site knowledge and computer graphics to guide synthesis of novel pesticides to be tested. Target sites could be selected that have low opportunity for change and retention of adequate fitness for the pest. A highly conserved gene such as the quinone-binding protein that is inhibited by the *s*-triazines is an example of a target site with limited opportunity for change and retention of adequate fitness. Additional critical catalytic sites unique to pests need to be identified.

Natural pesticides may circumvent synthetic pesticide resistance. For example, biocontrol organisms could be genetically engineered to produce natural pesticides. Beneficial organisms such as plants could be genetically engineered for endogenous production of natural pesticides (Schneiderman, 1984). In both, methods for timed bioproduction of the pesticide would be needed, since continuous production would facilitate the development of pesticide resistance. Agriregulators, as described later in this subsection, may be developed for temporal control of biopesticide biosynthesis.

Synergists may also circumvent pesticide resistance. These molecules are inactive as pesticides, but they synergize the activity of pesticides. As such they may decrease metabolic detoxification by inhibiting the detoxification system. Genetic, biochemical, and chemical biotechniques may improve our understanding, so that scientists can design synergists or produce quantities of the cloned detoxification system for use as in vitro screens for potential synergists. Genetic engineering may produce naturally occurring synergists, and biotechniques may synthesize modified synergists. Biotechnology techniques have been applied to several cytochrome P_{450} systems but not to any involved in pesticide detoxification.

Synthetic compounds that regulate gene expression will be major opportunities for agrichemicals and pharmaceuticals. One or more model examples already exist. The genes for biological nitrogen fixation are not expressed when N_2-fixing organisms are grown in an environment containing adequate fixed nitrogen or ammonia. A synthetic molecule, methionine sulfoxamine (MS), causes the expression of the biological N_2-fixing genes in the presence of adequate ammonia. Synthetic compounds such as MS will become important useful future agriregulator agrichemicals. They will be discovered by empirical synthesis screening and by designed synthesis as our knowledge of gene expression increases.

The opportunity for agriregulators is expected to be large. Plants may already contain the genetic information for natural pesticides, or genetic engineering will introduce the genetic information into crop plants. Agriregulators will be used to turn on the expression at the time of need. The idea that many of the detoxification systems for insecticides are regulated by a common genetic system suggests a major opportunity for an agriregulator that inhibits the genetic system regulating detoxification genes.

Resistance Exploitation

Xenobiotic resistance genes are and will be useful selectable markers to enable researchers to track and select organisms containing genetic constructions. Antibiotic resistance is the most common example, but pesticide-resistant genes will be increasingly used. Herbicide resistance may be used to follow genetic introductions into higher plants.

Introducing herbicide-resistant genes into crop plants to increase tolerance and enable crop rotation and the use of herbicides on a broader group of crops is being pursued aggressively and may be the first major practical example of genetic engineering in crop agriculture. Similar approaches may be used to introduce rodenticide and insecticide resistance genes into pets and food animals and insecticide resistance genes into beneficial insects such as bees.

With a dynamically expanding base of understanding of basic biological processes, researchers should be able to identify many exploitable targets, not only in agriculture (such as pest control and yield and quality improvement) but also in health care, food, energy, pollution control, and chemicals.

Application to Pesticides Other than Herbicides

Examples of the comprehensive application of biotechnology to fungicide, insecticide, or rodenticide resistance do not exist. An outline for such a study follows, using the rodenticide, warfarin, as the example.

Model studies would use microbes to develop warfarin resistance, with emphasis on identifying resistance in a microbe for which the biochemical and genetic information is greatest. The type of resistance(s) and the resistance genes and gene products would be identified as previously described for the sulfonylurea resistance microbes. Such resistances for warfarin may involve the biosynthetic pathway for vitamin K. The resistant microbes may provide useful screens to evaluate members of this class of rodenticides for ability to circumvent resistance. Diagnostic approaches such as MABs, DNA probes, and restriction maps may be developed to identify each type of resistance.

Information and diagnostics from these model studies should facilitate studies of resistance in the more complex rodent pests. The rodent resistance genes and gene products would be identified. Diagnostics would be developed to identify each type of resistance. The rodenticide structure/resistance relationships could be measured in vitro, eliminating effects of the nontarget components in the rodent. Rodenticides may be designed to circumvent or minimize the resistance using in vitro tests. Resistant animals such as pets could be developed by genetic engineering so as to decrease effects of rodenticides on nontarget animals. Natural rodenticides may be produced by biotechnology. Synergists may be developed on the basis of the understanding generated by biotechnology. Similar approaches could be used for fungicides such as benomyl, where an altered β-tubulin is the site of resistance, or for insecticides, where in many cases detoxification by cytochrome P_{450} systems generates resistance.

CONCLUSION

Biotechnologies have been used very little in pesticide resistance research and development. Biotechnology has tremendous potential in almost all phases of pesticide resistance investigations and applications, as shown in the sulfonylurea herbicide example. Biotechnology research and development with this and other herbicides has been useful in resistance development, understanding, and exploitation. If desirable, biotechnology would also be useful in pesticide resistance management and circumvention. The most successful biotechnology efforts in pesticide resistance, as in almost all other areas, will integrate a multiplicity of biotechnologies by a group of multidisciplinarians.

REFERENCES

Arntzen, C. J., and J. H. Duesing. 1983. Chloroplast-encoded herbicide resistance. Pp. 273–294 *in* Advances in Gene Technology: Molecular Genetics of Plants and Animals, K. Downey, R. W. Voellmy, F. Ahmand, and J. Schultz, eds. New York: Academic Press.

Brown, A. W. A. 1971. Pesticide resistance to pesticides. Pp. 457–552 *in* Pesticides in the Environment, Vol. 1, Part II, R. H. White-Stevens, ed. New York: Marcel Dekker.

Chaleff, R. S., and T. B. Ray. 1984. Herbicide resistant mutants from tobacco cell cultures. Science 223:1148.

Comai, L., L. D. Sen, and D. M. Stalken. 1983. An altered aroA gene product confers resistance to the herbicide glyphosate. Science 221:370.

Dekker, J., and S. G. Georgopoulos. 1982. Fungicide Resistance in Crop Protection. Wageningen, Netherlands: Centre for Agricultural Publishing and Documentation.

Georghiou, G. P., and R. B. Mellon. 1983. Pesticide resistance in time and space. Pp. 1–46 *in* Pest Resistance to Pesticides, G. P. Georghiou and T. Saito, eds. New York: Plenum.

Georghiou, G. P., and T. Saito, eds. 1983. Pesticide Resistance to Pesticides. New York: Plenum.

Hardy, R. W. F., and R. T. Giaquinta. 1984. Molecular biology of herbicides. BioEssays 1:152.

LaRossa, R. A., and J. V. Schloss. 1984. The sulfonylurea herbicide sulfometuron methyl is an extremely potent and selective inhibitor of acetolactate synthase in *Salmonella typhimurium*. J. Biol. Chem. 259:8753.

Ray, T. B. 1984. Site of action of chlorsulfuron inhibition of valine and isoleucine biosynthesis in plants. Plant Physiol. 75:827.

Schneiderman, H. A. 1984. What entomology has in store for biotechnology. Bull. Entomol. Soc. Am. 1984:55–62.

Sigal, I., B. G. Harwood, and R. Arentzen. 1982. Thiol β-lactamase: Replacement of the active-site serine of RTEM β-lactamase by a cysteine residue. Proc. Natl. Acad. Sci. 79:7157.

3

Population Biology of Pesticide Resistance: Bridging the Gap Between Theory and Practical Applications

WERE THE EVOLUTION OF PESTICIDE RESISTANCE not of grave concern to human health and well-being, it would have still been important as a major example of the power and potential of adaptive evolution. Surprisingly, population geneticists and ecologists have paid little attention to it. Similarly, relatively few investigators involved in management of resistance have directly applied the tools and theoretical concepts of academic population biology.

In this chapter we describe current attempts at bridging the gap between academic and applied population biology, discuss aspects of the genetics and population biology of resistance critical to developing resistance management programs, recommend future work needed in this area, and describe major impediments to developing and implementing programs to manage resistance.

A HEURISTIC MODEL OF MANAGING RESISTANCE

We present here a simplistic, idealized model of the resistance cycle resulting from pesticide use, solely for heuristic purposes (as a "thought experiment"), not as a realistic model for the long-term management of resistance. The model assumes that resistant genotypes arise in the pest population and, as a result of selection imposed by pesticide use, field control fails because these genotypes attain high frequencies. The model assumes that stopping use of the pesticide will result in a continuous decline in the frequency of resistant genotypes and, in a reasonable amount of time, the frequency of susceptible genotypes will become sufficiently high for the population to be effectively controlled by that pesticide once again (see Figure 1). Neither assumption is a necessary outcome.

The time period between initial use of the pesticide and control failure is the *resistance onset interval*, $TR(i)$. Stopping treatment with pesticide i results in relaxation of selection pressure for resistance to i and a decline in the frequency of resistant genotypes. The time between the end of treatment with i and a decline in the frequency of resistant genotypes low enough to resume effective control with compound i is the *susceptibility recovery interval*, $TS(i)$.

In theory, pest control is possible indefinitely by cycling through an array of compounds, *as long as resistance to each of them is independent of resistance to every other*. The total number of pesticides required for this cycling depends solely on the lengths of the resistance onset and susceptibility recovery intervals (Figure 1).

In this model, the goal of resistance management is to maximize the resistance onset intervals and minimize the susceptibility recovery intervals. The effect of this strategy would be to minimize the number of independent compounds needed for effective long-term control.

USE OF POPULATION BIOLOGY THEORY AND CONCEPTS IN RESISTANCE MANAGEMENT

To date, population biology theory has contributed to resistance management primarily in identifying the factors contributing to the rise of resistance, and to some extent in interpreting factors responsible for resistance in specific populations. We are unaware of any pesticide-use programs that have been entirely planned and executed in a manner prescribed from theoretical and empirical considerations of population genetics of resistance and the ecology of the organisms and ecosystem under treatment.

Elements of population biology theory have, however, been applied to some aspects of pest management. For example, the theory of the population biology of infectious disease played a role in the development of the successful multiline cultivar procedure used to reduce fungicide use in barley cultivation (Wolfe and Barrett, this volume). This theory has also been useful in a retrospective manner. Analyses of resistance cycles are generally consistent with those anticipated from population biology theory and laboratory experiments (Gutierrez et al., 1976; Comins, 1977b; Taylor et al., 1983; Tabashnik, this volume).

Nevertheless, we are unaware of any cases where a high-dose regime or any other tactic has been actually put into practice based solely upon considerations of population genetic theory, even though several theoretical investigations are directly relevant. For example, MacDonald (1959) noted that resistance would develop more slowly if it was recessive. Davidson and

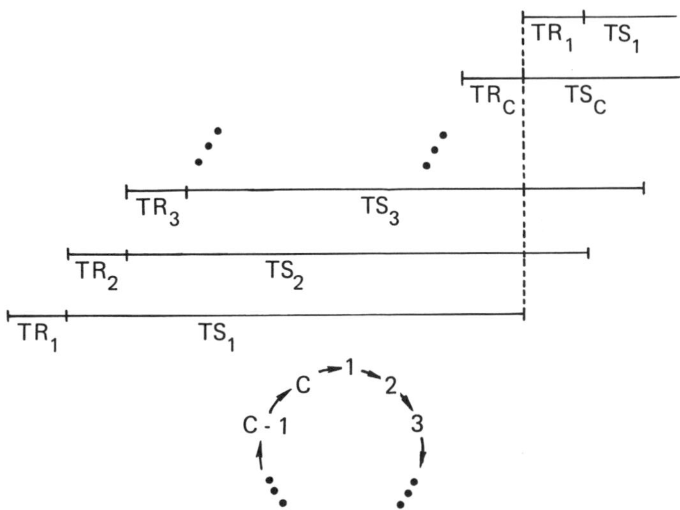

FIGURE 1 The Pesticide Resistance Cycle. TR(i) is the time period from the first use of a pesticide, i, to the time resistance precludes its use, the *resistance onset interval*. TS(i) is the time period between termination of use of compound i to the time the frequency of resistant pests is sufficiently low to maintain effective population control with that compound, the *susceptibility recovery interval*. C is the total number of compounds required for indefinite control.

Pollard (1958) found that higher doses of gamma-BHC (lindane) would kill heterozygotes and indicated that this would slow resistance development. More recently, the high-dose approach has been the subject of much theoretical work (Tabashnik and Croft, 1982). By and large, management of resistance to pesticides has made little direct use of population and community ecology theory. Earlier recognition by pesticide users of the "Volterra" principle (*when predators and their prey are both killed, prey populations will increase*) would have highlighted the danger of indiscriminate use of pesticides on populations where some control of prey species (pests) was achieved by natural enemies (predators).

GENETIC AND ECOLOGICAL INFORMATION REQUIRED FOR MODELS OF THE POPULATION BIOLOGY OF RESISTANCE

Even though specific resistance management programs should be designed on a case-by-case basis, the following general classes of information are required to develop realistic models of the population biology of pesticide resistance, and thus to design resistance management programs:

• *Mode of Inheritance*. Knowing the mode of inheritance of the resistant phenotype is critical to developing any model of pesticide resistance. Although it sounds formidable, a relatively modest amount of genetic information would actually be needed for models of the population genetics of resistance. Particularly critical for these models is whether resistance is inherited as a discrete character (involving one or two major genes) or acts as a continuously distributed (quantitative, or polygenic) character, because different classes of theoretical models are applicable to single-gene and polygenic resistance (Via, Uyenoyama, this volume). In the former we must know the number of alleles at the resistance-determining loci and the dominance relationship among these alleles (as a function of pesticide concentration) (Curtis et al., 1978). It would also be valuable to know the nature of the interactions (epistasis) between the genes determining resistance as well as other, modifying loci (Uyenoyama, this volume). Where resistance acts as a quantitative character, it is particularly critical to know the mean levels of resistance, the phenotypic variances, the additive and nonadditive genetic variance in these levels of resistance, and the genetic covariance in the tolerances to different pesticides (Via, this volume). We recognize that these cannot be known until resistance has evolved, but some generalities on inheritance of resistance are emerging (Chapter 2).

• *Fitness Relationships*. Estimating genotypic fitness is difficult, even in a well-controlled experiment. Nevertheless, at least rough estimates of the relative reproductive and survival rates of resistant and susceptible genotypes are necessary to consider their rates of increase, frequencies after pesticide treatment, and rate of decline when treatment is stopped (i.e., when selection is relaxed). *It is not sufficient to assume that fitness is simply a matter of the kill rate or that resistant genotypes will have a selective disadvantage in the absence of pesticides*. These fitness estimates have to be obtained for resistant and susceptible genotypes as functions of stage in the life cycle and concentrations of pesticides. Fitness should not be assumed to be a constant. In obtaining these estimates, it is necessary to control for a variety of other environmental and genetic factors such as temperature, season, physiological state, population density, and genetic background. Again, this information is not available until after resistance has evolved. In the case of insects and rodents, behavioral considerations should also be taken into account (Gould, 1984).

• *Population Structure*. Some details of intrinsic genetic structure of the target population and its spatial and temporal distribution are critical to developing a realistic model, especially: (1) whether generations are discrete or overlapping, (2) the nature of the alternation of haploid and diploid phases of the life cycle, (3) the relative lengths of sexual and asexual stages, and (4) the duration of the whole life cycle and its various stages. The lengths of both the resistance onset and susceptibility recovery intervals depend in part on how isolated the treated population is. A high rate of migration (gene flow) from susceptible populations would both delay an increase in the proportion of resistant genotypes and increase their rate of decline when treatment is stopped. Migrants from resistant populations, rather than independent mutants, could be the primary reason for the spread of resistance. To determine this physical component of population structure, the nature and timing of migration, as well as its absolute rate, should be considered. When considering gene flow, the frequency of resistant genotypes in the untreated population may also be important if that reservoir population is relatively small. In studying migration, an attempt should be made to estimate the genetically effective component, not just movement (Comins 1977b; Roush and Croft, May and Dobson, this volume).

• *Population Regulation*. Pest population growth is not necessarily exponential and unregulated in the absence of treatment. Interspecific and intraspecific competition, predation, and parasitism may help limit the rate of growth and densities of pest populations. The nature and importance of the population-regulating mechanisms have to be known and considered in the population biology of resistance. The Volterra principle suggests that pesticide use could exacerbate situations where the pest population is normally limited by parasites or predators that are susceptible to the controlling pesticide. The intensity of selection for and against resistant genotypes could be greatly affected by the nature of the trade-off between density-dependent and density-independent mortality and morbidity factors. Where there is substantial intraspecific competition, sublethal doses of a pesticide could have a strong selection effect by weakening the competitive abilities of susceptible individuals, even when it does not control the density of the population (McKenzie et al., 1982).

• *Refuges*. Reservoirs of susceptible genotypes within the treated area could result from pesticide dose variation in space or time. As is the case for weed seeds, these refuges could be quite substantial and play a significant role in augmenting the resistance onset interval (Gressel and Segel, 1978).

GENERAL AND SPECIFIC MODELS

It is possible to construct general models of the population biology of resistance with few—possibly no—data from natural sources. Models of this type have been used to identify the factors contributing to the rise of resistance and evaluate their relative importance (Comins, 1977a; Taylor, 1983; May and Dobson, this volume). These general models may be the only ones that can be constructed when little population biology information is available, and they can have considerable value. Finally, these models can be used to distinguish between the factors that are really important and those that play minor roles in the rise of resistance, thus playing a critical role in deciding which empirical studies should be conducted.

Where extensive information is available, more detailed applied models can be constructed and analyzed with analytical and computer simulation procedures (Tabashnik, this volume). Although more specific models can provide more quantitatively accurate predictions than general models, we see no justification to postpone developing resistance management programs based on models until all the data are available.

SOURCE OF DATA FOR MODEL CONSTRUCTION AND EVALUATION

• *The Roles of Laboratory and Field Studies*. Studies with pesticide-resistant mutants generated in the laboratory and fitness experiments performed with laboratory-selected strains may provide some information about the nature of the alleles conferring resistance and their anticipated fate in populations. Whenever possible, however, these investigations should use susceptible and resistant genotypes isolated from natural sources and perform fitness studies under natural conditions. The genetics of resistance in natural populations are probably different from those generated in the laboratory, because, for example, selection pressure under natural conditions might be different from that in short-term laboratory studies (McKenzie et al., 1982; Uyenoyama, this volume). Laboratory studies indicate that fitness differences are likely to exist in natural situations but do not provide accurate estimates of fitness differentials in the field. On the other hand, laboratory studies could provide reliable estimates of toxicological dominance, if they were performed under conditions that approximate field exposure to pesticides.

• *Extrapolating from Existing Genetic Information and Molecular Procedures*. To a great extent the high rate of progress in academic genetics can be attributed to the common use of relatively few species (model systems) that are particularly convenient to study. Unfortunately, real pest organisms are seldom ideal experimental organisms, so genetic information often has to be acquired by extrapolation from related organisms.

Using DNA and RNA probes to determine the physical location of genes and to ascertain whether homologous genes are responsible for the same phenotype in different species considerably broadens the range of organisms amenable to genetic analysis. Only limited use has been made of in vitro genetic procedures to investigate the genetics of pesticide resistance (see Georgopoulos, Gressel, Hardy, Hammock and Soderlund, MacNicoll, Plapp, Chapter 2, this volume). Obtaining DNA and RNA probes is not easy when the gene product is not known or known and present in low quantities, or when the physical location on the gene of the model organisms is not known, but molecular techniques should be considered for determining modes of inheritance for population studies of pesticide resistance.

It is both convenient and traditional to focus on phenotypes that are (or seem to be) discrete characters determined by one of two genes, but it is critical to consider that specific cases of resistance may be determined by multiple alleles and that resistance behaves as a quantitative character. There are well-developed procedures to analyze inheritance of quantitative characters and model the behavior of these characters under selection (Via, this volume).

EVALUATING MODELS AND PROGRAMS FOR PESTICIDE RESISTANCE MANAGEMENT

While we may believe that existing studies of the fit between theory and empirical observation justify the use of population biology theory to develop pesticide use and resistance management programs, a final demonstration of their utility remains necessary. In order to demonstrate the utility of mathematical and numerical modeling, the programs developed using them must: (1) maintain the required level of pest control, (2) be economically competitive, (3) yield lower levels of resistance than would be anticipated for alternative programs employing the same compound(s), and (4) be safe from both an environmental and health perspective.

While not sufficient in a formal sense, the a posteriori fit between observation and prediction should certainly be considered partial demonstration of the validity of models. Properly controlled pilot studies could provide further evidence, if they were run under field conditions using a few "model" systems with properties similar to those of the intended target species and communities. In cases where the pesticide is already in use, field data could serve as control. These studies should make the evaluation in the minimum time possible, and some acceleration could be achieved by using procedures to detect resistant organisms when they are rare and possibly heterozygous (for one- or two-gene resistance), or when resistance levels are low (for polygenic resistance).

The models and data will be quantitative, but fit will have to be evaluated somewhat qualitatively. The extraordinary number of interactions between the biotic and physical factors in a field study cannot all be controlled. On the other hand, if a program is effective, one would anticipate the desired level of pest control and significantly lower rates of increase of resistant genotypes in the experimental populations.

FOLKLORE, DOGMA, AND AD HOC PRACTICES

There are a number of current pesticide use practices and assumptions about their consequences for resistance management that seem to have little or no base in population biology theory.

• *Return to Pesticide Susceptibility.* While only occasionally stated explicitly, there seems to be a general belief that a decline in the frequency of resistant genotypes will necessarily follow when use of a compound is stopped. While we expect this to be true in the long run, the length of the susceptibility recovery interval may be effectively indefinite in many cases. In the absence of pesticide use, the selective differential between susceptible and resistant genotypes may be quite small. Even if the original resistant genotypes had a marked disadvantage in the absence of the pesticide, there may be selection for modifier genes that improve the fitness of resistant genotypes.

The limited empirical results on the fate of alleles conferring resistance following termination of pesticide use support a mixed view of the fate of resistance genotypes in the absence of pesticide selection. In some cases, the frequencies of alleles conferring resistance declined relatively rapidly (Greaves et al., 1977; Partridge, 1979; McKenzie et al., 1982; also see Greaves, Georgopoulos, this volume). In other cases, there was little change in the frequency of these alleles following the relaxation of selection (Whitehead et al., 1985; Georgopoulos, Roush and Croft, this volume).

Even in cases where the resistant genotypes have a clear selective disadvantage relative to sensitive genotypes, the intervals for susceptibility recovery will still be substantially longer than for the corresponding resistance onset. The intensity of selection favoring resistance during pesticide use will certainly be much greater than that favoring susceptibility following the termination of treatment. For a pesticide to be biologically effective for a period as long as that during its first use, the frequency of resistant genotypes in the recovered population would have to be similar to that prior to first use (see May and Dobson, this volume).

This conclusion has a number of immediate implications. First, the simplistic scheme depicted in our heuristic model is unlikely to be a realistic long-term solution to the problem of pesticide resistance. The recovery period following the rise of resistance could be extremely long and, for practical purposes, too long for individual pesticides to be used more than once. Thus, long-term control by pesticides alone would require an almost infinite supply of independent compounds. In a short-term view, the factors affecting evolutionary rates also illustrate the utility of (1) terminating pesticide use before the frequency of resistance is high; (2) developing procedures that increase the selection pressures favoring susceptible genotypes; and (3) programs that increase rates of gene flow from sensitive populations.

• *Pesticide Mixing and Cycling.* A current controversy is whether pesticides should be in rotations or mixtures before their target pest(s) become resistant. The answer is equivocal. Models can be constructed in which pesticide cycling or mixing either increases or decreases the resistance onset interval. The outcome depends critically on the way the different pesticides interact in determining the fitness of resistant and sensitive genotypes. Also important are: modes of inheritance of resistance; frequency of mutations for resistance; rates of recombination between the loci involved; and population dynamics of pest growth, refuges, migration and pesticide action.

These qualifications emphasize the need for considering tactics on a case by case basis with validation prior to implementation. The population biology of each type of pesticide use regime can be readily modeled, and the relative merits and liabilities of these pesticide use regimes can be assessed a priori.

- *Directed Evolution of Resistance*. A fundamental premise of evolutionary theory is that mutations occur at random; their incidence and nature are independent of specific selection pressure. Starting with the classical fluctuation test experiment of Luria and Delbruck (1943), there have been a number of lines of evidence in support of this interpretation (Crow, 1957). There have been suggestions, nevertheless, that pesticides will promote the generation of resistant organisms (as well as select for increase in their frequency) or that resistance to one compound will increase the rate of mutation to a second compound (Wallace and MacSwiney, 1976).

While it may be easy to discount these (or any) neolamarckian interpretations, we believe that the hypothesis that the rate and nature of *mutation* is influenced by selection for that mutation is interesting from both an academic and applied perspective and certainly worth testing. We can speculate on mechanisms that make mutations appear to be directed. In nonlethal doses, pesticides could cause "genomic shocks" that increase frequencies of transposition of chromosome pieces. If pesticide resistance is the result of inserting movable elements of chromosomes, then conceivably the initial transposition could increase the future rates of transposition. In cases where resistance to specific pesticides requires two mutations, one in a gene that is common to resistances to different compounds and one that is unique to each, mutation could appear to be directed.

IMPEDIMENTS

Implicit in this discussion is the assumption that the pesticide resistance problem is amenable to a technical solution. There is some justification for this assumption; for specific agricultural or clinical situations, programs using combinations of chemical and biological agents could be developed to prolong the useful life of compounds. On the other hand, we see little justification in maintaining the polite fiction that pesticide resistance is solely a technical problem and therefore solvable with the right tools. The design, execution, monitoring, and evaluation of pesticide-use programs and their ultimate implementation are major endeavors, even for single agricultural or clinical situations. Development and testing require cooperation of investigators in a variety of fields: chemistry, genetics, population biology, toxicology, botany, microbiology, zoology, epidemiology, and medicine. These activities have to be coordinated with people actually running and monitoring the program in the field or clinic. Pesticide-producing companies, primary users of these compounds (growers, physicians, veterinarians, and public health personnel) and government agencies regulating their use will have to participate.

- *The Dilemma of Interdisciplinary Programs*. Pesticide-use programs are interdisciplinary, yet universities, research institutes, and funding organizations responsible for their development and support are rigidly structured along traditional, disciplinary lines. In universities in the United States, academic and applied biology departments are almost always separate, both geographically and administratively, and have been maintained that way for 50 years or more. Most evolutionary geneticists, ecologists, and population biologists are in academic departments while biologists directly involved in pesticide use and management are in agricultural, clinical, and other more applied departments. Academic and applied biologists primarily publish in different journals and receive funding from different sources. As a result, there is relatively little intellectual intercourse between investigators in these two types of biology departments and often considerable xenophobia. While there are many situations where these administrative and geographic barriers have been breached (e.g., a number of papers in the bibliographies of the population biology papers in this volume and cited here), these are rare exceptions. More extensive breakdown of the traditional separation between applied and academic biology would be a major step toward the solution to the pesticide resistance problem as well as other biological-technical problems.

We see no easy general solution to this problem. While lip service is frequently given to the value of interdisciplinary programs, their active development has been limited at best, and this situation is likely to persist as long as universities, research institutes, and funding agencies are administratively partitioned into academic and applied areas. As long as these separate administrative units have primary control over personal rewards (salary, promotion, tenure), and as long as the kudos (invitations, travel, awards, and other recognition) are generated along disciplinary lines, from a purely careerist perspective, there is little positive incentive for individuals to engage in interdisciplinary projects; in some cases, there is pressure to avoid doing so. Funding may well be the greatest impediment to jointly applied and theoretical research. As long as research is funded either explicitly or implicitly (via the peer review system) along disciplinary lines, interdisciplinary projects will be at a disadvantage.

In the long run academic and applied biology could be somewhat unified, despite existing administrative barricades, with a more ecumenical approach to teaching. The genetics and population biology of pesticide resistance are certainly interesting applied problems that merit investigation even from the perspective of the most basic biology. Many other applied examples could replace more traditional model systems or natural populations used as examples in genetics and population biology courses.

RECOMMENDATIONS

RECOMMENDATION 1. **Pesticide use practices based on considerations of the population biology of pesticide resistance should be developed and implemented.**

Although the theory and observations of academic population biology have been used to explain past resistance episodes, at this juncture there have not been significant pesticide use programs developed and implemented from considerations of the principles of population biology.

RECOMMENDATION 2. **General models of the population biology of resistance can be used to develop pesticide-use practices, as long as the basic premises of these models can be empirically justified.**

While it may be a long-term goal to develop precise analogs of specific pesticide-use situations, population biology theory may be applied to develop pesticide-use regimes before specific models are developed. The fact that general population biology theory has been successful in a retrospective manner, by providing mechanistic explanations for past resistance episodes, justifies the use of this theory in a prospective manner.

RECOMMENDATION 3. **While general models may have broad utility, it remains necessary to gather the genetic and ecological information needed to construct specific models.**

In cases where general models prove inadequate, it will be necessary to employ specific and precise analogs of the populations and pesticides under consideration.

RECOMMENDATION 4. **The continuous monitoring of resistance frequencies should be an integral part of all programs to manage resistance.**

If the models are realistic analogs of the effects of the pesticide use regime on the genetic structure of the target population, there should be a good correspondence between the observed and predicted resistance frequencies and changes in those frequencies.

RECOMMENDATION 5. **Population biology theory should be used to examine current pesticide-use practices and controversies.**

There are a variety of ad hoc pesticide-use practices, e.g., alternating and mixing pesticides to extend the useful life of compounds, which may or may not be justifiable. Mathematical models of the population biology of pesticide use represent an efficient way to evaluate these practices in a prospective manner.

RECOMMENDATION 6. **An extensive effort should be made to encourage both research on pesticide use and resistance by academic biologists and the study of the population biology by applied biologists involved in pesticide use.**

Pesticide resistance is a long-term problem that will require the coordinated efforts of investigators representing several disciplines that currently suffer from a lack of interdisciplinary communication. While unlikely to be sufficient as a unique solution to the problem of coordinating efforts, some funds specifically earmarked for joint basic and applied research on the population biology of pesticide resistance may help surmount some of the institutional impediments to this type of interdisciplinary activity.

RECOMMENDATION 7. **A considerable effort should be put into developing pest-control measures that do not rely on the use of chemical pesticides.**

The continuous control of pest populations by cycling through novel chemical pesticides is unlikely to be a viable long-term strategy. There is no biological or evolutionary justification for the assumptions that (1) pest populations will return to sensitive states relatively quickly following the termination of the use of specific pesticides, or (2) an adequate supply of novel and safe pesticides can be developed and made available continuously to replace compounds that have lost their effectiveness due to resistance.

ACKNOWLEDGMENT

We would like to thank Ralph V. Evans for his comments on this manuscript.

REFERENCES

Comins, H. N. 1977a. The management of pesticide resistance. J. Theor. Biol. 65:399–420.

Comins, H. N. 1977b. The development of insecticide resistance in the presence of migration. J. Theor. Biol. 64:177–197.

Crow, J. F. 1957. Genetics of insect resistance to chemicals. Annu. Rev. Entomol. 2:227–246.

Curtis, C. F., L. M. Cook, and R. J. Wood. 1978. Selection for and against insecticide resistance and possible methods for inhibiting the evolution of resistance in mosquitoes. Ecol. Entomol. 3:515–522.

Davidson, G. and D. G. Pollard. 1958. Effect of simulated field deposits of gamma-BHC and

Dieldrin on susceptible hybrid and resistant strains of *Anopheles gambiae* Giles. Nature 182:739–740.

Gould, F. 1984. The role of behavior in the evolution of insect adaptation to insecticides and resistant host plants. Bull. Entomol. Soc. Am. 30:34–41.

Greaves, J. H., R. Redfern, P. B. Ayers, and J. E. Gill. 1977. Warfarin resistance: a balanced polymorphism in the Norway rat. Genet. Res. Camb. 89:295–301.

Gressel, J., and L. A. Segel. 1978. The paucity of plants evolving genetic resistance to herbicides: possible reasons and implications. J. Theor. Biol. 75:349–371.

Gutierrez, A. P., U. Regev, and C. G. Summers. 1976. Computer model aids in weevil control. Calif. Agr. April:8–18.

Luria, S. E., and M. Delbruck. 1943. Mutations of bacteria from virus sensitivity to virus resistance. Genetics 28:491–511.

MacDonald, G. 1959. The dynamics of resistance to insecticides by anophelines. Revista di Parassitologia 20:305.

McKenzie, J. A., M. J. Whitten, and M. A. Adena. 1982. The effect of genetic background on the fitness of diazon resistance genotypes of the Australian sheep blowfly, *Lucilia cuprina*. Heredity 19:1–19.

Partridge, G. G. 1979. Relative fitness of genotypes in a population of *Rattus norvegicus* polymorphic for warfarin resistance. Heredity 43:239–246.

Tabashnik, B. E., and B. A. Croft. 1982. Managing pesticide resistance in crop arthropod complexes: interactions between biological and operational factors. Environ. Entomol. 11:1137–1144.

Taylor, C. E. 1983. Evolution of resistance to insecticides: the role of mathematical models and computer simulations. Pp. 163–173 in Pest Resistance to Pesticides, G. P. Georghiou and T. Saito, eds. New York: Plenum.

Taylor, C. E., F. Quaglia, and G. P. Georghiou. 1983. Evolution of resistance to insecticides: a cage study on the influence of migration and insecticide decay rates. J. Econ. Entomol. 76:704–707.

Wallace, M. E., and F. MacSwiney. 1976. A major gene controlling warfarin resistance in the house mouse. J. Hyg. Camb. 76:173–181.

Whitehead, J. R., R. T. Roush, and B. R. Norment. 1985. Resistance stability and co adaptation in diazinon-resistant house flies (Diptera:Muscidae). J. Econ. Entomol. 78:25–29.

WORKSHOP PARTICIPANTS

Population Biology of Pesticide Resistance. Bridging the Gap Between Theory and Practical Applications

BRUCE R. LEVIN (*Leader*), University of Massachusetts
J. A. BARRETT, Cambridge University
ELINOR C. CRUZE, National Research Council
ANDREW P. DOBSON, Princeton University
FRED GOULD, North Carolina State University
JOHN H. GREAVES, Ministry of Agriculture, Fisheries and Food, Great Britain
DAVID HECKEL, Clemson University
ROBERT M. MAY, Princeton University
HAROLD T. REYNOLDS, University of California, Riverside
RICHARD T. ROUSH, Mississippi State University

BRUCE E. TABASHNIK, University of Hawaii
MARCY UYENOYAMA, Duke University
SARA VIA, University of Iowa
MAX J. WHITTEN, Commonwealth Scientific and Industrial Research Organization
M. S. WOLFE, Plant Breeding Institute, Cambridge

Pesticide Resistance: Strategies and Tactics for Management.
1986. National Academy Press, Washington, D.C.

Factors Influencing the Evolution of Resistance

GEORGE P. GEORGHIOU and CHARLES E. TAYLOR

Any attempt to devise management strategies for delaying or forestalling the evolution of pesticide resistance requires a thorough understanding of the parameters influencing the selection process. The parameters known to influence this process in pest populations are presented systematically under three categories—genetic, biological/ecological, and operational—and their relative importance is discussed with reference to available case histories.

INTRODUCTION

More than 447 species of arthropods have now developed resistance to insecticides (Georghiou, this volume). The main weapon for countering this resistance has been the use of alternative chemicals with structures that are unaffected by cross-resistance. The gradual depletion of available chemicals as resistance to them developed has revealed the limitations of this practice and emphasized the need for maximizing the ''useful life'' of new chemicals through their application under conditions that delay or prevent the development of resistance. To achieve this goal it is essential to understand the parameters influencing the selection process.

It is well established that resistance does not evolve at the same rate for all organisms that come under selection pressure. Resistance may develop rapidly in one species, more slowly in another, and not at all in a third. For example, despite enormous selection pressure during many years of intensive DDT treatment in the corn belt of the United States, the corn borer showed no evidence of resistance. Yet house flies in many areas developed resistance

TABLE 1 Known or Suggested Factors Influencing the Selection of Resistance to Insecticides in Field Populations

- A. Genetic
    - a. Frequency of R alleles
    - b. Number of R alleles
    - c. Dominance of R alleles
    - d. Penetrance, expressivity, interactions of R alleles
    - e. Past selection by other chemicals
    - f. Extent of integration of R genome with fitness factors
- B. Biological/Ecological
    - 1. Biotic
        - a. Generation turnover
        - b. Offspring per generation
        - c. Monogamy/polygamy, parthenogenesis
    - 2. Behavioral/Ecological
        - a. Isolation, mobility, migration
        - b. Monophagy/polyphagy
        - c. Fortuitous survival, refugia
- C. Operational
    - 1. The chemical
        - a. Chemical nature of pesticide
        - b. Relationship to earlier-used chemicals
        - c. Persistence of residues, formulation
    - 2. The application
        - a. Application threshold
        - b. Selection threshold
        - c. Life stage(s) selected
        - d. Mode of application
        - e. Space-limited selection
        - f. Alternating selection

SOURCE: Adapted from Georghiou and Taylor (1976).

within two to three years under selection pressure by this insecticide. Even within a species, resistance may develop more rapidly in one population than in another. The Colorado potato beetle, for example, showed far greater propensity for resistance on Long Island than on the mainland (Forgash, 1981, 1984).

There are many factors that can influence the rate at which this evolution proceeds. One effort to systematize them is shown in Table 1, modified slightly from a classification we proposed and discussed earlier (Georghiou and Taylor, 1976, 1977a,b). The factors are grouped into three categories, depending on whether they concern the genetics of resistance, the biology/ecology of the pest, or the control operations used. Most factors in the first two categories cannot be controlled, and the importance of some may not even be determined until resistance has already developed. Only through hindsight, for example, can one obtain any idea about the initial frequency of the alleles conferring resistance. Nor is it usually possible to measure dominance until one isolates such alleles and makes the appropriate crosses. In some cases these issues may be addressed in laboratory studies where resistant strains can be developed by selection on large, recently colonized populations. Nonetheless, some factors that influence the evolution of resistance are under man's control, especially those related to the timing and dose of insecticide application (Operational Factors, Table 1). The problem is to identify them and determine how their manipulation under the existing genetic and biological/ecological constraints may retard the evolution of resistance.

During the past few years, important contributions have been made by workers in a handful of laboratories, mainly in the United States, the United Kingdom, and Australia (Comins, 1977a,b, 1979a,b; Georghiou and Taylor, 1977a,b; Haile and Weidhaas, 1977; Curtis et al., 1978; Conway and Comins, 1979; Sutherst and Comins, 1979; Sutherst et al., 1979; Taylor and Georghiou, 1979, 1982; Gressel and Segel, 1982; Muggleton, 1982; Tabashnik and Croft, 1982; Levy et al., 1983; McPhee and Nestmann, 1983; Taylor et al., 1983; Wood and Cook, 1983; Knipling and Klassen, 1984; Mani and Wood, 1984; McKenzie and Whitten, 1984). Some of these contributions are examined in other papers in this symposium. We shall confine ourselves to a discussion of how, in a historical perspective, the accumulated knowledge on the occurrence and dynamics of resistance leads to the recognition of these factors (Table 1) as important.

GENETIC FACTORS IN RESISTANCE

Evolutionists frequently assume that organisms have the capacity to evolve nearly any type of resistance. From this follow many of the "optimization" arguments and the "adaptationist program" (Lewontin and Gould, 1979). This assumption is not warranted for insecticide resistance. Some populations obviously do not have the capacity to come up with the necessary resistant alleles in the first place, despite what would seem to be an obvious advantage for doing so. The corn borer is one species that did not. The paucity of cases of resistance to arsenicals in insects and to copper fungicides in plant pathogens are other examples. It has been speculated that herbivorous species, which have frequently evolved the capacity to deal with plant alkaloids, are in some sense preadapted to dealing with the problems posed by dangerous chemicals in their environment (Croft and Brown, 1975).

Related to this is the fact that there may be many ways to achieve resistance—by detoxifying the chemicals, altering site specificity, reducing penetration, behavioral avoidance of residues, to name a few. When more avenues are open it would be expected that resistance would evolve more easily.

Once alleles conferring resistance are present in the population, the frequency at which they occur may be important. There are several reasons for this. Obviously if the initial frequency is higher, then resistance has a head start. There may, however, be an Allee effect, so if the population is reduced to a sufficiently low level, the resulting population size is too small to sustain positive growth, perhaps by failure to find mates. More important, the selection pressures and immigration rates may impose an unstable equilibrium of gene frequencies, below which resistance alleles decrease in fitness and above which they increase (Haldane, 1930). In this case the initial frequency is especially important.

In practice the importance of many factors for resistance seems related to this unstable equilibrium. In the simplest instance this equilibrium depends largely on initial gene frequency, dominance, and immigration. These factors in turn may depend on others. Imagine a population with resistant allele, R, at a low frequency. Homozygous RR individuals may occur if the population is large enough, but will be very few in number. If the resistance is recessive or can be made recessive by application of an appropriately high dose of insecticide (Taylor and Georghiou, 1979), then following insecticide use all of the susceptible homozygotes (SS) and heterozygotes (RS) will be eliminated, leaving only the very few RR. If now there is an inflow of largely susceptible migrants, then those few RR will mate with SS homozygote immigrants, and the offspring for the next generation will be almost all SS and RS. These can be killed with another application of insecticide, keeping the population under control. It is possible to study this result mathematically and describe precisely when it should be observed (Comins, 1977a; Curtis et al., 1978; Taylor and Georghiou, 1979).

It is generally thought that resistance alleles are mildly deleterious prior to insecticide use, so that they are present initially at some sort of mutation-selection balance. This would typically be at an allele frequency of 10^{-2} to 10^{-4}, with the RR homozygotes present at 10^{-4} to 10^{-8}. Of course if two loci are required or if more than one nucleotide change is necessary then the frequency may be substantially less (Whitten and McKenzie, 1982).

McDonald (1959) proposed that dieldrin resistance, being more dominant than DDT resistance in Anopheline mosquitoes, would evolve at a faster rate. In theory there should be little difference between rates of evolution of dominant and recessive alleles in the absence of immigrants. But, in fact, McDonald's prediction has been more-or-less realized. The reason for this is probably related to the unstable equilibrium described above, which exists only when the resistant allele is recessive.

Dominance typically depends on the dose applied. Figure 1 shows the dosage-response curves for three genotypes of a mosquito, *Culex quinquefasciatus*, exposed to a pyrethroid insecticide. When a small dose, D_S, is applied, the heterozygotes survive, but with a larger dose, D_L, they do not. Thus, with D_S, the resistance is functionally dominant, but with D_L, it is

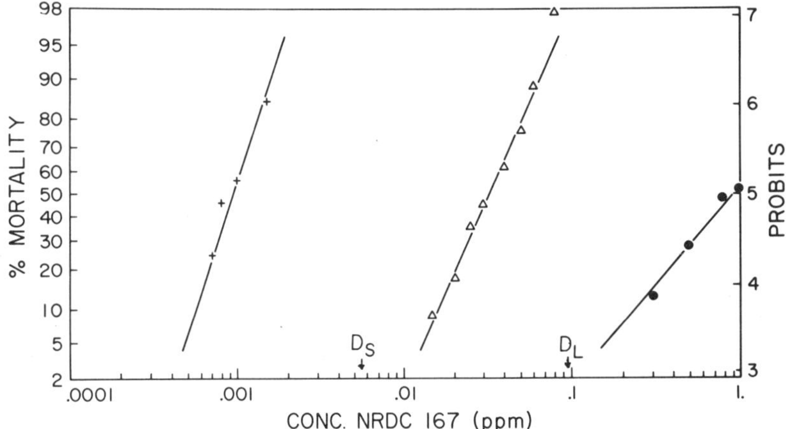

FIGURE 1 Dosage-response lines for larvae of *Culex quinquefasciatus* susceptible, heterozygous, and resistant tested with permethrin. The dominance is seen to depend on dose: with a small dose (D_S), resistance is functionally dominant, whereas with a large dose (D_L) it is functionally recessive.

functionally recessive. Modifier genes are known to change the location of the heterozygote line, typically moving it to the right.

Modifier genes may be important in other ways as well, most notably by helping to integrate the resistance allele into the rest of the genome to produce a "harmoniously coadapted genome" in the sense of Mayr (1963) or Dobzhansky (1970). There may be many pleiotropic effects from the substitution of a resistant allele for its wild-type alternative. Many of these are likely to be detrimental, so the resistant allele is initially mildly deleterious (Ferrari and Georghiou, 1981). Later, when there has been an opportunity for the modifiers to be selected and the pleiotropic side effects have been compensated for, such a disadvantage diminishes or disappears.

With few exceptions resistant populations demonstrate lower fitness than their susceptible counterparts. Continued selection may improve fitness through coadaptation of the resistant genome, resulting in more stable resistance. A dramatic illustration of this is a laboratory experiment of Abedi and Brown (1960). They selected for resistance, then released selection, then selected, and so forth. After several cycles resistance evolved much more rapidly and was more stable than initially. Almost certainly, modifier genes were the cause.

Instability of resistance may not necessarily be due entirely to differences in fitness, however. For example, genes for resistance to an organophosphate (temephos), a pyrethroid (permethrin), and a carbamate (propoxur) were introduced into a susceptible strain of *Culex quinquefasciatus* through a system of backcrosses. The resulting synthetic was subsequently divided into substrains and selected by these insecticides. Tests showed that the stability of resistance in each strain differed considerably: Organophosphate resistance regressed rapidly, pyrethroid resistance moderately, but resistance to the carbamate showed considerable persistence (Georghiou et al., 1983). It is, therefore, likely that the mechanism of resistance involved in each case may influence its persistence in populations.

Past selection by insecticides may facilitate evolution of resistance to new insecticides because of cross-resistance. Certain mechanisms of resistance have been found to confer resistance not only within an insecticide class but across classes as well. A classic example of this is the *kdr* gene. Both DDT and pyrethroids interfere with sodium gates along the axons of nerve cells. The *kdr* allele, by altering properties of the axonal membrane, makes it less receptive to binding. Thus, it confers resistance to pyrethroids in populations that had been selected earlier by DDT and vice versa (Priester and Georghiou, 1978; Omer et al., 1980).

Recently, Sawicki et al. (1984) showed that an esterase, E.O.33, selected in house flies by the organophosphates malathion and trichlorphon, confers mild cross-resistance to pyrethroids as well. By itself the esterase is of no consequence in the control of house flies with pyrethroids because the doses used in practice are strong enough to overcome the mild resistance it confers. In some populations, however, *kdr* is also present, albeit at low frequencies, probably as a result of previous use of DDT for control of flies. In these populations the introduction of pyrethroids led to the simultaneous selection of *kdr*, as well as the esterase, and to rapid control failure of pyrethroids. Thus, the earlier, sequential use of two different groups of insecticides, organophosphates and DDT, contributed to the rapid failure of a third group of compounds, the pyrethroids, through the selection of common resistance mechanisms.

The Colorado potato beetle also provides a pertinent example. On Long Island the population of this species required seven years to develop resistance to DDT, the first synthetic insecticide with which it was selected. The same population has required progressively less time to develop resistance to the subsequently used chemicals: five years for resistance to azinphosmethyl, two for carbofuran, two for pyrethroids, and one for pyrethroids with a synergist (Georghiou, this volume).

BIOLOGICAL/ENVIRONMENTAL FACTORS IN RESISTANCE

Ecology and life histories may dramatically alter the responsiveness to the selection that leads to resistance. Most obvious, of course, is that the larger the number of generations per year, the faster the evolution of resistance. The fruit tree mite *Panonychus ulmi*, which has as many as 10 generations

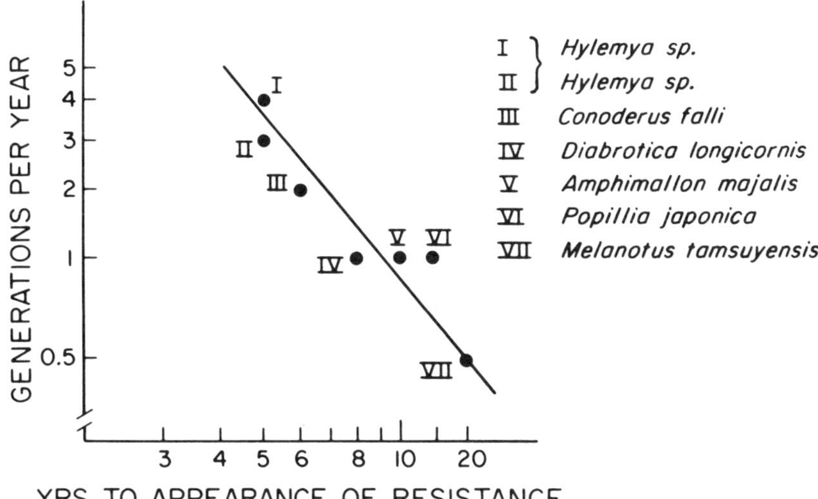

FIGURE 2 Relationship between generations per year and appearance of resistance in species selected by soil applications of aldrin/dieldrin.

per year, has developed resistance rapidly to many groups of insecticides. But another fruit tree mite *Bryobia rubrioculus*, which has only two generations per year, has yet to be reported as resistant (Georghiou, 1981).

Figure 2 illustrates the relation between generation turnover in various soil-inhabiting pest species and the number of years it has taken them to manifest resistance to soil applications of aldrin/dieldrin (Georghiou, 1980). It can be seen that root maggots (*Hylemya* spp.), which complete three to four generations per year, evolved resistance after five years of exposure, while *Conoderus falli*, with two generations per year, evolved resistance in six years. *Diabrotica longicornis*, *Amphimallon majalis*, and *Popillia japonica*, each with one generation per year, have required 8 to 14 years for resistance development, while the sugarcane wireworm (*Melanotus tamsuyensis*) in Taiwan, with a two-year life cycle, has taken 20 years to develop resistance. A similar correlation between generation turnover and rate of evolution of resistance is reported for apple tree pests by Tabashnik and Croft (1985).

All else being equal, populations with a higher reproductive potential are able to withstand a higher substitutional load, that is, they can tolerate a higher intensity of selection. Consequently one would expect to see a positive correlation between the rate of evolution of resistance and fertility. We are not aware of generalizations regarding this, however; nor are we aware of generalizations regarding monogamy/polygamy or mode of reproduction. Because of the unstable equilibrium discussed above, immigration may have a decisive role in retarding evolution. It is essential, however, that the few surviving RR homozygotes mate with SS immigrants. One might then expect polygamous species to evolve more slowly. Related to this is the importance of sexual selection and evolution of sex. It is thought that the principal advantage conferred by sexual systems over asexual ones is the ability to respond to environmental challenges, especially if the challenges are offered in rapid succession (the red queen hypothesis, as detailed in Maynard-Smith, 1978). There is clearly an opportunity for much interesting research here.

Polyphagous insect pests tend to develop resistance more slowly than monophagous ones. Two factors may contribute to this: A smaller part of polyphagous species are likely to be exposed, hence the selection is less intense on these species; because some of the insects would be in untreated refugia, they would provide a reservoir from which untreated, susceptible migrants could come. This may be the reason that resistance in ticks of livestock in South Africa appeared first in one-host species and only later in species that attack two or three hosts (Whitehead and Baker, 1961; Wharton and Roulston, 1970). Similarly, among aphids the spotted alfalfa aphid in California was one of the first to develop resistance, but the lettuce aphid, which moves to poplars during part of the year, has been controlled without evidence of resistance.

It is interesting that on strictly biochemical criteria polyphagy may enhance the potential of a species to develop resistance. Krieger et al. (1971) have provided evidence that in lepidopterous larvae the insecticide-metabolizing activity of microsomal oxidases is higher in polyphagous than in monophagous species. It is possible that a similar mechanism is involved in the tendency of plant-feeding insects to evolve resistance before their parasitoids do (Croft, 1972; Georghiou, 1972), although it should be apparent that the parasitoids can survive only after their hosts have become resistant, giving an evident bias in sampling.

We have suggested that one of the most important features of an insect's ecology, insofar as resistance is concerned, is the amount of immigration of susceptible individuals (Georghiou and Taylor, 1977a). After treatment with insecticides only a few RR individuals will usually survive (if a large enough dose, D_L, is used to make the resistance functionally recessive). If, then, enough SS immigrants arrive and mate with them, for all practical purposes the offspring will consist only of RS heterozygotes and SS homozygotes, both of which can be killed with subsequent treatment. If, however, there are no immigrants, or if they are too few, then substantial numbers of RR individuals will be produced and the population will be on its way to evolving resistance. This gives the unstable equilibrium alluded to above. The critical issues here are the numbers of RR survivors and SS immigrants. Low population densities contribute to fewer RRs, and immigration rates, refugia, polyphagy, and polygamy all contribute to this process.

As an illustration of the adverse effect of isolation, or absence of immigration, it may be noted that the highest resistance of house flies in California was found in populations breeding inside poultry houses. These houses had been screened, ostensibly for the purpose of excluding flies from entering. Ironically, prevention of immigrants has probably contributed to even higher levels of resistance.

In normal pest control all surviving individuals have not necessarily been reached by chemical treatment. Depending on the biological and behavioral characteristics of a species, a proportion may be present in refugia at the time of treatment, thus escaping selection. Refugia may consist of plant tissues, distorted foliage, growth buds, erineum, and the like, or they may represent a physiological state of lower susceptibility, such as diapause or pupation in soil. Whatever the reason, such refugia may be very important in providing a source of susceptible immigrants, thus retarding evolution (Georghiou and Taylor, 1976). The eriophyid mite *Aceria sheldoni*, which inhabits citrus buds, has been controlled for several years with chlorobenzilate and has yet to develop resistance. The citrus rust mite, however, also an eriophyid but feeding on leaf surfaces, has been reported as resistant.

Refugia may often be an important mechanism for delaying the buildup of resistance. Relative to the inward flux of migrants from the outside, they are less subject to the vagaries of weather, breeding sites, and other factors that may influence the timing or intensity of immigration from the outside. Further, we have suggested that refugia may be created artificially by intentionally excluding from treatment some segment of the population and it can thus be an operational factor in resistance management (Georghiou and Taylor, 1977b). Even with refugia, however, some inflow of migrants is necessary for an unstable equilibrium to exist.

OPERATIONAL FACTORS IN RESISTANCE

Operational factors in resistance are those related to the application of pesticides and are thought of as being under man's control. Most obviously these include the timing, dose, and formulation of pesticides used. But, in a way, effective dominance, refugia, and immigration may also be under some degree of control if conditions of application are made more-or-less favorable to them. For example, as indicated above refugia may be created by deliberately excluding some part of the population from treatment. The efficacy of this has been explored by Denholm et al. (1983), using house flies that had already been partially selected for resistance to a long-residual, synthetic pyrethroid, permethrin. Within three weeks after a single application of this persistent insecticide, to which virtually all flies were exposed, they became very resistant. But when a closely related pesticide, bioresmethrin, was applied as a space spray at two-week intervals, no buildup of resistance was observed. This difference was attributed to the fact that bioresmethrin exerted only an immediate toxic effect on the adult flies directly exposed to it. The many flies not in the adult stage, and thus in refugia, became part of the breeding population when they later emerged.

Timing of insecticide use may often be important. For an unstable equilibrium to exist there must be very few RR survivors following the initial treatment. This will occur if the R allele frequency is low, and also when the total population size is low. All else being equal, it is desirable to treat the population before its numbers become too large.

Pesticide dosage has been discussed above as an important determinant of dominance. Related to this are the formulation and rate of pesticide decay. After initial application the concentration of pesticide effectively decreases, because of breakdown, dilution and so forth. If this occurs rapidly then the population can be thought of as effectively receiving either a large dose, D_L, or none at all. With a persistent pesticide this occurs slowly, however, and for some time there is an effectively small dose, D_S, that may be very favorable for resistance development. A persistent pesticide may also kill susceptible immigrants and thus effectively prevent immigration.

Computer simulations have indicated that the timing and economic thresholds of application make little difference in the absence of migration. This is because selection is usually so intense that the selection coefficients are virtually the same in all these circumstances.

Of course the choice of insecticide is very important. Usually there is some degree of cross-resistance to other pesticides within the same class. Depending on the mechanism of resistance, there may also be cross-resistance among classes. Especially notable are cross-resistance between DDT and pyrethroids due to the gene *kdr* and between carbamates and organophosphates due to selection of ''insensitive'' acetylcholinesterase (Hama, 1983).

Whether insecticides are best used in combinations or sequentially is at present unclear. There are some suggestions that combinations may be more effective if there is much dominance and immigration in the system (Mani, in press; C. F. Curtis, London School of Hygiene and Tropical Medicine, personal communication, 1985). Our simulations, using quantitative genetic models, indicate that there is little difference if one works under the constraint of a constant selection differential. The available experimental evidence also suggests that there is little difference. Georghiou et al. (1983) selected mosquitoes by various combinations or sequences of temephos, permethrin, and propoxur, representatives of the three major classes of insecticides. The populations responded more-or-less the same. They observed, however, that there was some negative cross-resistance, in that strains that were more resistant to the organophosphate tended to be more susceptible to the pyrethroid. Just how this can be put to best use in an operational sense is still unclear. There is certainly a need for more experimental and theoretical work on this important problem.

CONCLUSION

Because insecticide resistance has become such a serious problem in recent years, it is abundantly clear that merely switching to new insecticides when the current one is no longer effective cannot continue. Integrated pest management, which will almost always involve some use of pesticides, is now regarded as essential. Recognizing and manipulating those factors that may help retard resistance should be an integral part of any such program. Throughout the preceding discussion we have emphasized the effects of pesticides on the target population alone. No mention has been made of the effects on competitors, parasites, or predators. These should be a part of the deliberation of which strategy to use, especially when considering the use of several insecticides in combinations. In any practical problem there are bound to be many unknowns, even surprises. There is a need for better knowledge of the factors influencing the evolution of resistance, enabling us to better assess the risk of resistance developing in each individual case and thus to formulate more realistic management practices for delaying or forestalling its evolution.

REFERENCES

Abedi, Z. H., and A. W. A. Brown. 1960. Development and reversion of DDT-resistance in *Aedes aegypti*. Can. J. Genet. Cytol. 2:252–261.

Comins, H. N. 1977a. The development of insecticide resistance in the presence of migration. J. Theor. Biol. 64:177–197.

Comins, H. N. 1977b. The management of pesticide resistance. J. Theor. Biol. 65:399–420.

Comins, H. N. 1979a. Analytical methods for the management of pesticide resistance. J. Theor. Biol. 77:171–188.

Comins, H. N. 1979b. The management of pesticide resistance: Models. Pp. 55–69 *in* Genetics in Relation to Insect Management, M. A. Hoy and J. J. McKelvey, eds. New York: The Rockefeller Foundation.

Conway, G. R., and H. N. Comins. 1979. Resistance to pesticides: Lessons in strategy from mathematical models. Span 22:53–55.

Croft, B. A. 1972. Resistant natural enemies in pest management systems. Span 15:19–22.

Croft, B. A., and A. W. A. Brown. 1975. Response of arthropod natural enemies to insecticides. Annu. Rev. Entomol. 20:285–335.

Curtis, C. F., L. M. Cook, and R. J. Wood. 1978. Selection for and against insecticide resistance and possible methods of inhibiting the evolution of resistance in mosquitoes. Ecol. Entomol. 3:273–287.

Denholm, I., A. W. Farnham, K. O'Dell, and R. M. Sawicki. 1983. Factors affecting resistance to insecticides in house flies, *Musca domestica* L. (Diptera: Muscidae). I. Long-term control with bioresmethrin of flies with strong pyrethroid-resistance potential. Bull. Entomol. Res. 73:481–489.

Dobzhansky, T. 1970. Genetics and the Evolutionary Process. New York: Columbia University Press.

Ferrari, J. A., and G. P. Georghiou. 1981. Effects of insecticidal selection and treatment on reproductive potential of resistant, susceptible, and heterozygous strains of the southern house mosquito. J. Econ. Entomol. 74:323–327.

Forgash, A. J. 1981. Insecticide resistance of the Colorado potato beetle, *Leptinotarsa decemlineata* (Say). Pp. 34–46 *in* Advances in Potato Pest Management, J. H. Lashomb and R. Casagrande, eds. Stroudsburg, Pa.: Hutchinson Ross.

Forgash, A. J. 1984. Insecticide resistance of the Colorado potato beetle, *Leptinotarsa decemlineata* (Say). Paper presented at 17th Int. Congr. Entomol., Hamburg, Federal Republic of Germany, August 1984.

Georghiou, G. P. 1972. The evolution of resistance to pesticides. Annu. Rev. Ecol. Syst. 3:133–168.

Georghiou, G. P. 1980. Insecticide resistance and prospects for its management. Residue Rev. 76:131–145.

Georghiou, G. P. 1981. The occurrence of resistance to pesticides in arthropods: an index of cases reported through 1980. Rome: Food and Agriculture Organization of the United Nations.

Georghiou, G. P., and C. E. Taylor. 1976. Pesticide resistance as an evolutionary phenomenon. Pp. 759–785 *in* Proc. 15th Int. Congr. Entomol., Washington, D.C. College Park, Md.: Entomological Society of America.

Georghiou, G. P., and C. E. Taylor. 1977a. Genetic and biological influences in the evolution of insecticide resistance. J. Econ. Entomol. 70:319–323.

Georghiou, G. P., and C. E. Taylor. 1977b. Operational influences in the evolution of insecticide resistance. J. Econ. Entomol. 70:653–658.

Georghiou, G. P., A. Lagunes, and J. D. Baker. 1983. Effect of insecticide rotations on evolution of resistance. Pp. 183–189 *in* IUPAC Pesticide Chemistry, Human Welfare and the Environment, J. Miyamoto, ed. Oxford: Pergamon.

Gressel, J., and L. A. Segel. 1982. Interrelating factors controlling the rate of appearance of resistance. The outlook for the future. Pp. 325–348 *in* Herbicide Resistance in Plants, H. M. LeBaron and J. Gressel, eds. New York: John Wiley and Sons.

Haile, D. G., and D. E. Weidhaas. 1977. Computer simulation of mosquito populations (*Anopheles albimanus*) for comparing the effectiveness of control technologies. J. Med. Entomol. 13:553–567.

Haldane, J. B. S. 1930. A mathematical theory of natural and artificial selection. VI. Isolation. Proc. Cambridge Philos. Soc. 26:220–230.

Hama, H. 1983. Resistance to insecticides due to reduced sensitivity of acetylcholinesterase. Pp. 229–331 *in* Pest Resistance to Pesticides, G. P. Georghiou and T. Saito, eds. New York: Plenum.

Knipling, E. F., and W. Klassen. 1984. Influence of insecticide use patterns on the development of resistance to insecticides: A theoretical study. Southwest. Entomol. 9:351–368.

Krieger, R. I., P. P. Feeny, and C. F. Wilkinson. 1971. Detoxication enzymes in the guts of caterpillars: An evolutionary answer to plant defenses? Science 172:579–581.

Levy, Y., R. Levi, and Y. Cohen. 1983. Buildup of a pathogen subpopulation resistant to a systemic fungicide under various control strategies: A flexible simulation model. Phytopathology 73:1475–1480.

Lewontin, R. C., and S. J. Gould. 1979. The spandrels of San Marco and the Panglossian paradigm; a critique of the adaptationist programme. Proc. R. Soc. London Ser. B 205:581–598.

MacDonald, G. 1959. The dynamics of resistance to insecticides by anophelines. Riv. Parassitol. 20:305–315.

Mani, G. S. In press. Evolution of resistance in the presence of two insecticides. Genetics.

Mani, G. S., and R. J. Wood. 1984. Persistence and frequency of application of an insecticide in relation to the rate of evolution of resistance. Pestic. Sci. 15:325–336.

Maynard-Smith, J. 1978. The Evolution of Sex. New York: Cambridge University Press.

Mayr, E. 1963. Animal Species and Evolution. Cambridge, Mass.: Harvard University Press.

McKenzie, J. A., and M. J. Whitten. 1984. Estimation of the relative viabilities of insecticide resistance genotypes of the Australian sheep blowfly, *Lucilia cuprina*. Aust. J. Sci. 37:45–52.

McPhee, W. J., and E. R. Nestmann. 1983. Predicting potential fungicide resistance in fungal populations by using a continuous culturing technique. Phytopathology 73:1230–1233.

Muggleton, J. 1982. A model for the elimination of insecticide resistance using heterozygous disadvantage. Heredity 49:247–251.

Omer, S. M., G. P. Georghiou, and S. N. Irving. 1980. DDT/pyrethroid resistance interrelationships in *Anopheles stephensi*. Mosq. News 40:200–209.

Priester, T. M., and G. P. Georghiou. 1978. Induction of high resistance to permethrin in *Culex pipiens quinquefasciatus*. J. Econ. Entomol. 71:197–200.

Sawicki, R. M., A. L. Devonshire, A. W. Farnham, K. E. O'Dell, G. D. Moores, and I. Denholm. 1984. Factors affecting resistance to insecticides in house flies, *Musca domestica* L. (Diptera: Muscidae). II. Close linkage on autosome 2 between an esterase and resistance to trichlorphon and pyrethroids. Bull. Entomol. Res. 74:197–206.

Sutherst, R. W., and H. N. Comins. 1979. The management of acaricide resistance in the cattle tick, *Boophilus microplus* (Canestrini) (Acari: Ixodidae), in Australia. Bull. Entomol. Res. 69:519–537.

Sutherst, R. W., G. A. Norton, N. D. Barlow, G. R. Conway, M. Birley, and H. N. Comins. 1979. An analysis of management strategies for cattle tick (*Boophilus microplus*) control in Australia. J. Appl. Ecol. 16:359–382.

Tabashnik, B. E., and B. A. Croft. 1982. Managing pesticide resistance in crop-arthropod complexes: Interactions between biological and operational factors. Environ. Entomol. 11:1137–1144.

Tabashnik, B. E., and B. A. Croft. 1985. Evolution of pesticide resistance in apple pests and their natural enemies. Entomophaga 30:37–49.

Taylor, C. E., and G. P. Georghiou. 1979. Suppression of insecticide resistance by alteration of gene dominance and migration. J. Econ. Entomol. 72:105–109.

Taylor, C. E., and G. P. Georghiou. 1982. Influence of pesticide persistence in the evolution of resistance. Environ. Entomol. 11:746–750.

Taylor, C. E., F. Quaglia, and G. P. Georghiou. 1983. Evolution of resistance to insecticides: A cage study on the influence of migration and insecticide decay rates. J. Econ. Entomol. 76:704–707.

Wharton, R. H., and W. J. Roulston. 1970. Resistance of ticks to chemicals. Annu. Rev. Entomol. 15:381–404.

Whitehead, G. B., and J. A. F. Baker. 1961. Acaricide resistance in the red tick, *Rhipicephalus evertsi*: Neuman. Bull. Entomol. Res. 51:755–764.

Whitten, M. J., and J. A. McKenzie. 1982. The genetic basis for pesticide resistance. Pp. 1–16 *in* Proc. 3rd Australas. Conf. Grassl. Invert. Ecol., K. E. Lee, ed. Adelaide, Australia: S.A. Government Printer.

Wood, R. J., and L. M. Cook. 1983. A note on estimating selection pressures on insecticide resistance genes. Bull. W.H.O. 61:129–134.

Pesticide Resistance: Strategies and Tactics for Management.
1986. National Academy Press, Washington, D.C.

Population Dynamics and the Rate of Evolution of Pesticide Resistance

ROBERT M. MAY and ANDREW P. DOBSON

For a wide range of organisms exposed to insecticides or the like, the number of generations taken for a significant degree of resistance to appear exhibits a relatively small range of variation, typically being around 5 to 50 generations; we indicate an explanation, and also seek to explain some of the systematic trends within these patterns. We review the effects of insect migration to and from untreated regions and of density-dependent aspects of the population dynamics of the target species. Combining population dynamics with gene flow considerations, we review ways in which the evolution of resistance may be speeded or slowed; in particular, we contrast the rate of evolution of resistance in pest species with that in their natural enemies. We conclude by emphasizing that purely biological aspects of pesticide resistance must ultimately be woven together with economic and social factors, and we show how the appearance of pesticide resistance can be incorporated as an economic cost (along with the more familiar costs of pest damage to crops and pesticide application).

INTRODUCTION

During the 1940s, around 7 percent of the annual crop in the United States was lost to insects (Table 1). Over the past two decades, this figure has risen to hold steady at around 13 percent. Much detail and some success stories are masked by the overall numbers in Table 1, but the essential message is clear: increasing expenditure on pesticides and the increasing application of pesticides have, on average, been accompanied by increased incidence of

TABLE 1 Agricultural Losses to Pests in the United States

Year	Percentage of Annual Crop Lost to			
	Insects	Diseases	Weeds	Total
1942–1950 (average)	7	11	14	32
1951–1960 (average)	13	12	9	34
1974	13	12	8	33
1984	13	12	12	37

SOURCE: Modified from Pimentel (1976) and May (1977).

resistance, with the net result being an increased fraction of crops lost to insects. Indeed, the fraction of all crops lost to pests in the United States today has changed little from that in medieval Europe, where it was said that of every three grains grown, one was lost to pests or in storage (leaving one for next year's seed and one to eat).

Beyond these practical worries, the appearance of resistance to pesticides illustrates basic themes in evolutionary biology. The standard example of microevolution in the current generation of introductory biology texts is industrial melanism in the peppered moth. This tired tale could well be replaced by any one of a number of field or laboratory studies of the evolution of pesticide resistance that would show in detail how selective forces, genetic variability, gene flow (migration), and life history can interact to produce changes in gene frequency. We believe such intrusion of agricultural or public health practicalities into the introductory biology classroom may help to show that evolution is not some scholarly abstraction, but rather is a reality that has undermined, and will continue to undermine, any control program that fails to take account of evolutionary processes.

In what follows, our focus is mainly on broad generalities. This paper complements Tabashnik's (this volume), which deals with many of the same issues in a very concrete way, giving numerical studies of models for the evolution of resistance to pesticides by orchard pests.

CHARACTERISTIC TIME TO EVOLVE RESISTANCE

The discussion in this paper is restricted to situations where the genetics of resistance involves only one locus with two alleles, in a diploid insect. This is the simplest assumption to begin with. It does, moreover, appear to be a realistic assumption in the majority of existing instances where detailed understanding of the mechanisms of resistance is available. The stimulating papers by Uyenoyama and Via in this volume indicate some of the important complications that may arise when two or many loci, respectively, are involved in determining resistance. We further restrict this discussion to a closed population, in which the selective effects of a pesticide act homogeneously in space; this assumption will be relaxed in later sections.

Following customary usage, we denote the original, susceptible allele by S, and the resistance allele by R; in generation t, the gene frequencies of R and S are p_t and q_t, respectively (with $p + q = 1$). The gentoype RR is resistant, SS is susceptible, and the heterozygotes RS in general are of intermediate fitness (but see below for discussion of exceptions). In the presence of an application of pesticide of specified intensity, the fitnesses of the three genotypes are denoted w_{RR}, w_{RS}, w_{SS}: we assume $w_{RR} \geq w_{RS} \geq w_{SS}$.

The equation relating the gene frequencies of R in successive generations is then the standard expression (Crow and Kimura, 1970):

$$p_{t+1} = \frac{w_{RR}p_t^2 + w_{RS}p_tq_t}{w_{RR}p_t^2 + 2w_{RS}p_tq_t + w_{SS}q_t^2} . \tag{1}$$

In the early stages of pesticide application, the resistant allele will usually be very rare, so that $p_t << 1$ and $q_t \simeq 1$. The initial ratio p_t/q_t will, indeed, usually be significantly smaller than the ratio w_{RS}/w_{RR} or w_{SS}/w_{RS}, so that to a good approximation equation 1 reduces to

$$p_{t+1}/p_t \simeq w_{RS}/w_{SS}. \tag{2}$$

Suppose the allele R is present in the pristine population at frequency p_0. By compounding equation 2, we see that the number of generations, n, that must elapse before a significant degree of resistance appears (that is, before p attains the value $p_f \simeq 1/2$, for example) is given roughly by

$$(p_f/p_0) \simeq (w_{RS}/w_{SS})^n. \tag{3}$$

We define T_R to be the absolute time taken for a significant degree of resistance to appear, and T_g to be the cohort generation time (Krebs, 1978) of the insect species in question. Then $n = T_R/T_g$, and the approximate relation of equation 3 may be rewritten as

$$T_R \simeq T_g \ln(p_f/p_0)/\ln(w_{RS}/w_{SS}). \tag{4}$$

It is to be emphasized that equation 4 is a rough approximation. In particular, if R is perfectly recessive, we have $w_{RS} = w_{SS}$, and equation 2 is an inadequate approximation to equation 1; even here, however, equation 4 is telling us something sensible, namely, that T_R is very long when R is perfectly recessive (taken literally, equation 4 gives $T_R \to \infty$).

Equation 4 shows that T_R depends directly on the organism's generation time T_g, but only logarithmically on other factors. In particular, T_R depends only logarithmically on (1) the initial frequency of the resistance allele, p_0;

(2) the choice of the threshold at which resistance is recognized, p_f; and (3) the selection strength, w_{RS}/w_{SS}, which in turn is determined by dosage levels and by the degree of dominance of R. Elsewhere in this volume, Roush suggests that p_0 values may range from 10^{-2} to 10^{-13}; this enormous range, however, collapses to a mere factor of six separating highest from lowest when logarithms are taken. Likewise, ratios of w_{SS}/w_{RS} ranging from 10^{-1} to 10^{-4} or less all make similar contributions to the denominator in equation 4, which involves only the logarithm of this ratio.

Table 2 sets out values of T_R for a variety of organisms (insects, and parasites of vertebrates), under the selective forces exerted by various insecticides or other chemotherapeutic agents. Table 3 (see p. 188) attempts a rough summary of the general trends exhibited in Table 2: we see that for the great diversity of animal life embraced by Table 2, T_R lies in the surprisingly narrow range of around 5 to 100 *generations*. We argue that such relative constancy of T_R, despite enormous variability in p_0 and w_{RS}/w_{SS}, is because T_R depends on all these factors (except T_g) only logarithmically. We will return to the systematic trends exhibited in Table 2 and crudely summarized in Table 3, after the discussions of migration, density dependence, and other miscellaneous factors.

The approximate expression for T_R in equation 4 mixes factors that are intrinsic to the genetic system underlying the resistance phenomenon (such as T_g, p_0, and the degree of dominance of R) with factors that are under the direct control of the manager (such as dosage levels). Comins (1977a) suggests a useful partitioning of these two kinds of factors. First, define the relative fitnesses of the genotypes RR, RS, SS, to be 1: $w^{1-\beta}$:w. Here w is the fitness of the susceptible homozygotes relative to the resistant homozygotes; w essentially measures the relative survivial of wild-type insects (high dosage of pesticide implies low w). The parameter β measures the degree of dominance of R: if R is perfectly dominant, $\beta = 1$; if R is perfectly recessive, $\beta = 0$; and in general, β will take some numerical value intermediate between 0 and 1. Equation 4 can now be rewritten as

$$T_R = T_0/\ln(1/w). \tag{5}$$

This separates the parameter w (which measures the selection strength as determined by the dosage level) from the parameter T_0 (which conflates intrinsic genetic factors). The quantity T_0 is defined as

$$T_0 = T_g \ln(p_f/p_0)/\beta. \tag{6}$$

Parameters such as p_0 or β usually cannot be estimated, and T_0 should be thought of as a phenomenological constant, to be determined empirically in the laboratory or in the field (Comins, 1977a).

Beyond explaining the general trends exhibited in Table 2 and other similar compilations, equations 4 or 5 (or more refined versions of them) may be

TABLE 2 Characteristic Times for the Appearance of Resistance, T_R, in Some Specific Systems

Species	Control Agent	Time to Resistance		Genetic Mechanism[2]
		Generations[1]	Years	
Avian Coccidia (Chapman, 1984)				
Eimeria tenella	Buquinolate	6 [<6]	1	Mutation
	Glycarbylamide	11 [9]	< 1	
	Nitrofurazone	12 [5]	7	
	Clopidol	20 [9]	6	
	Robenicline	22 [16]	10	
	Amprolium	65 [20]	14	
	Zoalene	11 [7]	22	
	Nicarbazin	35 [17]	27	
Gut Nematodes in Sheep (LeJambre et al., 1979; Kates et al., 1973)				
Haemonchus contortus	Thiabendazole	3	< 1	Autosomal
	Cambendazole	[4]	< 1	Semi-dominant

Ticks on Sheep				
(Stone, 1972; Tahori, 1978)				
Boophilus microplus	DDT	32	4	*
	HCH-dieldrin	2	< 1	X
	sodium arsenite	*	40	*
Lice Selected in the Laboratory				
(Eddy et al. *in* Brown and Pal, 1971)				
Pediculus corporis	DDT	12 [25]	*	*
Cotton Boll Weevil				
(Brazzel and Shipp, 1962; Graves and Roussel, 1962)				
Antonomus grandis	Endrin	25	*	*
Sheep Blow Fly				
(Shanahan and Roxburgh, 1974)				
Lucilia cuprina	Diazinon	*	12	*
House Flies in Denmark				
(Keiding, 1976, 1977)				
Musca domestica	Pyrethrum	*	21	*
	Parathion	*	9	*
	Trichlorophon	*	11	*
	DDT	*	3	*

TABLE 2 Continued

Species	Control Agent	Time to Resistance		Genetic Mechanism[2]
		Generations[1]	Years	
Black Flies in Japan and Ghana (Brown and Pal, 1971)				
Simulium aokii	DDT + Lindane	*	6	*
S. damnosum	DDT	*	5	*
Anopheline Mosquitoes: Different Parts of the World (Brown and Pal, 1971)				
Anopheles sacharovi	DDT	*	4–6	partly
	Dieldrin	*	8	behavioral
An. maculipennis	DDT	*	5	only partial
An. stephansi	DDT	*	7	*
	Dieldrin	*	5	*
An. culicifacies	DDT	*	8–12	*
An. annuaris	DDT	*	3–4	*
An. sundaicus	DDT	*	3	*
	Dieldrin	*	1–3	*
An. quadrimaculatus	DDT	*	2–7	*
	Dieldrin	*	2–7	*
An. pseudopunctipennis	DDT	*	>20	*
	Dieldrin	*	18 wk	*

[1]In this column the figures give the number of generations before a majority (>50 percent) of the individuals in the population are resistant to the control agent. The figures in brackets give the number of generations before resistance is first observed (usually >5 percent of individuals resistant).

[2]In this column an X implies that the data are for cross-resistance following the application of the previously listed substance. An asterisk indicates that no data are available.

used to make predictions about the way T_R depends on pesticide dosage levels or on degree of pesticide persistance in specific laboratory studies. Some such work is discussed in the next section.

The above ideas also apply to the back selection or regression to population-level susceptibility that may appear once a particular pesticide is no longer used. As discussed elsewhere (Comins, 1984), it is possible in principle that a pesticide may have cycles of useful life: the gene frequency of R first increases under the selection pressure exerted by use of the pesticide; eventually R attains a frequency sufficiently high to produce a noticeable degree of resistance, and shortly thereafter the pesticide is discontinued as ineffective; in the absence of the pesticide, usually $w_{SS} > w_{RR}$, and selection will now cause the frequency of R to decrease. Applying equation 4, mutatis mutandis, to this back-selection process, we note that the time elapsed before the population is again effectively susceptible to the pesticide will depend on (1) the intrinsic fitness ratios $w_{RR}:w_{RS}:w_{SS}$, which measure the strength of back selection in the absence of pesticide; (2) the frequency of R when the pesticide is discontinued; and (3) how low a frequency of R is required before reuse of the pesticide becomes sensible.

For factor 1 it has been shown that significant back-selection effects can indeed occur (Georghiou et al., 1983; Ferrari and Georghiou, 1981); Roush, in this volume, estimates the rate-determining ratio w_{RS}/w_{SS} to be in the range 0.75 to 1.0 for untreated populations. Even when demonstrably present, however, such back selection in the absence of a pesticide is typically weaker than the corresponding strengths of selection for resistance under pesticide usage, so that the denominator in equation 4 is smaller. For this reason alone, "regression times" will tend to be longer than "resistance times," T_R.

The influence of factor 2 is that regression will be faster if pesticide application is discontinued before the frequency of R gets too high. The possible complications discussed by Uyenoyama in this volume are more likely to arise when p_R is relatively high, which gives an additional reason for prompt discontinuation of a pesticide to which resistance has appeared.

For factor 3 we observe that in pristine populations the frequency of R may typically be around 10^{-6} to 10^{-8} (Roush, this volume). After use of a particular pesticide is stopped, resistance will be unobservable and effectively unmeasureable long before it attains levels as low as these pristine ones; when the frequency of R is around 10^{-2}, the population could easily be considered to have regressed to effective susceptibility. Taking the above numbers as illustrative, we see that resistance to the recycled pesticide is likely to appear significantly more quickly than it did in the first instance (T_R depends on $\ln(1/p_0)$, so that T_R is three or four times faster for $p_0 = 10^{-2}$ than for $p_0 = 10^{-6}$ or 10^{-8}).

In short, all three factors suggest that a population will usually take longer to recover susceptibility than it did to acquire resistance, and also that resistance will probably reemerge significantly faster following reintroduction of the pesticide. These broad generalities need to be fleshed out by detailed studies of specific mathematical models, backed where possible by long-term laboratory studies of relevant pest-pesticide systems.

MIGRATION AND GENE FLOW

The above discussion assumed that pesticides would be applied uniformly to a closed population of pests. In the field, the next generation of pests will virtually always include some immigration from untreated (or more lightly treated) regions, and this flow of susceptible genes will work against the evolution of resistance. This is a particular instance of one of the central questions of evolutionary biology: under what circumstances will gene flow wash out the selective forces that are tending to adapt an organism to a particular local environment? Earlier thinking of a qualitative kind suggested that very small amounts of gene flow may be sufficient to prevent local differentiation, and that geographical isolation was usually necessary before local adaptation could lead to new races or species (Mayr, 1963). More recently, population geneticists have shown that the occurrence of local differentiation (or "clines" in gene frequency) depends on the balance between the strength and the steepness of the spatial gradient of selection versus the amount and spatial scale of migration (Slatkin, 1973; Endler, 1977; Nagylaki, 1977). May et al. (1975) gives a brief review of migration theory and data. One illuminating study contrasts two examples of industrial melanism: *Biston betularia* is relatively vagile and thus is predominantly in the melanic form over most of England's industrial midlands; individuals of *Gonodontis bidentata* move significantly less in each generation, leading to weaker gene flow and a patchy pattern of local adaptation with melanic forms predominating near cities and wild types predominating in the intervening countryside (Bishop and Cook, 1975).

This academic literature is directly relevant to the problem of the evolution of pesticide resistance in the presence of migration. Comins (1977b) has given an analytic study of the implications for pesticide management, and Taylor and Georghiou (1979, 1982; Georghiou and Taylor, 1977) have presented numerical studies of particular examples. What follows is an attempt to lay bare the essential mechanisms; the above references should be consulted for a more accurate and detailed discussion.

To begin, suppose there is an infinite reservoir of untreated pests; within this untreated reservoir the gene frequency of R will therefore remain constant at the pristine value, which we denote by $\bar{p}_R$. In the treated region the next generation of larval pests will come partly from the previous generation of adults that have survived treatment (which tends to select for resistance) and have not emigrated, and partly from those among the previous generation of

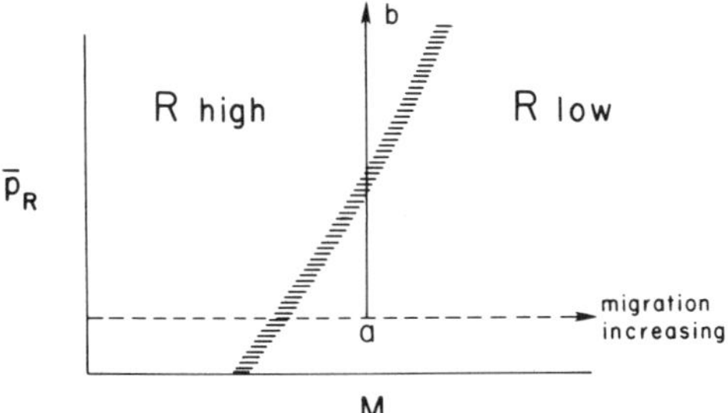

FIGURE 1 The degree of pesticide resistance that evolves in a treated region in the presence of immigration from untreated regions in each generation: $\bar{p}_R$ is the gene frequency of R in the untreated region, and m is a measure of the amount of migration (gene flow) as a ratio to the strength of selection. This figure abstracts the more complex and more detailed results of Comins (1977b), and is discussed more fully in the text.

untreated (and thus, largely susceptible) adults that have immigrated into the treated region. As discussed by Comins (1977b) and others, we assume it is the larval stage that damages the crops.

As shown in detail by Comins (1977b), the rate of evolution of resistance in the treated region will, under the above circumstances, depend on (1) the gene frequency of R in the untreated reservoir, $\bar{p}_R$; (2) the degree of dominance of R, as measured by the parameter β of equation 6 (actually, Comins uses a parameter h for arithmetically intermediate heterozygotes, rather than β for geometrically intermediate heterozygotes, but this is an unimportant detail); and (3) the magnitude of migration in relation to selection, as measured by a parameter m. Specifically, the migration/selection parameter m (Comins, 1977b) is defined as:

$$m = r/[(1 - r)(1 - w)]. \tag{7}$$

Here r is the migration rate (i.e., the fraction of adults in a given area that migrate rather than "staying at home"), and w measures the strength of selection ($w = w_{SS}/w_{RR}$, as in equation 5).

If β is low enough (R sufficiently recessive, corresponding very roughly to $\beta \lesssim 1/2$), the treated region will settle to a stable state in which the gene frequency of R remains *low*, providing migration is sufficiently high (m sufficiently large) (Comins, 1977b). Conversely, for relatively small m-values, selection overcomes gene flow and the system eventually settles to a resistant state (with $\bar{p}_R$ close to unity). This situation is illustrated schematically in Figure 1. In the treated region, the final steady state will be one of resistance or continued susceptibility, depending on the strength of migration relative to selection, as measured by m. There is a fairly sharp boundary between these two regions (indicated by the hatched line in Figure 1); the boundary depends weakly on the magnitude of $\bar{p}_R$, with slightly higher gene flow (higher *m*) being required to maintain susceptibility if $\bar{p}_R$ is higher. Comins shows that there can, in fact, be two alternative stable states for *m*-values close to the fuzzy boundary in Figure 1, but we suppress these elegant and rather fragile details in favor of the robust generalities shown schematically in Figure 1.

For β-values approaching unity (relatively dominant *R*), the treated regime will eventually become resistant no matter how large the gene flow. Even here, however, T_R can be very long if *m* is relatively large (Comins, 1977b).

More generally, the untreated region will be finite. The situation is now more symmetrical, with preponderately R genes migrating out from the treated regions into the untreated ones at the same time as preponderately S genes are flowing into the treated regions. The net outcome is that the gene frequency of R in the untreated regions, $\bar{p}_R$, will slowly increase. As indicated in Figure 1 (by the vertical trajectory from point *a* to point *b*), for any specified value of *m* such increase in $\bar{p}_R$ will in general eventually cause the treated region to move sharply from susceptibility (low R) to resistance (high R).

Thus, in the real world, resistance is always likely to appear in the long run. Its appearance can, however, be delayed by management strategies that keep *m* relatively high. Such strategies include maximizing the area of untreated regions or refugia, and keeping the dosage level as low as feasible in treated regions: both of these actions work toward higher *m*-values. In some situations it could pay to introduce susceptible adult males following treatment, which could enhance the gene frequency of S in the next generation without producing any additional pest larvae.

These analytic and numerical insights have been corroborated by laboratory experiments on *Musca domestica* exposed to dieldrin at various dosage levels and with various levels of influx of susceptibles (Taylor et al., 1983). As suggested by the mathematical models, the onset of resistance occurred sharply and at a time T_R that depended in a predictable way on dosage and immigration levels. It would be nice to have more laboratory studies of this kind. On the other hand, one should not place too much reliance on such laboratory studies, because they unavoidably fail to include many of the density-dependent mortality factors that are important in nature. This leads us into the next section.

DENSITY DEPENDENCE AND PEST POPULATION DYNAMICS

Density-dependent effects can enter at any stage in the life cycle of a pest. Such complications can be dissected with standard techniques, such as *k*-factor analysis (Varley et al., 1972). For simplicity the main density dependence is assumed to act on the adult population, N_t in generation *t*. Such nonlinearity, or density dependence, in the relationship between the popu-

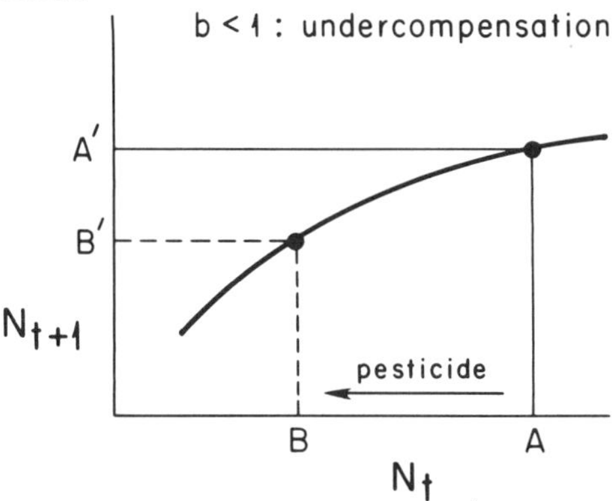

FIGURE 2 Undercompensating density dependence ($b < 1$ in equation 8). If the population in generation t, N_t, is displaced to a lower value (from A to B) by pesticide application or other effects, then the population in the next generation, N_{t+1}, will tend to be lower than it would otherwise have been (B′ rather than A′).

lation, N_t, in generation t and population, N_{t+1}, in the next generation may be characterized phenomenologically by a parameter b:

$$N_{t+1} = \lambda N_t (N_t)^{-b}. \tag{8}$$

Here λ is the intrinsic rate of increase (Krebs, 1978). This follows Haldane (1953) and Morris (1959); for a more complete discussion, see May et al. (1974).

The special case $b = 1$ gives "perfect" density dependence, with N_t tending to return immediately to the value λ in the next generation, following any disturbance. The case $b > 1$ is called overcompensating; if the population is perturbed below its long-term average or equilibrium value in one generation, it will tend to bounce back *above* this long-term value in the next generation. Conversely, $b < 1$ is called undercompensating; such populations will tend to recover steadily and monotonically following disturbance. As indicated in Figure 2, if a population with undercompensating density dependence ($b < 1$) is driven to low values in one generation (by pesticide application, for example), then in the next generation it will tend to remain at a lower value than would otherwise have been the case. But a population with overcompensating density dependence ($b > 1$) will tend to manifest a perverse response to pesticide application, as shown in Figure 3: if N_t is driven to a low value, then N_{t+1} will tend to be at a higher level than it would otherwise have been. These density-dependent factors may, of course, always be masked to a greater or lesser extent by superimposed density independent effects caused by the weather or other things; the underlying tendencies, however, remain.

What happens when we graft these considerations of population dynamics

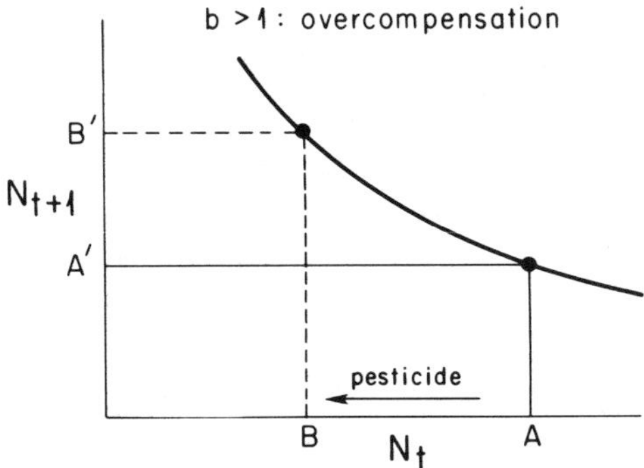

FIGURE 3 Overcompensating density dependence ($b > 1$ in equation 8). If N_t is perturbed to a lower value (from A to B), N_{t+1} tends to be bigger than would otherwise have been the case (B′ rather than A′).

onto the selective forces and gene flow of the previous section? With undercompensating density dependence ($b < 1$), the population densities of the next generation of pests on average will be lower in treated regions than in untreated ones. Consequently, the effects of migration from untreated regions will be more significant. In other words the *m*-value required to maintain susceptibility in treated regions will be lower for a pest population with $b < 1$ than for one with $b = 1$. Conversely, with overcompensation ($b > 1$) the next generation of pests cn average will be at higher density in treated regions than in untreated ones, whence higher *m*-values are required to maintain susceptibility. Figure 4 represents a generalization of the schematic Figure 1 to include now the complications arising from density dependence

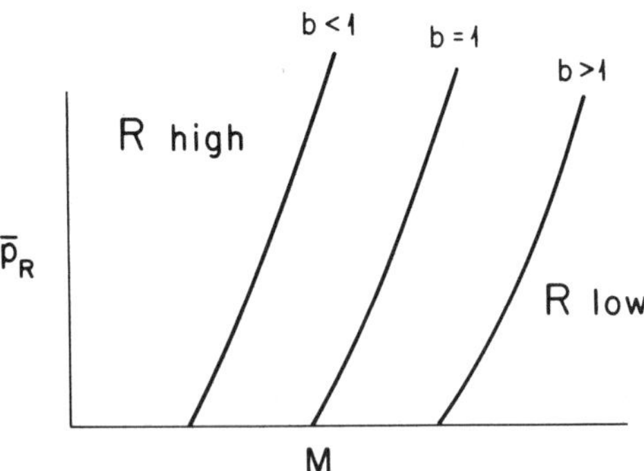

FIGURE 4 The results of Figure 1 are extended to show schematically how the population dynamics of the pest can affect the rate at which pesticide resistance evolves.

in the population dynamics of the pest. These ideas are developed more fully and more rigorously by Comins (1977b).

Another way of setting out the ideas encapsulated in Figure 4 is to observe that, other things being equal, resistance will appear more quickly in populations with overcompensating density dependence and more slowly in populations with undercompensating density dependence than in populations with perfect density dependence; that is, T_R increases as the density-dependence parameter b of equation 8 decreases.

Several studies have attempted to assess b-values of insect populations in the field and in the laboratory (Hassell et al., 1976; Stubbs, 1977; Bellows, 1981). (These studies all use more complex models than equation 8, but the distinction between overcompensating and undercompensating density dependence remains clear and valid). Most, although not all, populations that have been studied in the field show undercompensating density dependence. Among these studies the field population exhibiting the most pronounced degree of overcompensation is the Colorado potato beetle, which elsewhere in this volume (see Georghiou) is singled out as notorious for the speed with which it has developed resistance to a wide range of pesticides. In contrast to field populations, most laboratory populations in the above surveys show marked overcompensation. This difference between field and laboratory populations probably derives from the many natural mortality factors that commonly are not present in the laboratory; whatever the reason, this difference underlines the need for caution in extrapolating laboratory studies of the evolution of resistance into a field setting.

Comins (1977b) gives an interesting discussion of the detailed dependence of T_R on b and m. For $b = 1$, we simply have the results summarized in the preceding section. These amount to the rough estimate that, in the presence of a high level of migration,

$$T_R(m; b = 1) = T_R(0; b = 1)\,[\text{migration}/(1 - w)]. \qquad (9)$$

Here $T_R(0; b = 1)$ is the time for resistance to appear in a closed population, and $T_R(m; b = 1)$ is the time for it to appear in the presence of migration; w is the selection strength, as defined earlier (equation 5); and the factor labeled migration is a complicated term, involving m and other parameters, that measures the effects of migration. We see that $T_R(m; b = 1)$ will increase as selection becomes weaker (w larger), but that the dependence on w is more pronounced at low dosage ($T_R \rightarrow \infty$ as $w \rightarrow 1$) than at high dosage (T_R is roughly independent of w for $w << 1$).

For $b < 1$, the expression for $T_R(m; b)$ is more complicated than given in equation 9. Because undercompensating density dependence makes migration relatively more important, $T_R(m; b < 1)$ is always greater than $T_R(m; b = 1)$ for given values of m and w. At low levels of selection ($w \rightarrow 1$) the differences created by subsequent density-dependent effects are relatively

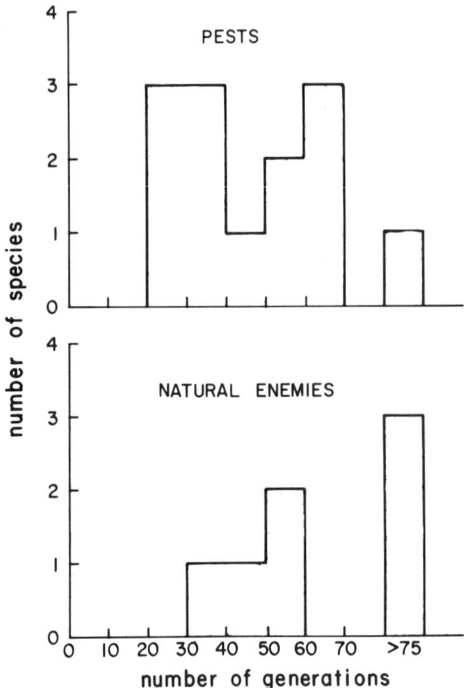

FIGURE 5 The number of generations taken for pesticide resistance to appear in species of orchard pests is contrasted with the corresponding patterns among their natural enemies (data from Tabashnik and Croft, 1985).

unimportant, but at high levels of selection ($w << 1$), density-dependent effects cause migration to assume increasing importance when $b < 1$. The result is that, for $b < 1$, T_R is longest at low and high selection levels, and shortest at intermediate values of w.

These theoretical insights of Comins (1977b) are concordant with the numerical simulations and laboratory experiments of Taylor et al. (1983) on flies with undercompensating density dependence. These authors found that (for a given level of immigration) resistance evolved fastest at intermediate dosage levels.

POPULATION DYNAMICS OF PESTS AND THEIR NATURAL ENEMIES

The propensity for pest species to evolve resistance more quickly than their natural enemies do has often been remarked (Tabashnik, this volume; Roush, this volume). Table 3 summarizes the trends for some groups of pests and their natural enemies, and Figure 5 presents detailed evidence for orchard crop pests and their predators. Clearly, such systematic differences in the rate of evolution of pesticide resistance can cause problems.

One reason for these differences might be that the coevolution between plants and phytophagous insects has preadapted the latter to the evolution of detoxifying mechanisms, whereas this is much less the case for the natural enemies of such insects. Laboratory studies show that there are in fact no simple, general patterns of this kind, and that under controlled conditions the rate of evolution of resistance in prey and in predator populations depends on the detailed molecular mechanisms underlying detoxification (Croft and Brown, 1975; Mullin et al., 1982). This in turn has prompted a search for pesticides that may be less lethal for natural enemies than for pests (Plapp and Vinson, 1977; Rock, 1979; Rajakulendran and Plapp, 1982; Roush and Plapp, 1982), or even the release of natural enemies that have been artificially selected for resistance to specific pesticides (Roush and Hoy, 1981).

An alternative explanation for the typically swifter evolution of resistance by pests than by their natural enemies lies in the population dynamics of prey-predator associations (Morse and Croft, 1981; Tabashnik and Croft, 1982; Tabashnik, this volume). Suppose a pesticide kills a large fraction of all prey and all predators in the treated region. For the surviving prey life is now relatively good (relatively free from predators), and the population is likely to increase rapidly. Conversely, for the surviving predators life is relatively bad (food is harder to find), and their population will tend to recover slowly. This argument can be supported by a standard phase plane analysis for Lotka-Volterra or other, more refined, prey-predator models. Such analysis shows that, in the aftermath of application of a pesticide that affects both prey and predator, prey populations will tend to exhibit overcompensating density-dependent effects (essentially with $b > 1$), while predator populations will tend to manifest undercompensation ($b < 1$). Returning to the arguments developed in the preceding section and illustrated schematically in Figure 4, we can now deduce that, for a given level of migration and pesticide application, pest species (which effectively have overcompensating density dependence) will tend to develop resistance faster than will their natural enemies (which effectively have undercompensating density dependence).

The detailed numerical studies of Tabashnik and Croft (1982) and Tabashnik (this volume) also make the above point, but in more detailed and specific settings. We think it is useful to buttress these concrete studies with the very general observation that pesticide resistance is likely to appear faster among pests than among their natural enemies, by virtue of the interplay between population dynamics and migration; in this sense, the phenomenon illustrates the general arguments made in the previous section.

Other work in this area includes the numerical studies by Gutierrez and collaborators on management of the alfalfa weevil, taking account of pest population dynamics, natural enemies, and the evolution of resistance (Gutierrez et al., 1976; Gutierrez et al., 1979), and Hassell's (in press) investigation of the dynamical behavior of pest species under the combined effects of pesticides and parasitoids. There is much scope for further work, both in the laboratory and with analytic or computer models.

MISCELLANEOUS TOPICS

This section comprises brief notes on a variety of factors that complicate the analyses presented above.

Life History Details

Throughout we have considered pests with deliberately oversimplified life cycles, in which pesticide application and density dependence acted only on one stage. Comins (1977a,b; 1979) indicates how the analysis can be extended, rather straightforwardly, to a life cycle with *n* distinct stages (pupae, several stages of larvae, adults). The numerical models of Tabashnik and of Gutierrez and collaborators also include such realistic complications.

High Dosage to Make R Effectively Recessive

As we noted earlier, if R is perfectly recessive, resistance will evolve much more slowly than is otherwise the case (Crow and Kimura, 1970). It has been argued that dosage levels high enough to kill essentially all heterozygotes may thus slow the evolution of resistance by making R, in effect, perfectly recessive. This strategy, however, will work only if pesticide dosage can be closely controlled in a closed population (Comins, 1984). This is roughly the case for acaricide dipping of cattle against ticks, for example (Sutherst and Comins, 1979). In general, lack of close control and/or the immigration of pests from untreated regions is likely to render such a strategy infeasible.

Heterozygote Superiority

There appear to be some instances among insects where the RS genotypes are more resistant to an insecticide than either RR or SS (Wood, 1981). The spotted root-maggot *Euxesta notada* may exhibit such heterozygous advantage in the presence of DDT or dieldrin (Hooper and Brown, 1965). Although familiar for rat resistance to warfarin, such heterozygous superiority raises questions that do not seem to have been discussed for pesticides directed at insects.

Pesticide Resistance Compared with Drug Resistance

Resistance to antibiotics and antihelminths poses growing problems in the control of infections among humans and other animals. Reviewing recent work, Peters (in press) concludes that both high dosage rates and the use of drug mixtures may tend to retard the evolution of resistance. Drug administration to humans and other animals often does permit close control in a closed population, such that these strategies have a chance to work (rather than be washed out by gene flow; see Life History Details, above).

Pesticide Resistance Compared with Herbicide Resistance

Herbicide resistance has usually been slower to evolve than pesticide resistance, even when the longer generation time of most weeds is taken into account (Gressel and Segel, 1978; Gressel, this volume). Gressel suggests that this is due to the presence of seed banks in the soil (corresponding, in effect, to gene flow over time instead of space) and to the lower reproductive fitness of resistant genotypes. Gressel and Segel's analysis (1978) leads to an expression tantamount to equation 4 for T_R, but with the denominator replaced by:

$$\ln[w_{RS}/w_{SS}] \rightarrow \ln[1 + (w_{RS}/w_{SS})(f_{RS})/f_{SS})(1/T_{soil})] \qquad (10)$$

Here f_{RS}/f_{SS} is the ratio of the reproductive success of the two genotypes, which may be 0.5 or less; T_{soil} represents the number of years that a typical seed spends in the seed bank, which can be 2 to 10 years. These two factors can diminish the RR/SS selective advantage by an order of magnitude, leading to significantly longer T_R.

The array of complications discussed above helps to explain several of the general trends set out in Table 3.

ECONOMIC COST OF PESTICIDE RESISTANCE

The foregoing discussion has dealt exclusively with biological aspects of the evolution of pesticide resistance. Such a discussion, however, only makes sense if embedded in a larger economic context.

Some broad insight into the economic costs of pesticide resistance can be obtained by the following modification of a more detailed analysis by Comins (1979). Agricultural costs associated with pests are of at least three kinds: the damage done to crops, the cost of pesticide application, and the more subtle costs arising from the need to develop new pesticides as the appearance of resistance retires old ones. To a crude approximation we may think of the parameter w (which measured the strength of selection in our previous analysis) as determining the fraction of the pest population surviving pesticide

TABLE 3 Some Possible Trends in the Way T_R/T_g (the number of generations that elapse before resistance is noticed, as cataloged in Table 2) Depends on the Biological and Environmental Setting

Generations to Resistance	Organism	Variation in Life History Parameter or Efficiency of Treatment
2		Large proportion of population treated
	Gut Coccidia	
5		High population densities and strong density-dependent effects
	Gut Nematodes	
10		Asexual reproduction and high mutation rates
	Mosquitoes House Flies	
20		Increasing mobility into and out of treated area
	Rats Cattle Ticks	
50		Increasing proportion of lifetime fecundity prior to treatment
	Phytophagous Insects Weeds	
		Seed Banks
	Entomophagous Insects	
100		Low population density and reduced contact rate between organisms and control agent
	Tsetse Fly	
	Mediterranean Fruit Fly	

NOTE: The first column sets a scale (measured logarithmically in generations); the second column places some organisms along this scale in a very approximate way; and the third column comments on some rough correlations between the time scale and life histories or treatment efficiencies.

application; the cost of insect damage to the crop may then be estimated as Aw. Comins (1979) argues that application costs are likely to be related logarithmically to the fraction killed, whence these costs may be estimated as $B \ln(1/w)$. A and B are proportionality constants that can be empirically determined. Finally we need to estimate the amount of money that must be set aside each year such that after T_R years, when resistance necessitates the introduction of a new pesticide, its development costs (C') will be met. If the set-aside money compounds at an annual interest rate δ, a standard calculation gives the average "cost of resistance" as $C' [\exp(\delta) - 1]/[\exp(\delta T_R) - 1]$. (This is a more realistic estimate of the cost than that used by Comins, 1979.) The total annual cost that pests pose to the farmer is thus

$$\text{Total cost} = Aw + B \ln(1/w) + (\delta T_0)C/[\exp(\delta T_0/\ln(1/w)) - 1]. \quad (11)$$

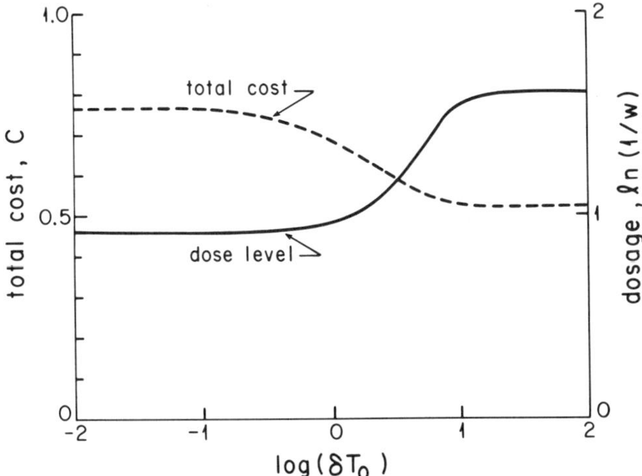

FIGURE 6 The solid curve shows the pesticide dosage (measured by $\ln(1/w)$) that minimizes the total economic costs associated with pests (crop damage, cost of pesticide application, cost of developing new pesticides as resistance renders old ones ineffective). The dashed line correspondingly shows the minimized total costs. The curves are based on equation 11, with the parameters A, B, C here having the representative values 1.0, 0.2, 0.2, respectively (in some arbitrary monetary units); the basic features of Figure 6 are not qualitatively dependent on these parameter values. Both dosage levels and total costs are shown as a function of the parameter combination δT_0, which is essentially the ratio between the intrinsic time scale associated with the evolution of resistance and the doubling time of invested money (at interest rate δ: for more precise definitions, see the text).

Here equation 5 has been used to express T_R in terms of the intrinsic time scale for resistance, T_0, and the selection strength, $1/w$. The cost constant C is defined as $C = C' [\exp(\delta) - 1]/(\delta T_0)$; in the limit $\delta \rightarrow 0$, C is essentially the insecticide development cost per year, $C = C'/T_0$.

In accord with common sense, equation 11 says that as dosage levels increase (that is, as w decreases), the cost associated with pest damage to the crop decreases, but the cost of pesticide application increases, as does the cost associated with developing new pesticides (because this task becomes more frequent). For any specific set of values of A, B, C, and δT_0, some intermediate level of w (between 0 and 1) will minimize the total cost. Figure 6 shows this optimal dosage level (solid line) and the associated total cost (dosage + application + pesticide development; dashed line) as a function of δT_0 for representative values of A, B, and C. For a combination of low interest rates and/or intrinsically short times to evolve resistance ($\delta T_0 << 1$), the optimum strategy suggests relatively low dosage rates (and the lowest possible total cost is necessarily relatively high). Conversely, if δT_0

$>> 1$, optimum dosage rates are relatively high (and total costs are relatively low).

In other words the right-hand side of Figure 6 corresponds to characteristic resistance times being longer than the time it takes for invested money to double (which is proportional to $1/\delta$); resistance is effectively far off, and optimal dosage can thus be high. The left-hand side of Figure 6 corresponds to characteristic resistance times being short compared with the doubling time of invested money; resistance looms, and therefore useful pesticide life should be extended by lower dosages.

An essential point, which is given little attention elsewhere in this volume, is that not all actors in this drama discount the future at the same rate. Pesticide manufacturers may often tend to inhabit the right-hand side of Figure 6, seeing money as fungible, and taking δ to be relatively high. Many farmers, however, may tend instead to inhabit the left-hand side of Figure 6, with assets tied up in their land, the future of which they would wish to discount slowly.

In short even with goodwill and a clear biological understanding of how best to manage pesticide resistance, different groups can come to different decisions. This is a particular case of a more general phenomenon, discussed lucidly by Clark (1976) for fishing, whaling, and logging.

CONCLUSION

Our aim has been to combine population biology with population genetics, to show how migration and density-dependent dynamics can affect the rate of evolution of resistance to pesticides. To advance this enterprise we need a better understanding of the detailed genetic mechanisms underlying resistance and more information about the population biology of pests and natural enemies in the laboratory and in the field. Insofar as the dynamical behavior of pest populations influences the rate of evolution of resistance, we must be wary of extrapolating the laboratory studies into field situations; it would be nice to see more control programs being designed with a view to acquiring a basic understanding at the same time as they serve practical ends.

If dosage levels, migration, refugia, natural enemies, and other factors are to be managed to slow down the evolution of pesticide resistance, efforts must be coordinated over large regions. Some crops lend themselves to this, and some do not. Often the best interests of individuals will differ from those of groups, leading to problems that are social and political rather than purely biological.

Beyond this, even with good biological understanding and coherent planning of group activities, it can be that different sectors—pesticide manufacturers, farmers, planners responsible for feeding people—have different aims stemming from different rates of discounting the future and the absence of a truly common coinage. Population biology can clarify these tensions, but it cannot resolve them.

ACKNOWLEDGMENTS

This work was supported in part by the National Science Foundation, under grant BSR83-03772 (RMM), and by the North Atlantic Treaty Organization Postdoctoral Fellowship Program (APD).

REFERENCES

Bellows, T. S., Jr. 1981. The descriptive properties of some models for density dependence. J. Anim. Ecol. 50:139–156.

Bishop, J. A., and L. M. Cook. 1975. Moths, melanism and clean air. Sci. Am. 232(1):90–99.

Brazzel, J. R., and O. E. Shipp. 1962. The status of boll weevil resistance to chlorinated hydrocarbon insecticides in Texas. J. Econ. Entomol. 55:941–944.

Brown, A. W. A., and R. Pal. 1971. Insecticide Resistance in Arthropods. Geneva: World Health Organization.

Chapman, H. D. 1984. Drug resistance in avian coccidia (a review). Vet. Parasitol. 15:11–27.

Clark, C. W. 1976. Mathematical Bioeconomics. New York: John Wiley and Sons.

Comins, H. N. 1977a. The management of pesticide resistance. J. Theor. Biol. 65:399–420.

Comins, H. N. 1977b. The development of insecticide resistance in the presence of migration. J. Theor. Biol. 64:177–197.

Comins, H. N. 1979. Analytic methods for the management of pesticide resistance. J. Theor. Biol. 77:171–188.

Comins, H. N. 1984. The mathematical evaluation of options for managing pesticide resistance. Pp. 454–469 in Pest and Pathogen Control: Strategic, Tactical and Policy Models, G. R. Conway, ed. New York: John Wiley and Sons.

Croft, B. A., and A. W. A. Brown. 1975. Responses of arthropod natural enemies to insecticides. Annu. Rev. Entomol. 20:285–335.

Crow, J. F., and M. Kimura. 1970. An Introduction to the Theory of Population Genetics. New York: Harper and Row.

Endler, J. A. 1977. Geographic Variation, Speciation and Clines. Princeton, N.J.: Princeton University Press.

Ferrari, J. A., and G. P. Georghiou. 1981. Effects of insecticidal selection and treatment on reproductive potential of resistant, susceptible, and heterozygous strains of the southern house mosquito. J. Econ. Entomol. 74:323–327.

Georghiou, G. P., and C. E. Taylor. 1977. Genetic and biological influences in the evolution of insecticide resistance. J. Econ. Entomol. 70:319–323.

Georghiou, G. P., A. Lagunes, and J. D. Baker. 1983. Effect of insecticide rotations on evolution of resistance. Pp. 183–189 *in* Pesticide Chemistry: Human Welfare and the Environment, J. Miyamoto, ed. New York: Pergamon.

Graves, J. B., and J. S. Roussel. 1962. Status of boll weevil resistance to insecticides in Louisiana during 1961. J. Econ. Entomol. 55:938–940.

Gressel, J., and L. A Segel. 1978. The paucity of plants evolving genetic resistance to herbicides: Possible reasons and implications. J. Theor. Biol. 75:349–371.

Gutierrez, A. P., U. Regev, and C. G. Summers. 1976. Computer model aids in weevil control. Calif. Agric. April:8–18.

Gutierrez, A. P., U. Regev, and H. Shalet. 1979. An economic optimization model of pesticide resistance: Alfalfa and Egyptian alfalfa weevil—An example. Environ. Entomol. 8:101–109.

Haldane, J. B. S. 1953. Animal populations and their regulation. New Biol. 15:9–24.

Hassell, M. P. In press. The dynamics of arthropod pest populations under the combined effects of parasitoids and pesticide application. J. Econ. Entomol.

Hassell, M. P., J. H. Lawton, and R. M. May. 1976. Patterns of dynamical behaviour in single-species populations. J. Anim. Ecol. 45:471–486.

Hooper, G. M. S., and A. W. A. Brown. 1965. Dieldrin resistant and DDT resistant strains of the spotted root maggot apparently restricted to heterozygotes for resistance. J. Econ. Entomol. 58:824–830.

Kates, K. C., M. L. Colglazier, and F. D. Enzie. 1973. Experimental development of a cambendazole-resistant strain of *Haemonchus contortus* in sheep. J. Parasitol. 59:169–174.

Keiding, J. 1976. Development of resistance to pyrethroids in field populations of Danish houseflies. Pestic. Sci. 7:283–291.

Keiding, J. 1977. Resistance in the housefly in Denmark and elsewhere. Pp. 261–302 *in* Pesticide Management and Insecticide Resistance, D. L. Watson and A. W. A. Brown, eds. New York: Academic Press.

Krebs, C. J. 1978. Ecology: The Experimental Analysis of Distribution and Abundance. New York: Harper and Row.

LeJambre, L. F., W. M. Royal, and P. J. Martin. 1979. The inheritance of thiabendazole resistance in *Haemonchus contortus*. Parasitology 78:107–119.

May, R. M. 1977. Food lost to pests. Nature (London) 267:669–670.

May, R. M., G. R. Conway, M. P. Hassell, and T. R. E. Southwood. 1974. Time delays, density dependence, and single species oscillations. J. Anim. Ecol. 43:747–770.

May, R. M., J. A. Endler, and R. E. McMurtrie. 1975. Gene frequency clines in the presence of selection opposed by gene flow. Am. Nat. 109:659–676.

Mayr, E. 1963. Animal Species and Evolution. Cambridge, Mass: Harvard University Press.

Morris, R. F. 1959. Single-factor analysis in population dynamics. Ecology 40:580–588.

Morse, J. G., and B. A. Croft. 1981. Developed resistance to azinphosmethyl in a predator-prey mite system in greenhouse experiments. Entomophaga 26:191–202.

Mullin, C. A., B. A. Croft, K. Strickler, F. Matsumura, and J. R. Miller. 1982. Detoxification enzyme differences between a herbivorous and predatory mite. Science 217:1270–1272.

Nagylaki, T. 1977. Selection in One- and Two-Locus Systems: Lecture Notes in Biomathematics, Vol. 15. New York: Springer-Verlag.

Peters, W. In press. Resistance to antiparasitic drugs and its prevention. Parasitology.

Pimentel, D. 1976. World food crisis: Energy and pests. Bull. Entomol. Soc. Am. 22:20–26.

Plapp, F. W., and S. B. Vinson. 1977. Comparative toxicities of some insecticides to the tobacco budworm and its ichneumonid parasite, *Campoletis sonorensis*. Environ. Entomol. 6:381–384.

Rajakulendran, S. V., and F. W. Plapp. 1982. Comparative toxicities of fire synthetic pyrethroids to the tobacco budworm, an ichneumonid parasite, and a predator. J. Econ. Entomol. 75:769–772.

Rock, G. C. 1979. Relative toxicity of two synthetic pyrethroids to a predator *Amblyseius fallacius* and its prey *Tetranychus urticae*. J. Econ. Entomol. 72:293–294.

Roush, R. T., and M. A. Hoy. 1981. Laboratory, glasshouse and field studies of artificially selected carbaryl resistance in *Metaseiulus occidentalis*. J. Econ. Entomol. 74:142–147.

Roush, R. T., and F. W. Plapp. 1982. Biochemical genetics of resistance to aryl carbamate insecticides in the predaceous mite, *Metaseiulus occidentalis*. J. Econ. Entomol. 75:304–307.

Shanahan, G. J., and N. A. Roxburgh. 1974. The sequential development of insecticide resistance problems in *Lucilia cuprina* in Australia. PANS 20:190–202.

Slatkin, M. 1973. Gene flow and selection in a cline. Genetics 75:733–756.

Stone, B. F. 1972. The genetics of resistance by ticks to acaricides. Aust. Vet. J. 48:345–350.

Stubbs, M. 1977. Density dependence in the life-cycles of animals and its importance in K- and r-strategies. J. Anim. Ecol. 46:677–688.

Sutherst, R. W., and H. N. Comins. 1979. The management of acaricide resistance in the cattle tick *Boophilus microplus* in Australia. Bull. Entomol. Soc. Am. 69:519–540.

Tabashnik, B. E., and B. A. Croft. 1982. Managing pesticide resistance in crop-arthropod complexes: Interactions between biological and operational factors. Environ. Entomol. 11:1137–1144.

Tabashnik, B. E., and B. A. Croft. 1985. Evolution of pesticide resistance in apple pests and their natural enemies. Entomophaga 30:37–49.

Tahori, A. S. 1978. Resistance of ticks to acaricides. Refu. Vet. 35:177–179.

Taylor, C. E., and G. P. Georghiou. 1979. Suppression of insecticide resistance by alteration of gene dominance and migration. J. Econ. Entomol. 72:105–109.

Taylor, C. E., and G. P. Georghiou. 1982. Influence of pesticide persistence in evolution of resistance. Environ. Entomol. 11:745–750.

Taylor, C. E., F. Quaglia, and G. P. Georghiou. 1983. Evolution of resistance to insecticides: A case study on the influence of migration and insecticide decay rates. J. Econ. Entomol. 76:704–707.

Varley, G. C., G. R. Gradwell, and M. P. Hassell. 1972. Insect Population Ecology. Oxford: Blackwell.

Wood, R. J. 1981. Strategies for conserving susceptibility to insecticides. Parasitology 82:69–80.

Pesticide Resistance: Strategies and Tactics for Management.
1986. National Academy Press, Washington, D.C.

Computer Simulation as a Tool for Pesticide Resistance Management

BRUCE E. TABASHNIK

Computer simulation may be useful for devising strategies to retard pesticide resistance in pests and to promote it in beneficials. This paper demonstrates the use of simulation to study interactions among factors influencing resistance development, describes efforts to test models of resistance development, and illustrates management applications of computer models. Suggested guidelines for future tests of resistance models are to (1) establish baseline data on susceptibility before populations are selected for resistance, (2) conduct tests under field conditions, (3) use experimental estimates of biological parameters in models, and (4) replicate treatments. Modelers of pesticide resistance must test models, explore the implications of polygenic resistance, and incorporate alternative controls such as biological control in models.

INTRODUCTION

Pest species have developed resistance to pesticides faster than beneficial organisms, limiting the integration of biological and chemical controls. Resistant strains of more than 400 insect and mite species have been recorded, but fewer than 10 percent are beneficial (Georghiou and Mellon, 1983; Croft and Strickler, 1983). The goals of resistance management are to retard resistance in pests and to promote it in beneficials. Models of pesticide resistance can be useful tools for working toward these goals. Various types of models have played an essential role in building a conceptual framework for resistance management (Table 1; Taylor, 1983). This paper emphasizes simulation modeling as a component of management and identifies future di-

TABLE 1 Modeling Studies of Pesticide Resistance

Studies	Factors Emphasized Biological	Operational	Economic
Analytical			
MacDonald, 1959	X		
Comins, 1977a	X		
Curtis et al., 1978	X	X	
Gressel and Segel, 1978	X	X	
Taylor and Georghiou, 1979	X		
Cook, 1981	X		
Skylakakis, 1981	X	X	
Wood and Mani, 1981	X	X	
Muggleton, 1982	X		
Simulation			
Georghiou and Taylor, 1977a,b	X	X	
Greever and Georghiou, 1979	X	X	
Plapp et al., 1979	X		
Kable and Jeffrey, 1980		X	
Curtis, 1981	X	X	
Taylor and Georghiou, 1982	X	X	
Tabashnik and Croft, 1982, 1985	X	X	
Levy et al., 1983		X	
Taylor et al., 1983	X	X	
Knipling and Klassen, 1984		X	
Dowd et al., 1984	X	X	
Optimization			
Hueth and Regev, 1974		X	X
Taylor and Headley, 1975		X	X
Guttierrez et al., 1976, 1979		X	X
Comins, 1977b, 1979		X	X
Shoemaker, 1982		X	X
Statistical/Empirical			
Georghiou, 1980	X		
Tabashnik and Croft, 1985	X		

SOURCE: The model classifications are based on Logan (1982) and Taylor (1983). The list of studies is expanded from Taylor (1983) but is not intended to be exhaustive.

rections for modeling that can increase its usefulness as a resistance management tool.

MODELING ASSUMPTIONS

The key assumptions of the models discussed in this paper (Tabashnik and Croft, 1982, 1985; Taylor and Georghiou, 1982; Taylor et al., 1983) are as follows:

1. Resistance is controlled primarily by a single-gene locus with two

alleles, R (resistant) and S (susceptible), with a fixed dose-mortality line for each genotype.

2. The dose-mortality line for RS heterozygotes is intermediate between the SS (susceptible) and RR (resistant) lines. At low pesticide doses RS heterozygotes are not killed, and the R gene is effectively dominant; at high doses RS heterozygotes are killed, and the R gene is effectively recessive.

3. The insect life cycle is divided into substages, with transition probabilities between substages determined by natural and pesticide mortalities.

4. Immigrants are primarily susceptible and have at least one day to mate and reproduce before being killed by a pesticide.

INTERACTIONS

There are four main classes of conditions for resistance development: (1) no immigration, low pesticide dose (R gene functionally dominant); (2) no immigration, high pesticide dose (R gene functionally codominant or recessive); (3) high immigration, low dose; and (4) high immigration, high dose. Initial modeling studies that focused on different subsets of these four main classes arrived at apparently conflicting results (e.g., contrast Georghiou and Taylor, 1977a,b, with Comins, 1977a, and Taylor and Georghiou, 1979). It was not clear whether contradictions arose from differences in modeling approaches or from differences in conditions among various studies.

Tabashnik and Croft (1982) examined the influence of various factors on rates of resistance development under all four main classes of conditions. Results showed that the way certain factors influence the rate of resistance evolution depends on which of the four classes of conditions are present. In other words the same factor may have a different influence under different background conditions.

One of the most striking examples of the interaction effect is the influence of pesticide dose on the time to develop resistance (Figure 1). Without immigration resistance developed faster as dose increased. With immigration there were two distinct phases. At low doses resistance developed faster as dose increased, paralleling the case without immigration. At high doses, however, resistance developed more slowly as dose increased. These results are consistent with Comins (1977a). Without immigration the rate of resistance development is determined primarily by the rate at which S genes are removed from the population. As dose increases, S genes are removed more rapidly; resistance develops faster. The situation with low doses and immigration is similar. With immigration and doses high enough to kill RS heterozygotes, however, pesticide mortality also removes R genes from the population. As dose increases in this range, more RS heterozygotes are killed, leaving relatively few resistant (RR) individuals. The RR survivors are ef-

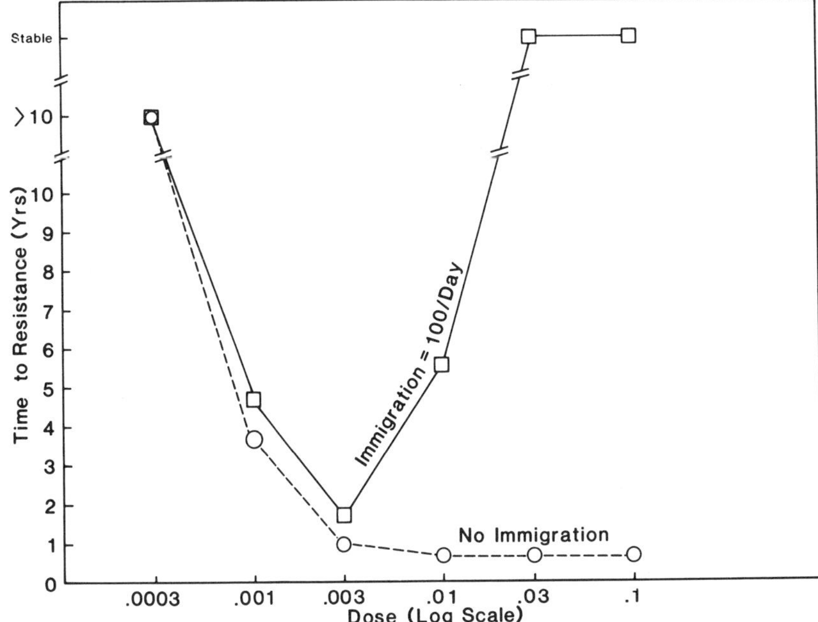

FIGURE 1 Effects of dose on the rate of evolution of resistance. Conditions: 0 or 100 immigrants daily, biweekly treatments of adults. Source: Tabashnik and Croft (1982).

fectively swamped out by susceptible immigrants, thereby retarding resistance development.

The simulation results suggest that one of the most important factors influencing the rate of resistance evolution is the number of generations per year. Under all four classes of conditions, resistance developed faster as the number of generations per year increased. Field observations of resistance development in soil and apple arthropods (Georghiou, 1980; Tabashnik and Croft, 1985) are consistent with this prediction.

A summary of the influence of various factors on resistance development (Table 2) highlights the interactions among factors. Increases in the operational factors (dose, spray frequency, and fraction of the life cycle exposed to pesticide) made resistance develop faster when there was no immigration (both low- and high-dose range) and when there was immigration and a low dose. The opposite occurred with immigration and a high dose. Some biological factors (fecundity, survival, and initial population size) had little effect in the absence of immigration, but increases in these factors made resistance evolve faster when there was immigration. Two biological factors (generations/year and immigration) had the same influence under all four classes of conditions.

TABLE 2 The Influence of Operational and Biological Factors on Resistance Development under Four Main Classes of Conditions

	No Immigration		High Immigration	
Factors	Low Dose[a]	High Dose[b]	Low Dose[a]	High Dose[b]
Operational				
Dose	+	+	+	–
Spray Frequency	+	+	+	–
Life Stages Exposed	+	+	+	–
Biological				
Generations per Year	+	+	+	+
Immigration	–	–	–	–
Fecundity	0	0	+	+
Survivorship	0	0	+	+
Initial Population Size	0	0	+	+
Initial R Gene Frequency	+	0	+	+
Reproductive Disadvantage	–	0	–	–
Dominance[c]	+	0	+	+

NOTE: + shows that increasing the listed factor speeds resistance development; – shows that increasing the listed factor slows resistance development; 0 shows little or no effect.

[a]Kills only SS, R gene functionally dominant.

[b]Kills SS and some RS, R gene functionally codominant or recessive.

[c]Based on Comins (1977a), Georghiou and Taylor (1977a), Wood and Mani (1981), and Tabashnik (unpublished).

SOURCE: Tabashnik and Croft (1982).

The most important conclusion from this simulation approach is that the influence of certain factors will depend on the presence or absence of immigration by susceptibles and on the functional dominance of the R gene (i.e., dose). Therefore, it is necessary to develop resistance management strategies that are appropriate for specific ecological and operational contexts.

TESTING MODELS

Experimental tests of pesticide resistance models are sorely needed (Taylor, 1983). There have been more than 25 papers describing resistance models during the past 10 years (Table 1), but only two studies explicitly test such models (Taylor et al., 1983; Tabashnik and Croft, 1985). These two studies represent opposite types of validation. The following discussion summarizes results of the studies and suggests how elements of both approaches can be combined to produce an especially powerful test of resistance models.

Taylor et al. (1983) used laboratory house fly (*Musca domestica*) populations to test a model of evolution of resistance to dieldrin, an organochlorine insecticide. Resistance to dieldrin is due to a single gene, and three fly genotypes are distinguishable by bioassay (Georghiou et al., 1963). Taylor et al. (1983) simulated five different treatment regimes, then compared the predicted resistance gene frequencies and population sizes with those observed in five corresponding experimental cages.

All of the biological parameters used in the simulations were measured directly from laboratory fly populations. The initial conditions were alike for all cages (90 SS + 10 RS individuals of each sex per cage), and each cage received a different treatment: (A) control—no insecticide and no immigration, (B) slow insecticide decay and immigration, (C) fast decay and immigration, (D) no decay and no immigration, and (E) no decay and immigration. Immigration was achieved by adding 25 individuals (24 SS + 1 RS) to the appropriate cages three times weekly. Dieldrin was incorporated in the larval medium and acted only on larvae and newly eclosed adults. The initial dieldrin concentration (40 ppm) was the same in treatments B to E, but decay rates corresponding to insecticide half-lives of 1.0 and 0.5 days were mimicked by using decreasing dieldrin concentrations in successive treatments. Each cage was run for 57 days (about four generations).

The results showed a strong correlation between predicted and observed values for the final R gene frequency in each treatment (Figure 2). Both the simulations and experiments support earlier predictions that immigration by susceptibles can retard the evolution of resistance, especially when the ratio of immigrants to residents in the treated population is high (Comins, 1977a; Taylor and Georghiou, 1979; Tabashnik and Croft, 1982).

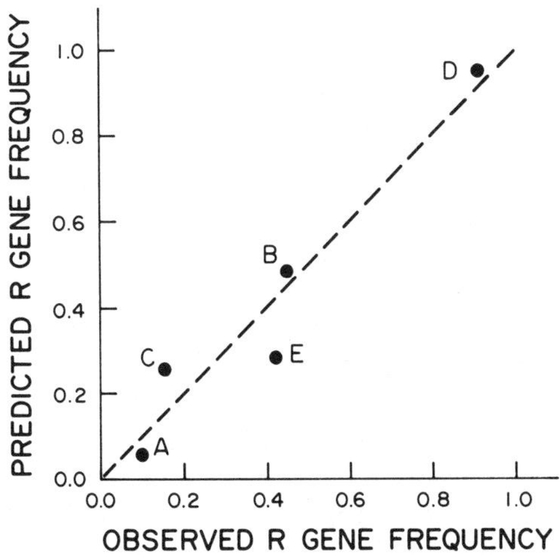

FIGURE 2 Predicted versus observed resistance (R) gene frequencies in caged house flies. Dashed line shows predicted = observed. Letters indicate treatments (see text) (Taylor et al., 1983).

This validation study shows that in a highly defined situation, model predictions may correspond well with reality. Because virtually all of the biological and operational parameters were either measured or controlled, the correspondence between predictions and observations is no accident. The model appears to incorporate the essential processes affecting evolution of resistance in the system studied. The system studied, however, was highly artificial, and its relationship to field systems is unclear. Validation in an artificial system probably cannot adequately address the question of whether model predictions apply to field situations.

Tabashnik and Croft (1985) tested a resistance model by comparing simulated times versus historically observed times to evolve resistance to azinphosmethyl in the field for 24 species of apple pests and natural enemies. Azinphosmethyl is an organophosphorous insecticide that has remained a major apple pest-control tool in North America for almost 30 years. The long-term patterns of evolution of resistance to azinphosmethyl among the diverse apple orchard insects and mites constitute a unique data set for testing predictions about resistance.

To represent 24 different apple arthropod species in the simulation, the following population ecology parameters were estimated independently for each species: generations/year, fecundity, immigration, natural (nonpesticide) mortality, initial population size, development rate, sex ratio, pesticide exposure in orchards, and percent of time spent in orchards by adults. Parameter values and historically observed times to evolve resistance for each species were based on a survey of 24 fruit entomologists (Croft, 1982).

Operational and genetic factors were held constant for all 24 species. All species were subjected to the same simulated pesticide dose, spray schedule, and pesticide half-life because all species were present in the same habitat and were exposed to a similar treatment regime in the field. The genetic basis of resistance, dose-mortality lines, and initial R gene frequency were assumed to be the same for all species because these parameters are virtually impossible to estimate for most species. Further, Tabashnik and Croft (1985) sought to determine how much of the variation in rates of evolution of resistance could be explained by differences among species in population ecology, with all other factors being constant.

The results show a significant rank correlation between predicted and historically observed times to evolve resistance for the 12 pest species and the 12 natural-enemy species (Figure 3). Thus, ecological differences among apple species are sufficient for explaining observed variation in rates of resistance development among pests and natural enemies.

There was no consistent bias in the predictions for pests, but predicted times were consistently less than observed times for natural enemies, suggesting that the original assumptions may omit factors that slow resistance development in natural enemies. The original assumptions about natural

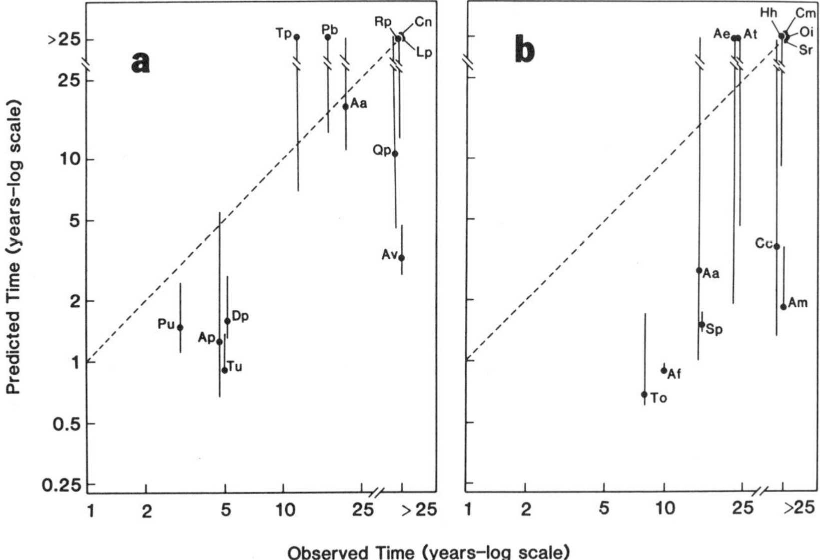

FIGURE 3 Predicted versus observed times to evolve resistance to azinphosmethyl for apple arthropods. Predicted time (●) = simulated time to evolve resistance using means of estimates of population ecology parameters. Observed time = years after 1955 (first widespread use of azinphosmethyl) to first report of resistance. Vertical bars show range of predicted times from sensitivity analysis. Dashed lines show predicted = observed. **A. Pests:** n = 12. Spearman's rank correlation coefficient, $r_s = 0.652$, $p < 0.05$. *Key:* Aa = *Archips argyrospilus*, Ap = *Aphis pomi*, Av = *Argyrotaenia velutinana*, Cn = *Conotrachelus nenuphar*, Dp = *Dysaphis plantaginea*, Lp = *Laspeyresia pomonella*, Pb = *Phyllonorcyter blancardella*, Pu = *Panonychus ulmi*, Qp = *Quadraspidiotus perniciousus*, Rp = *Rhagoletis pomonella*, Tp = *Typhlocyba pomaria*, Tu = *Tetranychus urticae* **B. Natural enemies:** n = 12. r_s = 0.692, $p < 0.025$. *Key:* Aa = *Aphidoletes aphidimyza*, Ae = *Anagrus epos*, Af = *Amblyseius fallacis*, Am = *Aphelinus mali*, At = *Aphelopus typhlocyba*, Cc = *Chrysopa carnea*, Cm = *Coleomegilla maculata lingi*, Hh = *Hyaliodes harti*, Oi = *Orius insidiosus*, Sp = *Stethorus punctum*, Sr = *Syrphus ribesii*, To = *Typhlodromus occidentalis*. *Source:* Tabashnik and Croft, 1985).

enemies were modified to incorporate the preadaptation and food-limitation hypotheses. Incorporating the preadaptation hypothesis (pests are preadapted to detoxify pesticides because they detoxify plant poisons, but natural enemies are less preadapted) (Croft and Morse, 1979; Mullin et al., 1982) did not substantially improve the correspondence between predicted and observed times. Adding the food-limitation hypothesis (a natural enemy evolves resistance only after its prey/host is resistant, because pesticides drastically reduce food for natural enemies by eliminating susceptible prey/hosts) (Huf-

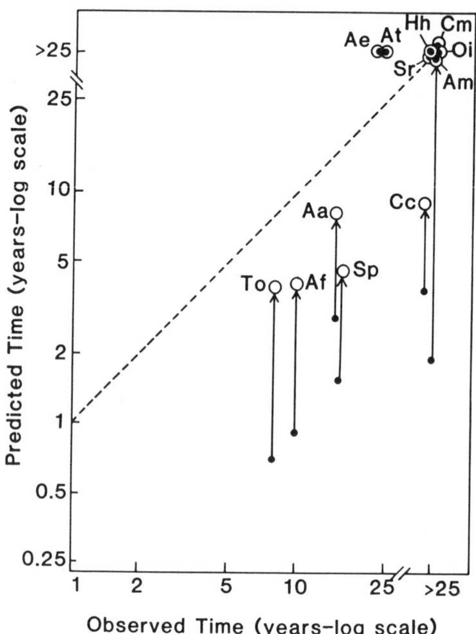

FIGURE 4 Effects of the food-limitation hypothesis on predicted times for natural enemies to evolve resistance. $n = 12$. $r_s = 0.806$, $p < 0.005$ (see Figure 3 for key to species names). Open circles indicate predictions with the food-limitation hypothesis incorporated; dark circles indicate predictions under initial assumptions. Arrows show change in predictions due to food-limitation hypothesis.

faker, 1971), however, substantially improved the correspondence between predicted and observed times for all six natural enemies that were initially predicted to evolve resistance too fast (Figure 4).

These results suggest that food limitation following pesticide applications may be an important factor in retarding evolution of resistance in natural enemies. If this is so it may be possible to promote resistance development in natural enemies by ensuring them an adequate food supply following sprays—either by reducing mortality to their prey/hosts or by providing an alternate food source when prey/hosts are scarce.

The validation study of Tabashnik and Croft (1985) provides encouragement that model results can be applied to field situations. That study, however, relies on estimated values for many important parameters. Tabashnik and Croft (1985) address this problem in part by a sensitivity analysis demonstrating that many of the model's predictions were minimally affected by substantial variation in some key parameters that are difficult to estimate, but that are potentially influential (immigration, initial population size, and fecundity; see sensitivity bars in Figure 3).

TABLE 3 Predicted Time (years) for the European Red Mite (Panonychus ulmi) to Evolve Pesticide Resistance under Different Pesticide Doses and Application Frequencies

Pesticide Dose[a]	Initial Mortality	Application Frequency (Sprays/Year) 6	3	1	1/2[b]
0.01	93%	1.5	1.7	2.6	5.7
0.002	73%	1.6	1.9	6.5	19.6
0.001	50%	1.5	2.2	13.6	>25

[a]Arbitrary units
[b]One spray every 2 years

SOURCE: Tabashnik and Croft (1985).

It seems that a powerful approach to testing resistance models can be developed by combining elements from both of the studies described above. Guidelines are as follows:

- Establish baseline data on susceptibility before populations are selected for pesticide resistance. Rates of resistance development can be measured only if initial susceptibility is known.
- Conduct tests under field conditions or conditions similar to the field. It may be especially important to use large initial population sizes if genes conferring resistance are rare.
- Obtain experimental estimates of basic biological parameters (e.g., fecundity) required for modeling.
- Replicate treatments.

Field experiments that might promote rapid evolution of new resistances in pests should not be performed. Although experimental selection for resistance is costly and time-consuming (Taylor, 1983), unintentional selection for resistance is widespread. Extremely valuable data bases on resistance could be developed by concomitant monitoring of field treatment regimes and susceptibility levels in field populations. Such data would provide a sound basis for evaluating management tactics as well as models of pesticide resistance.

MANAGEMENT APPLICATIONS

Computer simulations can be used to project the consequences of alternative control strategies. For example, Tabashnik and Croft (1985) simulated resistance development by the European red mite (*Panonychus ulmi*) under 12 management schemes based on three pesticide doses and four application schedules (Table 3). Resistance was predicted to occur within three years when intermediate to high acaricide doses (causing 50 to 93 percent initial mortality) and frequent applications (three to six per season) were simulated. If both dose and application frequency are reduced, resistance in the European red mite is predicted to be delayed from 7 to more than 25 years.

The projected times for resistance development in the European red mite are consistent with observed patterns of resistance to the acaricide cyhexatin in the United States. Since cyhexatin was introduced in 1970, resistance has not occurred in apple orchards, where it has been used judiciously in conjunction with biological control by predators. Cyhexatin resistance has occurred rapidly, however, in pear-apple interplants, where biological control is difficult and acaricide use is more intensive (Croft and Bode, 1983).

CONCLUSION

Modelers of pesticide resistance face three major challenges in the immediate future. First, and most important, models of pesticide resistance must be tested. Second, the implications of polygenically based pesticide resistance need to be explored. With few exceptions models of pesticide resistance assume one locus-two allele genetics, but many resistances may be polygenic (Plapp et al., 1979). Two of the papers in this volume take important steps toward addressing this challenge (Uyenoyama, Via). Third, alternative control methods such as biological control should be incorporated into models of pesticide resistance. The most promising way to retard resistance is to reduce pesticide use by integrating pesticides with other controls, yet current models generally assume that pesticides are the sole control method. If these challenges are addressed, modeling will play an increasingly important role in managing pesticide resistance.

ACKNOWLEDGMENTS

Special thanks to B. A. Croft for his assistance and encouragement. R. T. Roush and R. M. May provided valuable comments. Support was provided by the Research and Training Fund, University of Hawaii and USDA-HAW00947H. Paper Number 2919 of the Hawaii Institute of Tropical Agriculture and Human Resources journal series.

REFERENCES

Comins, H. N. 1977a. The development of insecticide resistance in the presence of migration. J. Theor. Biol. 64:177–197.

Comins, H. N. 1977b. The management of pesticide resistance. J. Theor. Biol. 65:399–420.

Comins, H. N. 1979. The control of adaptable pests. Pp. 217–266 *in* Pest Management, Proceedings of an International Conference, G. A. Norton and C. S. Holling, eds. Oxford: Pergamon.

Cook, L. M. 1981. The ecological factor in assessment of resistance in pest populations. Pestic. Sci. 12:582–586.

Croft, B. A. 1982. Arthropod resistance to insecticides: A key to pest control failures and successes in North American apple orchards. Entomol. Exp. Appl. 3:88–110.

Croft, B. A., and W. M. Bode. 1983. Tactics for deciduous fruit IPM. Pp. 270–291 *in* Integrated Management of Insect Pests of Pome and Stone Fruits, B. A. Croft and S. C. Hoyt, eds. New York: Interscience.

Croft, B. A., and J. G. Morse. 1979. Recent advances on pesticide resistance in natural enemies. Entomophaga 24:3–11.

Croft, B. A., and K. Strickler. 1983. Natural enemy resistance to pesticides: Documentation, characterization, theory and application. Pp. 669–702 *in* Pest Resistance to Pesticides, G. P. Georghiou and T. Saito, eds. New York: Plenum.

Curtis, C. F. 1981. Possible methods of inhibiting or reversing the evolution of insecticide resistance in mosquitoes. Pestic. Sci. 12:557–564.

Curtis, C. F., L. M. Cook, and R. J. Wood. 1978. Selection for and against insecticide resistance and possible methods of inhibiting the evolution of resistance in mosquitoes. Ecol. Entomol. 3:273–287.

Dowd, P. F., T. C. Sparks, and F. L. Mitchell. 1984. A microcomputer simulation program for demonstrating the development of insecticide resistance. Bull. Entomol. Soc. Am. 30:37–41.

Georghiou, G. P. 1980. Insecticide resistance and prospects for its management. Residue Rev. 76:131–145.

Georghiou, G. P., R. B. March, and G. E. Printy. 1963. A study on genetics of dieldrin resistance in the house fly (*Musca domestica* L.). Bull. W. H. O. 29:155–165.

Georghiou, G. P., and R. B. Mellon. 1983. Pesticide resistance in time and space. Pp. 1–46 *in* Pest Resistance to Pesticides, G. P. Georghiou and T. Saito, eds. New York: Plenum.

Georghiou, G. P., and C. E. Taylor. 1977a. Genetic and biological influences in the evolution of insecticide resistance. J. Econ. Entomol. 70:319–323.

Georghiou, G. P., and C. E. Taylor. 1977b. Operational influences in the evolution of insecticide resistance. J. Econ. Entomol. 70:653–658.

Greever, J., and G. P. Georghiou. 1979. Computer simulations of control strategies for *Culex tarsalis* (Diptera: Culicidae). J. Med. Entomol. 16:180–188.

Gressel, J., and L. A. Segel. 1978. The paucity of plants evolving genetic resistance to herbicides: Possible reasons and implications. J. Theor. Biol. 75:349–371.

Guttierez, A. P., U. Regev, and C. G. Summers. 1976. Computer model aids in weevil control. Calif. Agric. April:8–9.

Guttierez, A. P., U. Regev, and H. Shalet. 1979. An economic optimization model of pesticide resistance: Alfalfa and Egyptian alfalfa weevil—an example. Environ. Entomol. 8:101–109.

Hueth, D., and U. Regev. 1974. Optimal agricultural pest management with increasing pest resistance. Am. J. Agric. Econ. 56:543–552.

Huffaker, C. B. 1971. The ecology of pesticide interference with insect populations. Pp. 92–107 *in* Agricultural Chemicals—Harmony or Discord for Food, People, and the Environment, J. E. Swift, ed. Berkeley: University of California, Division of Agricultural Science.

Kable, P. F., and H. Jeffery. 1980. Selection for tolerance in organisms exposed to sprays of biocide mixtures: A theoretical model. Phytopathology 70:8–12.

Knipling, E. F., and W. Klassen. 1984. Influence of insecticide use patterns on the development of resistance to insecticides: A theoretical study. Southwest. Entomol. 9:351–368.

Levy, Y., R. Levi, and Y. Cohen. 1983. Buildup of a pathogen subpopulation resistant to a systemic fungicide under various control strategies: A flexible simulation model. Phytopathology 73:1475–1480.

Logan, J. A. 1982. Recent advances and new directions in phytoseiid population models. Pp. 49–71 *in* Recent Advances in Knowledge of the Phytoseiidae, M. A. Hoy, ed. Publ. 3284. Berkeley, Calif.: Agricultural Sciences Publications (Publ. No. 3284.)

MacDonald, G. 1959. The dynamics of resistance to insecticides by anophelines. Riv. di Parassitol. 20:305–315.

Muggleton, J. 1982. A model for the elimination of insecticide resistance using heterozygous disadvantage. Heredity 49:247–251.

Mullin, C. A., B. A. Croft, K. Strickler, F. Matsumura, and J. R. Miller. 1982. Detoxification enzyme differences between an herbivorous and predatory mite. Science 217:1270–1272.

Plapp, F. W., Jr., C. R. Browning, and P. J. H. Sharpe. 1979. Analysis of rate of development of insecticide resistance based on simulation of a genetic model. Environ. Entomol. 8:494–500.

Shoemaker, C. A. 1982. Optimal integrated control of univoltine pest populations with age structure. Oper. Res. 30:40–61.

Skylakakis, G. 1981. Effects of alternating and mixing pesticides on the buildup of fungal resistance. Phytopathology 71:1119–1121.

Tabashnik, B. E., and B. A. Croft. 1982. Managing pesticide resistance in crop-arthropod complexes: Interactions between biological and operational factors. Environ. Entomol. 11:1137–1144.

Tabashnik, B. E., and B. A. Croft. 1985. Evolution of pesticide resistance in apple pests and their natural enemies. Entomophaga 30:37–49.

Taylor, C. E. 1983. Evolution of resistance to insecticides: The role of mathematical models and computer simulations. Pp. 163–173 *in* Pest Resistance to Pesticides, G. P. Georghiou and T. Saito, eds. New York: Plenum.

Taylor, C. E., and G. P. Georghiou. 1979. Suppression of insecticide resistance by alteration of gene dominance and migration. J. Econ. Entomol. 72:105–109.

Taylor, C. E., and G. P. Georghiou. 1982. Influence of pesticide persistence in evolution of resistance. Environ. Entomol. 11:746–750.

Taylor, C. E., F. Quaglia, and G. P. Georghiou. 1983. Evolution of resistance to insecticides: A cage study on the influence of migration and insecticide decay rates. J. Econ. Entomol. 76:704–707.

Taylor, C. R., and J. C. Headley. 1975. Insecticide resistance and the evolution of control strategies for an insect population. Can. Entomol. 107:237–242.

Wood, R. J., and G. S. Mani. 1981. The effective dominance of resistance genes in relation to the evolution of resistance. Pestic. Sci. 12:573–581.

Pesticide Resistance: Strategies and Tactics for Management.
1986. National Academy Press, Washington, D.C.

Pleiotropy and the Evolution of Genetic Systems Conferring Resistance to Pesticides

MARCY K. UYENOYAMA

The evolution of pesticide detoxification is portrayed as the response to extreme selection pressures by a genetic network of catabolic enzymes and their regulators. Empirical and theoretical studies necessary for the assessment of this view and the exploration of its implications are described.

INTRODUCTION

Effective strategies designed to oppose the evolution of pesticide resistance must address the problem of preventing or retarding the development of the full expression of resistance, as well as the problem of controlling the density of highly resistant individuals. Most of the extensive mathematical and numerical models reviewed by Taylor (1983) investigate only the latter question, the control of quantitative aspects of resistance, including the rate of increase of highly effective mechanisms of resistance within and among populations. In this paper I consider the evolutionary process at the earlier stage, in which qualitative improvement of the expression of resistance arises as an adaptation both to the pesticide and to natural selection.

In this discussion I consider pesticide resistance as an expression of an entire genetic system and examine the implications of this multilocus perspective with respect to the optimal conditions for its evolution. Pesticide resistance in insects and novel metabolic capabilities in microorganisms represent adaptations to selection of extreme intensity that are fashioned from elements of normal metabolism. Sewall Wright's shifting balance theory, which addresses the significance of population structure to the evolution of genetic networks, provides the theoretical framework of this discussion, which seeks to convey some sense of why answers to such questions are essential from an evolutionary perspective.

EVOLUTION OF NEW FUNCTION IN MICROORGANISMS

Biochemical and genetic analyses of new catabolic pathways in laboratory populations of bacteria have yielded a wealth of information on the assembly and integration of genetic networks (Clarke, 1978; Mortlock, 1982; Hall, 1983). The processes of adaptation occurring in microbes in the laboratory and in pests of commercial crops in the field share two characteristics: the extraordinary intensity of selection imposed and the sophistication of the genetic mechanisms for the coordinated induction and repression of catabolic enzymes that respond. Responses of modern microbes to laboratory selection may in fact reveal more about the evolution of pesticide resistance than the evolution of primitive microorganisms.

Selection Procedures

Two major strategies for selecting mutants that possess extended metabolic capabilities have been adopted: one approach challenges populations to subsist on a novel substrate and the other requires the restoration of a known function by strains in which the structural locus that normally performs the function has been deleted. Investigators using the first approach focus on the identification of the regulatory and structural loci that participate in the new pathways. For example, *Klebsiella* and *Escherichia* populations presented with sugars one or several biochemical steps removed from the normal substrates constructed new metabolic pathways by borrowing enzymes from existing pathways (Mortlock, 1982). Clarke (1978) reviews experiments on *Pseudomonas* that used a variant of this first approach: altered regulation and activity of a specific amidase was selected by challenging populations with analogues of the normal substrate (acetamide). Investigators using the second approach focus on the execution of a specific task by a specific operon; they study the re-evolution of a key link in a known pathway rather than the formation of entire pathways. Selection has been imposed on *Escherichia coli* strains carrying deletions of the lacZ (β-galactosidase) gene from the lac operon to obtain lines in which β-galactosidase activity has been restored. The mutations of the regulatory and structural loci of the EBG (evolved β-galactosidase) operon, from which a well-regulated, high-activity response was eventually fashioned, are reviewed by Hall (1983).

On the molecular level the appearance de novo of a new functional locus, with appropriate sequences for initiating transcription, directing the processing of the mRNA, initiating translation, and terminating translation, represents an extraordinary macromutation. In every case the response that permitted survival involved existing enzymes having the fortuitous ability to metabolize the substrate. Regulatory mutations that induced the production of these enzymes in the absence of their normal substrates played key roles. Hall and Hartl (1974) obtained mutants characterized by hyperinducibility of the EBG operon by lactose, as well as constitutive mutants. In other experiments the key catabolic enzyme was induced by a substance in the selective medium (Clarke, 1978).

Costs Associated with Pleiotropy

If the modification of normal regulation or specificity of the key enzyme favored under artificial selection interferes with its original function, then the mutant form may suffer a disadvantage relative to the wild type in the absence of artificial selection. This disadvantage under natural selection may be regarded as the cost of pleiotropy. The EBG operon, possibly "an evolutionary remnant" (Clarke, 1978) of a relict lactose utilization pathway, may represent an exception to this generalization because it does not appear to perform any essential metabolic function in wild-type cells. Even in this case constitutive synthesis may reduce fitness under natural selection through wasteful overproduction of an enzyme (Hall, 1983; Clarke, 1978). Further, metabolism of possibly toxic analogues of the new substrate may inhibit the growth of organisms with nonspecific induction mechanisms (Hall, 1983).

Disruption of normal regulation may contribute to pleiotropic costs through imbalances of catabolites and catabolic repression (Mortlock, 1982). Clarke (1978, Table III) lists a number of amides whose catabolism can provide carbon and nitrogen but inhibits growth. Scangos and Reiner (1978) demonstrated that the inhibition (by compounds to which the wild type was insensitive) of *E. coli* strains capable of growing on the novel substrate (xylitol) was due to the activity of an enzyme whose derepression permitted use of xylitol. Further, inhibition by the novel substrate itself was relieved only at the expense of the ability to metabolize the normal substrate.

Further evolution of microbial populations with extended metabolic capabilities likely involves improved effectiveness and specificity of the response to the substrate (Mortlock, 1982; Hall, 1983). Wu et al. (1968) obtained a structural locus mutation that improved the rate of catalysis of xylitol and halved the doubling time of constitutive *Klebsiella* populations. A second mutation improved xylitol uptake and permitted another 50 percent reduction in doubling time. A sequence of four mutations in the regulatory and structural loci of the EBG operon was required for the formation of a well-regulated lactose utilization operon, in which lactose induced the synthesis of a modified EBG enzyme whose catalytic activity converted lactose into an inducer of the lactose transport system.

These examples support the view that prolonged selection in the new environment results in the refinement of the response that permits survival in that environment. Inducibility, higher rates of activity, greater specificity, and even modification of the catalyzed conversion improve the operation of the new pathway. Further, if the population repeatedly encounters both the original and the novel environments, then adaptation entails the ability to respond to both selection regimes (Clarke, 1978; Mortlock, 1982). Independent regulation of the old and new functions, which permits the expression of genetic loci primarily in response to the selective regime under which they evolved, requires the release of the elements of the new pathway from the control of the old pathway (Mortlock, 1982). Reduction in pleiotropic costs associated with new functions permits adaptation by the population to both environments.

MECHANISMS OF PESTICIDE RESISTANCE

The effective, highly evolved mechanisms for tolerating or detoxifying pesticides possessed by laboratory strains derived from resistant populations are not very likely to be representative of the rudimentary resistance mechanisms that were marshaled on initial exposure to the pesticides. Inferences regarding aspects of the resistance mechanism (including its specificity, the type of mutations involved, and the magnitude of pleiotropic costs) made on the basis of comparisons among inbred laboratory strains are relevant to questions surrounding the initial stages of the evolution of resistance only to the extent that differences among such strains reflect variation that was present in the natural populations in which resistance evolved. This caveat applies with particular force to the assessment of pleiotropic costs, because such costs may themselves evolve toward lower values as regulation of the resistance mechanism and its integration into the genome proceeds. In this section I draw analogies between the microbial evolution experiments and the evolution of pesticide resistance, while recognizing that any interpretations are open to question.

Specificity of the Response

Detoxification of certain classes of pesticides involves catabolic enzymes of low substrate specificity (Plapp and Wang, 1983). The primary function of the mixed-function oxidases that detoxify carbamate and organophosphate pesticides in the house fly and other insects appears to lie in normal metabolism (Georghiou, 1972). Resistant strains produce unusually high concentrations of microsomal oxidases that differ from the oxidases of susceptible strains with respect to substrate specificity and other properties (Plapp, 1976). Resistance to juvenile hormone analogues may also involve these broad-spectrum oxidases (Plapp, 1976; Tsukamoto, 1983). Nonspecific resistance to a variety of pesticides may involve mechanical rather than catabolic defenses. A reduction in rates of absorption of pesticides contributes to resistance in diverse organisms (Georghiou, 1972; Plapp, 1976). Such mechanisms of reduced penetration confer limited resistance and are most effective in combination with detoxification.

Specific structural changes have also been implicated in mechanisms of resistance. The shift in substrate specificity of certain mixed-function oxidases cited above indicates that structural as well as regulatory mutations are involved. Plapp (1976) describes qualitative differences in acetylcholinesterase and carboxylesterase activity that improve tolerance to or detoxification of organophosphate and carbamate insecticides. Loci controlling specific modifications of acetylcholinesterase and sensitivity of neurons to DDT reside on chromosomes II and III in the house fly (Tsukamoto, 1983).

The Evolution of Pleiotropic Costs

Crow (1957) demonstrated that the chromosomes contribute nonepistatically to the survival rate of *Drosophila melanogaster* exposed to DDT. He hypothesized that epistatic networks can evolve under close inbreeding or asexual reproduction, but that selection in outcrossing, genetically heterogeneous populations produces nonepistatic mechanisms of resistance. If elements of rudimentary resistance mechanisms evolving in nature contribute nonepistatically to fitness in *both* treated and untreated environments, then the characterization of resistance as the response of a genetic network is inappropriate. No direct evidence on this point is available; Keiding (1967) has suggested that reversion may be caused by elements whose deleterious effects reflect a lack of integration with the genetic background rather than inherent harmfulness.

Crow (1957) has discussed the potential for erroneously attributing correlations between resistance and other traits to pleiotropy in cases where those traits simply reflect differences between the particular strains representing the resistant and susceptible phenotypes. Lines et al. (1984) examined the F_2 progeny of resistant and susceptible strains in order to distinguish between effects due to strain differences per se and effects due to resistance loci (or closely linked loci). The question of pleiotropy is particularly sensitive to the general problem of choosing an appropriate control (susceptible) strain, because pleiotropic costs may evolve. With respect to the early stages of the evolution of resistance, the proper control should represent susceptible individuals of the same population, because it is in this context that the initial, rudimentary resistance mechanisms must be refined.

Apparent reversion of resistance during periods in which use of the pesticide had been suspended has been observed in field populations (Keiding,

1967; Georghiou, 1972). Curtis et al. (1978) estimated the pleiotropic costs associated with resistance by monitoring the decline of resistance in populations of *Anopheles*; they caution that such field studies may wrongly attribute declines due to migration of susceptibles to reversion. Perhaps the best demonstration that characters influencing fitness in the absence of insecticides evolve in treated populations comes from the work of McKenzie et al. (1982) on diazinon resistance in natural populations of the blow fly, *Lucilia cuprina*. In 1969–1970, population experiments indicated lower fitness in resistant flies relative to flies from a standard reference strain (McKenzie et al., 1982). In contrast resistant lines derived from a field population in 1979 suffered no disadvantage relative to the control strain, either in laboratory population cages or in field viability tests. Results resembling the earlier observations were obtained following placement of the major resistance gene on the control background by backcrossing. These results indicate that regardless of the appropriateness of the standard reference strain as a susceptible control, continued pesticide treatment in the field has modified characters that contribute to fitness in the absence of the pesticide: the pleiotropic costs have undergone evolution.

Evolution of Epistatic Resistance

The question of fashioning resistance to pesticides from the components of normal metabolism centers on the evolutionary process by which an integrated genetic network controlling normal metabolism transforms into another genetic network capable of responding to both treated and untreated environments. Known single-locus determinants of resistance may represent highly evolved mechanisms, the products of the evolutionary process discussed here. The evolutionary process under which genetic systems evolve differs fundamentally from the processes involving the independent evolution of single characters (Wright, 1960). Analysis of the process of the evolution of genetic networks may contribute toward the control of pesticide resistance by suggesting some means of retarding the development of effective mechanisms of resistance.

THE SHIFTING BALANCE THEORY

Genetic Systems as Sets of Interacting Loci

A complex developmental process integrating a myriad of internal and external influences is interposed between genes and characters of selective importance (Wright, 1934, 1960, 1968). Substitution of an allele at a given locus by another allele of different effect alters the entire developmental network, thereby inducing a response in several characters. Wright based this principle of "universal pleiotropy" (1968, Chapter V) on his extensive studies of inheritance in laboratory populations of guinea pigs, whose extraordinary diversity of morphology, vigor, and temperament derived from the interaction between various genetic factors and particular backgrounds (Wright, 1978).

Shifts Among Peaks in the Adaptive Topography

Wright (1932) characterized the possible genetic states of an individual as points in a gene frequency space whose dimensions correspond to loci, and associated with each point the adaptive value of individuals carrying the corresponding array of genes. Under pleiotropy and epistasis certain genetic combinations confer particularly high fitness, corresponding to peaks of this adaptive topography, and others confer low fitness, corresponding to valleys. In the imagery of the adaptive topography, populations ascend toward peaks. Having once attained a peak the population undergoes no further improvement except insofar as new mutations elevate the peak at which it resides or otherwise modifies the surrounding topography (Wright, 1942). Sustained advance requires some means of momentary release from convergence toward a peak to permit the population to explore other regions of the topography. Continual shifts to higher peaks constitute the essence of the shifting balance process.

Among the several mechanisms enumerated by Wright (1931, 1932, 1940, 1948, 1955, 1959) that can modulate the selective process that compels populations to proceed up gradients in the adaptive topography are genetic drift and qualitative changes in selection pressure. Genetic drift introduces an element of stochasticity into evolutionary changes in gene frequency and permits the nonadaptive passage of populations into and even through valleys of the adaptive topography. Variable selection pressures, especially in cases in which the direction of evolution undergoes periodic reversals, can trigger peak shifts (Wright, 1932, 1935, 1940, 1942, 1956). In the imagery of the adaptive topography, valleys may be temporarily uplifted, permitting the population to wander into the domain of attraction of a new peak by means of a wholly adaptive process.

THE EVOLUTION OF PESTICIDE RESISTANCE

In its simplest form the evolution of a rudimentary resistance mechanism and the reduction of pleiotropic costs through the separation of incipient detoxification pathways from metabolic pathways represents a peak shift under fluctuating selection. Alternation of treated and untreated generations requires the maintenance of adaptations to both selective regimes. Moderate levels of migration between treated and untreated populations may promote peak shifts in both regions.

Multiple Peaks in the Adaptive Topography

Upon initial exposure to the pesticide, rare individuals survive by virtue of regulatory mutations that induce sufficient production of an enzyme having the fortuitous ability to detoxify the compound in the absence of its normal substrate. All individuals possess the bifunctional structural locus; the sole genetic difference between susceptible and resistant individuals at this stage lies at the regulatory locus. Temporary suspension of pesticide treatments tends to reduce the level of resistance in the population by restoring the original selective regime, which favors a lower rate of production.

Distinct modifier loci contribute to the resistance mechanism by releasing the key enzyme from its original metabolic pathway. Such mutations are likely to induce deleterious effects in the absence of the pesticide by interfering with the regulation of the original metabolic pathway. Under pesticide treatment these mutations are favored by directional selection because any degree of separation between the two pathways permits the detoxification pathway to operate more efficiently.

Selection by pesticides favors maximal synthesis of the key enzyme and maximal separation of the pathways. Natural selection in the absence of the pesticide either favors moderate levels of synthesis of the enzyme if the pathways are not separated or is insensitive to the rate of synthesis if the pathways are entirely separated. Only one combination, maximal synthesis of the key enzyme and complete separation of the pathways, confers high fitness under both selective regimes. In the absence of the pesticide, however, this optimal combination is separated from the current position of the population by the disadvantage of incompletely separated pathways. The transfer of the population from its original state to the optimal state through the alternation of the two selective regimes represents a peak shift.

Effects of Migration Between Treated and Untreated Areas

Migration of susceptible individuals into areas under treatment by pesticides can delay the increase in density of individuals carrying well-developed, single-locus resistance mechanisms by inflating the frequency of the susceptible allele and ensuring that most resistance alleles are carried by heterozygotes (Georghiou and Taylor, 1977; Comins, 1977; Tabashnik and Croft, 1982). Comins (1977) showed that intermediate levels of migration promote the optimal balance between its positive effect (increasing the frequency of the susceptible allele in the treated deme) and its negative effect (increasing the frequency of the resistant allele in the untreated deme). If the untreated population is effectively infinite so that emigration from treated areas is negligible, the benefits of reducing the frequency of the resistant allele must be weighed against the damage inflicted by susceptible immigrants (Tabashnik and Croft, 1982). The deliberate increase of the frequency of susceptibles by the creation of untreated refugia or by the release of susceptible individuals has been suggested as a strategy of control (Georghiou and Taylor, 1977; Taylor and Georghiou, 1979). The effect of migration on the rate of refinement of resistance through the joint evolution of structural, regulatory, and modifier loci demands a full analytical treatment. In contrast with the conclusions drawn from single-locus models, migration may have a uniformly detrimental effect as a control strategy opposing the evolution of genetic networks because it promotes the evolution of modifiers of resistance by increasing the effective rate of mutation in the treated area and introducing a preadaptation for resistance into untreated areas.

Migration into the treated area may promote peak shifts by increasing the level of genetic variation and the effective population size in the treated area. Reductions in the pleiotropic costs associated with rudimentary resistance mechanisms await mutations at modifier loci that promote the separation of the detoxification pathway from the original metabolic pathways. Fisher (1958) described the dependence of the rate of production of advantageous mutations and their probabilities of extinction on the population size. Large populations contain more potential sites of mutation, and the probability of extinction of advantageous mutations in the first few generations after their appearance declines with increasing population size. Mutations that permit separation of the pathways are initially advantageous only under treatment by the pesticide; the suggestion that migration into treated areas promotes peak shifts may need qualification under alternating selective regimes.

Migration from treated areas into untreated populations promotes the spread of alleles that improve the separation of the pathways and contributes to preadaptation to the pesticide. Because such alleles are assumed to be deleterious until some minimal degree of separation is achieved, natural selection in untreated areas will oppose their introduction. They may nevertheless proceed to fixation under nonadaptive processes such as genetic drift. The introduction of these alleles by migration occurs at rates and in frequencies far greater than expected under mutation alone. Each fixation further increases the separation of the pathways and promotes more fixations. Walsh (1982) computed the probability of fixation of an allele, introduced into the population as a single gene, under the assumption of an arbitrary level of underdominance in fitness (Wright, 1941; Bengtsson and Bodmer, 1976; Lande, 1979). Sufficient separation of the pathways in the untreated population may form the basis of a preadaptation to the pesticide. Upon the introduction of the pesticide the population can respond without interfering with normal metabolism and evolve resistance without bearing the pleiotropic costs that opposed the rise of resistance in the first population.

A CALL FOR EMPIRICAL AND THEORETICAL WORK

This discussion and its conclusions draw upon a number of suppositions and assumptions: primitive resistance mechanisms redirect the activity of enzymes that normally participate in metabolism toward detoxification; such redirection entails pleiotropic costs that, in the absence of pesticide treatment, lower the fitness of resistant individuals relative to susceptible individuals; pleiotropic costs can be reduced through adaptation by a genetic network of modifiers; peak shifts of this kind occur under alternation of treated and untreated generations; and migration from treated areas promotes peak shifts that may form the basis of preadaptations to the pesticide. An informed assessment of this argument and the validity of any control strategies it may suggest requires empirical and theoretical investigation.

Empirical Studies of Rudimentary Resistance

Analysis of the genetic structure of primitive mechanisms of resistance and the direct assessment of pleiotropic costs associated with such mechanisms would provide empirical information of crucial importance for the prevention or retardation of the evolution of resistance. The highly successful strategy of the microbial evolution experiments could be modified for the study of rudimentary resistance mechanisms either by challenging organisms in the laboratory with new pesticides to which effective resistance has not yet evolved or by deleting a locus of major effect on resistance and monitoring the restoration of its function. The objectives would include (1) classification of the key mutations with respect to regulatory or structural function, (2) estimation of the relative importance of regulatory mutations causing constitutivity and hyperinducibility, and (3) assessment of the effects of the key mutations on normal metabolism.

Direct estimates of pleiotropic costs associated with poorly formed resistance mechanisms could be obtained by comparing the levels of additive genetic variance in fitness in experimental populations before and after exposure to a novel pesticide. Fitness in the absence of the pesticide may be regarded as a character which is correlated with the character of resistance and which is disrupted by the selection imposed by the pesticide (Falconer, 1953, 1981). Before pesticide application, the additive genetic variance of characters closely associated with fitness is expected to be low (Fisher, 1958; Falconer, 1981). After exposure the surviving individuals are likely to differ in a variety of characters from individuals that succumbed. If certain of those characters contribute to fitness in the absence of the pesticide, then the

TABLE 1 Relative Fitnesses in the Absence of Pesticide Treatment (Regime 1)

	BB	Bb	bb
AA	w_1	$w_1 - s$	t
Aa	w_2	$w_2 - s$	t
aa	w_3	$w_3 - s$	t

additive genetic variance in fitness is expected to increase after treatment. The magnitude of change in additive genetic variance in fitness reflects the magnitude of the pleiotropic costs associated with resistance. It is this component of variance that determines the rate of reversion of resistance in the absence of pesticide treatment (Falconer, 1981).

A Model of Epistatic Resistance

In its simplest form the peak shift required for the evolution of resistance mechanisms that incur low pleiotropic costs entails genetic changes at two loci: the regulatory locus controlling the level of synthesis at the key structural locus and a modifier locus permitting separation of the two pathways. The effects of migration and population size on the refinement of resistance in a population that exchanges migrants with untreated populations could be investigated through the analysis of the two-locus model described in this section.

In the absence of pesticide treatment, genetic variation at the regulatory locus is maintained by heterosis in fitness and the modifier locus is monomorphic. The introduction by mutation or migration of a new allele at the modifier locus results in the production of heterozygotes that suffer a reduction in fitness due to interference between the detoxification pathway and normal metabolism. In homozygotes for the new allele the pathways are independent, rendering variation at the regulatory locus, which now controls the production of an enzyme involved only in detoxification, selectively neutral. Regime 1 corresponds to natural selection in the absence of treatment by the pesticide.

Table 1 presents the fitness matrix associated with Regime 1. Locus A represents the regulatory locus at which variation is maintained by heterosis ($w_2 > w_1, w_3$). Locus B represents the modifier locus at which the heterozygote detracts from fitness ($s > 0$) and the homozygote improves fitness by causing the separation of the pathways ($t > w_i - s$ for all i). Because the new allele (b) at the modifier locus causes underdominance in fitness in combination with all genotypes at the regulatory locus, its introduction is uniformly opposed by natural selection.

Exposure to the pesticide favors maximal rates of synthesis of the key

TABLE 2 Relative Fitnesses Under Treatment by Pesticides (Regime 2)

	BB	Bb	bb
AA	x_1	x_1+u	x_1+v
Aa	x_2	x_2+u	x_2+v
aa	x_3	x_3+u	x_3+v

enzyme and any reduction in the interdependence of the two pathways. Table 2 presents the fitness matrix associated with Regime 2, which corresponds to pesticide treatment. Selection at locus A, which was balancing under Regime 1, now becomes directional ($x_1 > x_2 > x_3$). Selection at locus B, which was underdominant under Regime 1, now also becomes directional, favoring the new allele ($v > u > 0$).

In treated areas Regime 1 alternates with Regime 2 at a frequency determined by the generation time of the pest relative to the interval between treatments. Evolution in untreated populations is governed solely by Regime 1. Migration is represented by an exchange of genes between the treated population and one or more unexposed populations.

The key objectives of the theoretical analysis of this system include the description of evolution in treated and untreated regions separately and the influence of migration between these regions. Such studies should explore the effect of relative population sizes in treated and untreated areas, the migration rate, the frequency of treatment, and the intensity of selection on the rate of introduction of the new allele (b) and the probability and rate of fixation of the optimal combination in treated populations. Numerical and mathematical analyses of the model could be used to explore the process of formation of preadaptations to the pesticide in untreated areas by studying the effect of migration rate and population size on the rate of introduction of the new modifier allele (b) through the barrier of underdominance in fitness.

CONCLUSION

The central concern of this discussion has been to suggest that empirical and theoretical investigation be directed toward the elucidation of the process under which primitive responses to pesticides develop into highly effective mechanisms of resistance. The bifunctionality of components of primitive resistance mechanisms suggests that in the early evolutionary stages the defense against pesticides involves some disruption of normal physiological processes. Direct empirical investigations of primitive responses to new pesticides would provide crucial evidence to support or refute the hypothesis that primitive mechanisms of resistance incur substantial pleiotropic costs.

The evolution of genetic systems entails changes at several genetic loci under epistatic selection. Taylor (1983) cites only one paper (Plapp et al., 1979) that addresses multilocus models of resistance. The multilocus approach permits the study of qualitatively new phenomena which have no representation in one-locus models: epistasis deriving from pleiotropy, the central issue of this discussion, requires a multilocus approach. In the preceding section, a simple two-locus model was proposed that incorporates migration within subdivided populations and loci that contribute to both detoxification and normal metabolism. Of particular relevance to the development of effective control policies is the question of whether migration between treated and untreated regions promotes the reduction of pleiotropic costs and the rate of preadaptation to the pesticide by untreated populations.

The confrontation of theoretical population genetics with the practical problems of the control of pesticide resistance enriches both fields by revealing new perspectives on old problems and by provoking the development of new questions. While the establishment of improved channels for dialogue can hardly be expected to produce panaceas, the clear necessity of effective policies governing the control and management of pest populations demands the best efforts of a variety of disciplines.

ACKNOWLEDGMENTS

I thank Bruce E. Tabashnik and Richard T. Roush whose insight and knowledge of the literature served as my introduction to the study of pesticide resistance. John A. McKenzie, on very short notice, graciously forwarded preprints and offered suggestions that improved the paper. This study was supported by PHS Grant HD-17925.

REFERENCES

Bengtsson, B. O., and W. F. Bodmer. 1976. On the increase of chromosome mutations under random mating. Theor. Popul. Biol. 9:260–281.

Clarke, P. H. 1978. Experiments in microbial evolution. Pp. 137–218 *in* The Bacteria, T. C. Gunsalos, ed. New York: Academic Press.

Comins, H. N. 1977. The development of insecticide resistance in the presence of migration. J. Theor. Biol. 64:177–197.

Crow, J. F. 1957. Genetics of insect resistance to chemicals. Annu. Rev. Entomol. 2:227–246.

Curtis, C. F., L. M. Cook, and R. J. Wood. 1978. Selection for and against insecticide resistance and possible methods of inhibiting the evolution of resistance in mosquitoes. Ecol. Entomol. 3:273–287.

Falconer, D. S. 1953. Selection for large and small size in mice. J. Genet. 51:470–501.

Falconer, D. S. 1981. Introduction to Quantitative Genetics, 2nd ed. London: Longman.

Fisher, R. A. 1958. The Genetical Theory of Natural Selection, 2nd ed. New York: Dover.

Georghiou, G. P. 1972. The evolution of resistance to pesticides. Annu. Rev. Ecol. Syst. 3:133–168.

Georghiou, G. P., and C. E. Taylor. 1977. Operational influences in the evolution of insecticide resistance. J. Econ. Entomol. 70:653–658.

Hall, B. G. 1983. Evolution of new metabolic functions in laboratory organisms. Pp. 234–257 *in* Evolution of Genes and Proteins, M. Nei and R. K. Koehn, eds. Sunderland, England: Sinauer.

Hall, B. G., and D. L. Hartl. 1974. Regulation of newly evolved enzymes. I. Selection of a novel lactase regulated by lactose in *Escherichia coli*. Genetics 76:391–400.

Keiding, J. 1967. Persistence of resistant populations after the relaxation of the selection pressure. World Rev. Pest Control 6:115–130.

Lande, R. 1979. Effective deme sizes during long-term evolution estimated from rates of chromosomal rearrangement. Evolution 33:234–251.

Lines, J. D., M. A. E. Ahmed, and C. F. Curtis. 1984. Genetic studies of malathion resistance in *Anopheles arabiensis*. Bull. Entomol. Res. 74:317–325.

McKenzie, J. A., M. J. Whitten, and M. A. Adena. 1982. The effect of genetic background on the fitness of the diazinon resistance genotypes of the Australian sheep blowfly, *Lucilia cuprina*. Heredity 49:1–9.

Mortlock, R. P. 1982. Regulatory mutations and the development of new metabolic pathways by bacteria. Evol. Biol. 14:205–268.

Plapp, F. W., Jr. 1976. Biochemical genetics of insecticide resistance. Annu. Rev. Entomol. 21:179–197.

Plapp, F. W., Jr., C. R. Browning, and P. J. H. Sharpe. 1979. Analysis of rate of development of insecticide resistance based on simulation of a genetic model. Environ. Entomol. 8:494–500.

Plapp, F. W., Jr. and T. C. Wang. 1983. Genetic origins of insecticide resistance. Pp. 47–70 *in* Pest Resistance to Pesticides, G. P. Georghiou and T. Saito, eds. New York: Plenum.

Scangos, G. A., and A. M. Reiner. 1978. Acquisition of ability to utilize xylitol: Disadvantages of a constitutive catabolic pathway in *Escherichia coli*. J. Bacteriol. 134:501–505.

Tabashnik, B. E., and B. A. Croft. 1982. Managing pesticide resistance in crop-arthropod complexes: Interactions between biological and operational factors. Environ. Entomol. 11:1137–1144.

Taylor, C. E. 1983. Evolution of resistance to insecticides: The role of mathematical models and computer simulations. Pp. 163–173 *in* Pest Resistance to Pesticides, G. P. Georghiou and T. Saito, eds. New York: Plenum.

Taylor, C. E., and G. P. Georghiou. 1979. Suppression of insecticide resistance by alteration of gene dominance and migration. J. Econ. Entomol. 72:105–109.

Tsukamoto, M. 1983. Methods of genetic analysis of insecticide resistance. Pp. 71–98 *in* Pest Resistance to Pesticides, G. P. Georghiou and T. Saito, eds. New York: Plenum.

Walsh, J. B. 1982. Rate of accumulation of reproductive isolation by chromosomal rearrangements. Am. Nat. 120:510–532.

Wright, S. 1931. Evolution in Mendelian populations. Genetics 16:97–159.

Wright, S. 1932. The roles of mutation, inbreeding, crossbreeding, and selection in evolution. Proc. 6th Int. Congr. Genet. 1:356–366.

Wright, S. 1934. Physiological and evolutionary theories of dominance. Am. Nat. 68:24–53.

Wright, S. 1935. Evolution in populations in approximate equilibrium. J. Genet. 30:257–266.

Wright, S. 1940. The statistical consequences of Mendelian heredity in relation to speciation. Pp. 161–183 *in* The New Systematics, J. Huxley, ed. Oxford: Clarendon.

Wright, S. 1941. On the probability of fixation of reciprocal translocations. Am. Nat. 75:513–522.

Wright, S. 1942. Statistical genetics and evolution. Bull. Am. Math. Soc. 48:223–246.

Wright, S. 1948. On the roles of directed and random changes in gene frequency in the genetics of populations. Evolution 2:279–294.

Wright, S. 1955. Classification of the factors of evolution. Cold Spring Harbor Symp. Quant. Biol. 20:16–24.

Wright, S. 1956. Modes of selection. Am. Nat. 90:5–24.

Wright, S. 1959. Physiological genetics, ecology of populations, and natural selection. Perspect. Biol. Med. 3:107–151.

Wright, S. 1960. Genetics and twentieth century Darwinism: A review and discussion. Am. J. Hum. Genet. 12:365–372.

Wright, S. 1968. Evolution and the Genetics of Populations. Genetic and Biometric Foundations, Vol. I. Chicago, Ill.: University of Chicago Press.

Wright, S. 1978. The relation of livestock breeding to theories of evolution. J. Anim. Sci. 46:1192–1200.

Wu, T. T., E. C. C. Lin, and S. Tanaka. 1968. Mutants of *Aerobacter aerogenes* capable of utilizing xylitol as a novel carbon. J. Bacteriol. 96:447–456.

Pesticide Resistance: Strategies and Tactics for Management.
1986. National Academy Press, Washington, D.C.

Quantitative Genetic Models and the Evolution of Pesticide Resistance

SARA VIA

When tolerance to pesticides varies continuously among individuals, a quantitative genetic approach to resistance evolution is more useful than is the usual single-locus view. Relative characteristics of polygenic and single-gene resistance are described; then the evolution of polygenic resistance is discussed in terms of basic quantitative genetics principles. Finally, polygenic models that use the quantitative genetic analog of negative cross-resistance (genetic correlation) are described. These models suggest that the joint application of selected compounds in some spatial array may be a useful means of retarding the evolution of polygenic resistance. Further refinements of the models and ways to validate them with experimental data are considered. Estimates of genetic parameters and selection intensities are essential to assess the validity of the suggestions presented here. These models are discussed primarily as heuristic tools that may provide a new conceptual view on the problem of pesticide resistance; they do not as yet provide descriptions of particular cases of resistance evolution in real pest populations.

INTRODUCTION

The increasing frequency of pesticide resistance is an undeniable example of the process of evolution. Basic Darwinian principles assert that when genetic variation is available, populations under selection by some aspect of the environment will increase adaptation through evolutionary change. When pesticides are the agents of selection, the response will be some form of pesticide resistance, such as detoxification, physiological adaptation, or behavioral avoidance (Georghiou, 1972; Wood and Bishop, 1981).

Mathematical models have been instrumental in the identification and study of the genetic and environmental factors that influence the rate and direction of evolution. Because pesticides are agents of selection, pesticide resistance can be studied by using the same theoretical frameworks as have been applied to other types of evolutionary change.

Previous population genetic models have considered that resistance is determined by a single gene. These models are generally not immediately applicable when resistance is a quantitative (polygenic) trait, in which the underlying genes may not (and indeed need not) be identified individually. This paper describes how resistance can be studied from a polygenic perspective and suggests how models that were derived to describe the evolution of quantitative characters in different environments may be used to design genetically sound strategies of pesticide application to retard the evolution of pesticide resistance.

Cases of polygenic resistance are well known (Crow, 1954; King, 1954; Liu, 1982; Wood and Bishop, 1981). Although polygenic resistance in field situations may be less common than monogenic resistance, the potential for polygenic resistance may be more widespread than is currently recognized, because different populations exhibit different mechanisms of resistance (Thomas, 1966; Wood and Bishop, 1981) and mutations affecting resistance can be mapped to different loci (Wood and Bishop, 1981; Pluthero and Threlkeld, 1983). In fact the high frequency of major gene resistance in field populations may result more from the very strong selection imposed by current regimes of pesticide application (Lande, 1983; Roush, 1984) than from an inherent bias in genetic potential. The intent of new methods of pesticide application is to lower the effective intensity of selection (Taylor and Georghiou, 1982; Tabashnik and Croft, 1982). Such methods may increase the incidence of polygenic resistance.

POLYGENIC RESISTANCE

When pesticide resistance is polygenic (owing to effects at several gene loci), the resistance phenotype as expressed in the dose-response curve will be continuous (Figure 1B). The polygenic curve spans the range of the separate resistance classes seen in the single-locus case (Figure 1A). The range in dose response of a single genotype in the true one-locus case is due to environmental effects: if there were no environmental variation, all individuals of a given genotype would die at the same dose, and the dose-response curves in Figure 1A would be vertical lines. In this paper the effects of modifier genes on the dose-response curves for the major locus will be ignored. Such modifiers, however, will lower the slopes of

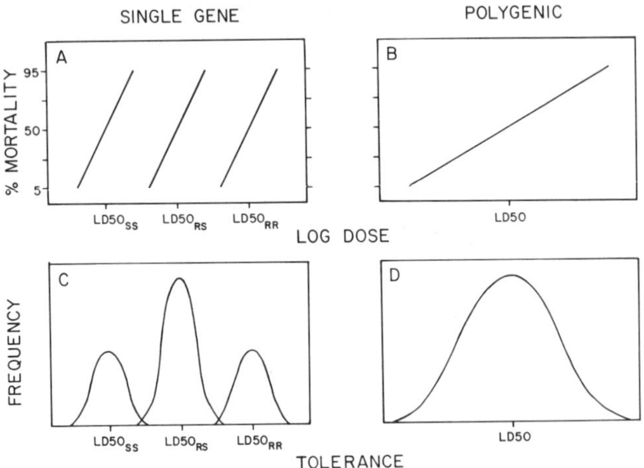

FIGURE 1 Comparison of dose-response curves (A,B) and tolerance distributions (C,D) for pesticide resistance with single-gene or polygenic genetic basis. A,B: Dose-reponse curves corresponding to the cumulative distribution of mortality with increasing dose on a log scale. C,D: Tolerance curves are probability density functions for the sensitivity to dose. (Redrawn from Via and Lande, 1985.)

the dose-response curves in Figure 1A, having the same effect as environmental variance.

In polygenic resistance a continuous dose-response relationship results from the combination of environmental and genetic factors. No distinct genotypic classes can be identified because classes overlap when several loci determine a trait; polygenic characters thus are also called "continuous characters" (Falconer, 1981). Because only the additive genetic variance in tolerance to a given compound (V_A) contributes to the evolution of resistance by individual selection, it is necessary to determine the fraction of the total phenotypic variance in tolerance to that pesticide (V_P) that is due to additive genetic causes. This is accomplished by partitioning V_P into its components,

$$V_P = V_A + V_E \tag{1}$$

where V_E includes the nonadditive genetic variance plus the microenvironmental variation in tolerance. Other more complete partitionings are also possible (Falconer, 1981).

The various partitionings of the phenotypic variance into its causal components rely on theory first developed by R. A. Fisher (1918). The theory of quantitative genetics is based on the fact that family members resemble one another because they share genes; variation among families can thus be used to estimate genetic variation. Experiments designed to determine the genetic components of variance for quantitative traits therefore rely heavily on breeding designs that generate family groupings with certain degrees of relatedness (Falconer, 1981). Variation in the phenotypic characters of interest (here, tolerance to certain pesticides) can then be estimated within and among families to derive the desired estimates of the genetic components of variance (Via, 1984a,b).

Selection for Tolerance

The dose-response curves in Figures 1A and 1B are cumulative distribution functions (Mood et al., 1963). They express the total fraction of the population that is dead by the time a pesticide has reached a certain dosage. In contrast Figures 1C and 1D are probability distribution functions (Mood et al., 1963) that express the proportion of individuals that die at a particular dosage. These probability distribution functions represent tolerance curves for the population. A normal distribution of tolerance means that a few individuals in the population are very sensitive to pesticide treatment, a few will survive until the dose is extremely high, and most will have an average degree of tolerance. Tolerance curves illustrate the proportion of the population that dies at a particular dose. Variation in tolerance for each curve in the single-locus case is presumed to be entirely environmental. In the polygenic case, variation is the sum of genetic and environmental components.

The mean tolerance in a population is the LD_{50} (Figure 1). In the presence of a pesticide, selection will act to increase the LD_{50}—individuals with high tolerance are favored. The selection response of a quantitative trait is the product of the proportion of variation in a character that is caused by additive genetic variation and the intensity of selection (Falconer, 1981). Using this result the dynamics of the evolution of tolerance when the population is exposed to a single pesticide can be described mathematically as

$$\Delta LD_{50} = (V_A/V_P)s \qquad (2)$$

where ΔLD_{50} is the change in the mean tolerance in every generation, and s is the difference in mean tolerance before and after selection (the selection differential). Equation 2 illustrates that the rate at which pesticide resistance (tolerance) evolves is proportional to the magnitude of the total variation in tolerance that is additive genetic and to the intensity of selection. Although the genetic parameters may change during selection, equation 2 will hold for several generations, after which the genetic parameters must be reestimated.

Genetic Correlations Among Traits

The univariate formulation presented in equation 2 applies only when selection acts on a single character, such as tolerance to a particular pesticide.

Usually many characters are under selection simultaneously. For example, natural selection on fertility and fecundity operates at the same time as selection for pesticide resistance. The disadvantage of individuals with major genes for insecticide resistance, with respect to natural selection on correlated traits, may account for some of the reversion of resistance seen in the absence of pesticides (Abedi and Brown, 1960; Curtis et al., 1978; McKenzie et al., 1982). The case considered here concerns simultaneous selection of tolerances to multiple pesticides and considers the effect of genetic correlations in tolerances on the evolution of resistance.

A study of the evolution of suites of characters must consider the degree to which the traits of interest have the same genetic basis. The genetic similarity of two traits can be estimated as the genetic correlation (Falconer, 1981). Genetic correlations result from the pleiotropic (multiple) effects of genes. Because pleiotropy is considered to be universal (Wright, 1968), significant genetic correlations among traits are common.

Genetic correlations affect the course of evolution; when selection impinges on any character in a correlated group, all traits that are influenced by the same genes will also show an evolutionary change in their phenotypes, even if they are not directly affected by selection. This is called correlated response to selection. These correlated changes are not necessarily in the direction that is adaptive for all characters. Correlated characters cannot evolve independently: if two traits are negatively correlated, selection for one to increase may result in a correlated decrease in the other—even if this is disadvantageous. Therefore, genetic correlations can constrain the evolution of the whole phenotype and can cause maladaptation of some traits within a correlated suite. This process may be a useful way to temporarily retard evolution in insect pest populations.

Genetic Correlations in Tolerance to Different Pesticides

The present model illustrates what may happen when different pesticides are sprayed in adjacent fields. The key feature of the model is an observation first made by Falconer (1952): a character expressed in two environments can be considered as two genetically correlated traits. Here, tolerance to two pesticides is considered to be two traits that may have a genetic correlation of less than +1 if different genes produce tolerance to each compound. For example, if different enzymes are required to detoxify two compounds or if different loci are involved in behavioral avoidance (Wood and Bishop, 1981), the genetic correlation in tolerance to the pair of compounds may be low. With this view the basic theory of evolution in correlated characters (Hazel, 1943; Lande, 1979) can be expanded to encompass genetic correlations across environments (Via and Lande, in press). Here, the correlations of interest are across pesticides.

The Model Consider tolerance to a particular pesticide to be a normally distributed character, as illustrated in Figure 1B. The phenotypic variation in tolerance may be decomposed into additive genetic and environmental components, as in equation 1, using the resemblances among relatives (parent-offspring regression or some other standard breeding design such as sibling analysis) (Falconer, 1981; Via, 1984a). From such an analysis the additive genetic variation in tolerance to each pesticide can be determined. Environmental effects influencing tolerance to a particular pesticide are assumed to follow a Gaussian (normal) distribution. When several loci of small effect influence the tolerance phenotype, the distribution of additive genetic effects on tolerance can also be assumed to be approximately Gaussian.

If one simultaneously measures the tolerances of family members to two pesticides by subjecting some siblings to each compound, the additive genetic correlation in tolerance to the two compounds can be estimated (Falconer, 1981; Via, 1984b). As discussed previously the genetic correlation between tolerances to the two pesticides is an estimate of the extent to which they have the same genetic basis.

The specific scenario modeled here concerns adjoining fields that are sprayed with different compounds. Individuals are assumed to assort at random into the fields with some probability (q into the fields with the first pesticide and $1-q$ into the fields sprayed with the other compound). The term q represents either some fixed preference for the different field types that is uniform among all individuals or denotes the proportional representation of each pesticide in the overall environment. In this model any given individual experiences only one pesticide.

This model is presented here primarily for its heuristic value; it is not ready for immediate application to field problems. The model is limited in its applicability for several reasons:

- The characters must be normally distributed (such as "tolerance" in Figure 1D), with independent mean and variance (Wright, 1968).
- The characters are assumed to be under stabilizing selection, that is, the fitness function has an intermediate optimum. The models use Gaussian (normal) fitness functions for selection on characters with intermediate optima. This approximation is most accurate when the population is near the optimum value of the character. Because an intermediate optimum is assumed, the model does not apply to characters like total fitness or survival, which are assumed to be under continual directional selection to increase. Pesticide tolerance may have an intermediate optimum: individuals with high membrane impermeability or excessive behavioral avoidance of chemicals that they could metabolize may be at a disadvantage relative to individuals with more intermediate values of the features that confer tolerance. The shape of the fitness function for individuals exposed to pesticides is an empirical

question. Estimates could be made by using a regression technique like that described in Lande and Arnold (1983), but to date no such estimates exist.

- The population is assumed to be panmictic: individuals subjected to each pesticide are assumed to mix in a mating pool and then to reassort into locations where the various pesticides are sprayed. This assumption makes the models more accurate for species that mate in a common place away from the site of exposure than for species that have several generations per season and mate at the site where selection occurs. Subdivided population models (Via and Lande, 1985) suggest that the retardation of evolution will not be as effective when migration is low among fields sprayed with different compounds as it is when there is complete panmixis.
- The models were originally formulated for weak selection. This maintains normality in the phenotypic distributions and allows genetic variation, which is depleted by selection, to be replenished by mutation (Lande, 1976; 1980). With strong selection, as is probable when pesticides are applied intensively, the approximate course and rate of evolution described by these models will be less accurate.

The extent to which the models discussed here will actually describe the course of evolution in laboratory or field populations remains to be determined: it is an empirical problem. The applicability of these and other genetic models must be tested by estimating genetic parameters and selection intensities. Until they are tested or proved, the models function primarily to introduce hypotheses about what can happen in the course of evolution of pesticide resistance.

The mode of selection that seems most realistic here is so-called hard selection, in which the contribution of each patch to the mating pool after selection is proportional to both q and to the relative mean fitness of individuals selected in that patch ($\overline{W}_i/\overline{W}$, where $W = q\overline{W}_1 + (1 - q)\overline{W}_2$). The relative mean fitness of a subpopulation (W_i) can qualitatively be considered to be proportional to its contribution to the total population; mean fitness is an indicator of population growth rate (Lande, 1983). In this case the expected changes in LD_{50}s (the tolerances to the two compounds) are

$$\begin{array}{lccc} & \textit{Direct Responses} & + & \textit{Correlated Responses} \\ \Delta LD_{50(1)} = & [q\overline{W}_1/\overline{W}]G_{11}P_{11}{}^{-1}s_1 & + & [(1 - q)\overline{W}_2/\overline{W}]G_{12}p_{22}{}^{-1}s_2 \\ \Delta LD_{50(2)} = & [(1 - q)\overline{W}_2/\overline{W}]G_{22}P_{22}{}^{-1}s_2 & + & [q\overline{W}_1/\overline{W}]G_{21}P_{11}{}^{-1}s_1 \end{array} \quad (3)$$

where G_{ii} is the additive genetic variance in tolerance to the ith compound, G_{ij} is the additive genetic covariance in tolerances ($i \neq j$), and $P_{ii}{}^{-1}s_i$ is the selection intensity on tolerance to the ith compound (Lande and Arnold, 1983).

The evolutionary effects of genetic correlation between tolerances to different compounds on the rate and direction of the evolution of pesticide resistance can be seen in equation 3: the responses to selection of correlated characters have two components. For example, in $LD_{50(1)}$ the direct response is the product of (1) the increase in tolerance to pesticide 1 resulting from direct selection on resistance to that compound ($P_{11}^{-1}s_1$), (2) the genetic variance of tolerance to that pesticide (G_{11}), and (3) a weighting factor ($q_1\overline{W}_1/\overline{W}$) that is required because only part of the population experiences compound 1. The correlated response is the product of (1) selection on the other pesticide ($P_{22}^{-1}s_2$), (2) the genetic covariance between tolerances to the two compounds (G_{12}), and (3) the weighting factor $[(1 - q)\overline{W}_2/\overline{W}]$.

Equation 3 illustrates that the magnitude and sign of the genetic covariance between tolerances to different pesticides can affect the rate of response of either of the tolerances viewed singly. If the genetic covariance for tolerance to different pesticides (G_{12}) is negative, and both characters are selected to increase ($s_1 > 0$ and $s_2 > 0$), the change in tolerance to pesticide 1 will be less than if G_{12} is positive. This is the obvious way that unfavorable genetic correlations in tolerance to different compounds can be used to retard evolution in pest populations. The same principle has been invoked in discussions of negative cross-resistance for the single-locus case (Dittrich, 1969; Curtis et al., 1978; Chapman and Penman, 1979). As will be shown later, however, a negative genetic correlation in tolerance to different compounds is not absolutely required for maladaptation to one of the compounds to occur.

Two scenarios follow that illustrate the models. For these examples, several simplifying assumptions were made:

- Genetic and phenotypic variances in tolerance to each compound are assumed to be equal.
- The width of the fitness function is the same for tolerance to each pesticide (resistance to each compound is assumed to be under equal strengths of stabilizing selection).
- Genetic variances are assumed to remain constant. This assumption is violated if selection is very strong, but it is otherwise correct (Via and Lande, 1985).

In example 1 the population has low tolerance to each of two compounds. One compound is used over a larger acreage than the other (70 percent of the total). When the correlation in tolerance to the two pesticides is positive, evolution of resistance to both will occur readily (Figure 2). If, however, the genetic correlation is low, evolution of resistance to the rarer compound will be slow to occur; most of the population experiences the other pesticide. For strongly negative genetic correlations, Figure 2 illustrates that tolerance to the rare compound can actually decrease as the evolution of resistance to the common pesticide occurs.

In example 2 a new compound is used in conjunction with a compound to which the pests have already become highly resistant. Here the pesticides are deployed in equal proportions in some spatial array in a local area. As evolution increases tolerance to the new compound, either a high positive or a large negative genetic correlation in tolerances will lead to maladaptation (decrease in tolerance) to the old pesticide (Figure 3). This example requires that an intermediate optimum tolerance actually exists, so that a positive genetic correlation in tolerances will cause an overshoot of the optimum tolerance to pesticide 1 and a corresponding decrease in mean fitness.

When maladaptation is occurring, mean fitness in the population will decrease. Thus, not only will resistance be less and less among the survivors, the population size and growth rate will be expected to decrease. Using pesticides in combinations that would create maladaptation to one of the pair could be an effective way to combat the nearly ubiquitous increases in pesticide resistance.

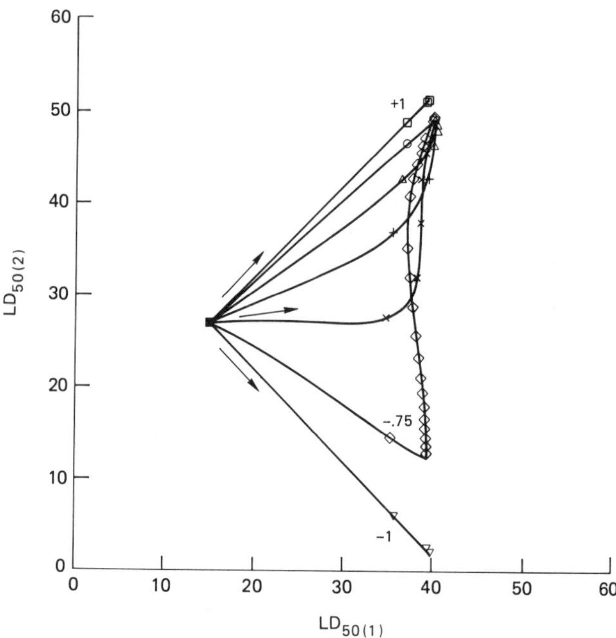

FIGURE 2 Expected evolutionary trajectories for populations with different additive genetic correlations in tolerance to two pesticides. Seventy percent of the total area is sprayed with compound 1. The joint optimum tolerance is the point at which most of the trajectories eventually converge (40,50). Values of the genetic correlations are +1 (□), +0.75 (○), +0.375 (△), 0 (+), −0.375 (×), −0.75 (◇), −1 (▽). Selected values are indicated on the graph near the corresponding trajectories. Evolution occurs in the direction of the arrows. Parameters are $q = 0.7$, $G_{11} = G_{22} = 10$, $P_{11} = P_{22} = 20$; width of both fitness functions = 200, $LD_{50(1)} = 27$, $LD_{50(2)} = 25$. (Redrawn from Via and Lande, 1985.)

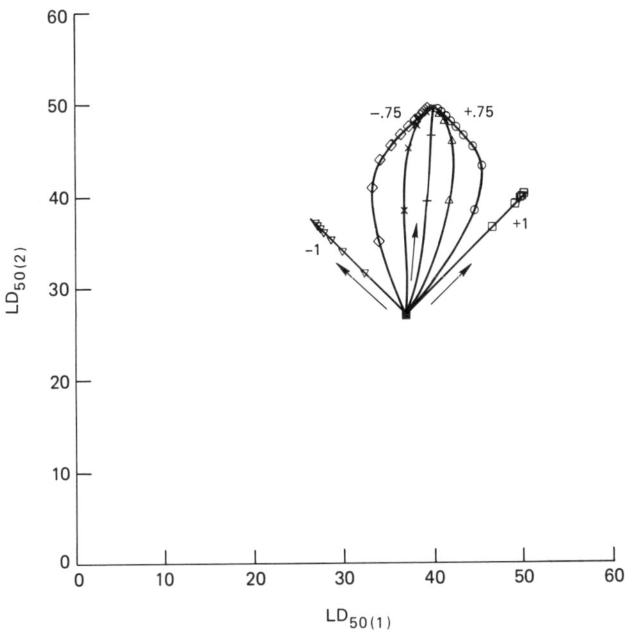

FIGURE 3 Expected evolutionary trajectories when resistance is high for compound 1 at the time a second compound is introduced. The two pesticides are then applied in a joint spraying regime. The joint optimum tolerance is the point where most of the trajectories converge (40,50). Values of the correlations and parameter values are the same as in Figure 2, except $q = 0.5$, and $LD_{50(1)} = 45$. (Redrawn from Via and Lande, 1985.)

Other Approaches

As seen in Figures 2 and 3, the effect of the genetic correlation in tolerance on resistance evolution depends on the initial mean tolerance to each compound relative to the optimum level of tolerance. Within the context of the basic model described here and its attendant assumptions, several alternative strategies of pesticide application could be investigated.

Simultaneous Application of Pesticides The suggestion has been made that mixtures of pesticides with different modes of action might prevent adaptation in pest populations with single-locus negative pleiotropic effects (negative cross-resistance) (Ogita, 1961a,b; Chapman and Penman, 1979; Gressel, in press). The simultaneous application of compounds means that all individuals experience both pesticides. In this case tolerance to compound 1 and tolerance to compound 2 are two genetically correlated characters that can be measured on the same individual (in the previous example each individual expressed tolerance to only *one* pesticide, owing to spatial separation of application). General models for evolution in correlated characters similar to equation 2, but including no weighting terms (Lande, 1979), could be used to investigate the implications of simultaneous application. Because all individuals experience both pesticides, the overall rate of evolution will probably be more rapid than in example 1, implying more rapid resistance evolution to one of the compounds, but perhaps also a more rapid correlated decrease in tolerance to the other. One drawback of simultaneous application is that it may radically increase the overall intensity of selection (Gressel, in press).

Alternating the Proportion of Acreage Sprayed with Different Compounds Maladaptation may occur to a pesticide that is even slightly rare (Figure 2, where 30 percent of the total population experienced compound 2). If one compound is "rare" for several years and then the other compound is made the rare one, the overall progress toward total resistance may be seriously retarded. If no alternation is made, resistance will evolve relatively quickly to the more common compound.

Temporal Alternation Resistance evolution may be retarded if individuals are selected for resistance to one compound and then a few years later are selected for resistance to another compound. This technique will be effective only if tolerance to the two compounds is negatively genetically correlated. The expected results in this case are the same as in the extreme case of the alternating frequency of compounds described above.

Use of More than Two Pesticides in a Given Area With a larger matrix of potentially antagonistic genetic correlations in tolerance, evolution may be retarded for even longer than in the two examples previously described. This approach, however, has two drawbacks: (1) resistance will evolve to many of the available compounds at once, decreasing reserves; and (2) with spatially patchy deployment a larger area would have to be involved, lessening the degree of panmixia and reducing the retarding effect of antagonistic correlations in tolerance, which work only with mixing of individuals with different selection (pesticide exposure) histories. Simultaneous application of multiple pesticides is not the answer, since it could cause an increase in selection intensity and thus would probably speed rather than retard evolution of resistance.

To improve the descriptive power of a quantitative genetic model of pesticide resistance, a model of directional selection that is not tied to the weak selection requirement is necessary. In such a model genetic variance for tolerance would be expected to be exhausted, and the response to selection would be a function of mutation. Such a model does not presently exist, although it is possible that a modification of Lande's (1983) treatment of the relative rates of spread of a single locus and polygenic characters under directional selection could provide a useful beginning.

CONCLUSION

These simple quantitative genetic models are only a first step toward a population-genetic and evolutionary approach to the problem of polygenic pesticide resistance. Problems in pest management must be addressed as evolutionary problems. The pests are evolving to become better adapted, not only to the use of toxic compounds but also to resistant plant varieties (Pathak and Heinrichs, 1982) and a host of other management practices. Pests, like every other class of organisms on earth, evolve by virtue of heritable genetic variation and selection by some environmental agent. Agroevolution differs from evolution in natural populations only in that humans impose selection in the form of various management strategies.

Understanding the processes that lead to certain evolutionary outcomes is the function of population genetic modeling. The applicability of particular models is an empirical issue that cannot be resolved without experimental estimates of critical parameters in the models.

Genetic variances and covariances (or correlations) in tolerance to different pesticides are virtually unknown. The quantitative genetic variance in tolerance can be estimated by breeding individuals to generate families and then exposing some siblings from each family to the different compounds in replicate groups. If one notes the dose at which each individual dies, then variation in tolerance within and among families can be estimated. The among-family variations can be used to derive an estimate of the genetic variance for tolerance.

Other parameters that require estimation are

- The intensity of selection attributable to different compounds (Lande and Arnold, 1983)
- The extent of migration among groups of individuals subjected to different pesticides
- The shape of the fitness functions for tolerance to different pesticides (Lande and Arnold, 1983): are they directional or stabilizing, and how well are they approximated by the usual exponential or Gaussian functions?

Empiricists have another role: to determine the validity of the models as descriptions of evolution. Experiments must be designed to produce observations of evolution in conjunction with models that can produce predictions based on parameters estimated before selection.

Empiricists and theoreticians must work together. With a better understanding of how pests evolve, improved strategies to retard that evolution can be developed.

ACKNOWLEDGMENTS

I thank R. Lande, F. Gould, and R. Roush for useful discussions. This work was supported by NIH Grant No. GM34523.

REFERENCES

Abedi, Z. H., and A. W. A. Brown. 1960. Development and reversion of DDT-resistance in *Aedes aegypti*. Can. J. Genet. Cytol. 2:252–261.

Chapman, R. B., and D. R. Penman. 1979. Negatively correlated cross-resistance to a synthetic pyrethroid in organophosphorous-resistant *Tetranychus urticae*. Nature (London) 218:298–299.

Crow, J. F. 1954. Analysis of a DDT-resistant strain of *Drosophila*. J. Econ. Entomol. 47:393–398.

Curtis, C. F., L. M. Cook, and R. J. Wood. 1978. Selection for and against insecticide resistance and possible methods of inhibiting the evolution of resistance in mosquitoes. Ecol. Entomol. 3:273–287.

Dittrich, V. 1969. Chlorphenamidine negatively correlated with OP resistance in a strain of two spotted spider mite. J. Econ. Entomol. 62:44–47.

Falconer, D. S. 1952. The problem of environment and selection. Am. Nat. 86:293–298.

Falconer, D. S. 1981. Introduction to Quantitative Genetics, 2nd ed. New York: Longman.

Fisher, R. A. 1918. The correlation between relatives on the supposition of Mendelian inheritance. Trans. Roy. Soc. Edinb. 52:399–433.

Georghiou, G. P. 1972. The evolution of resistance to pesticides. Annu. Rev. Ecol. Syst. 3:133–168.

Gressel, J. In press. Strategies for prevention of herbicide resistance in weeds. *In* Rational Pesticide Use, K. J. Brent, ed. Cambridge, England: Cambridge University Press.

Hazel, L. N. 1943. The genetic basis of constructing selection indices. Genetics 28:476–490.

King, J. C. 1954. The genetics of resistance to DDT in *Drosophila melanogaster*. J. Econ. Entomol. 47:387–393.

Lande, R. 1976. The maintenance of genetic variability by mutation in a polygenic character with linked loci. Genet. Res. 26:221–235.

Lande, R. 1979. Quantitative genetic analysis of multivariate evolution, applied to brain:body size allometry. Evolution 33:402–416.

Lande, R. 1980. The genetic covariance between characters maintained by pleiotropic mutation. Genetics 94:203–215.

Lande, R. 1983. The response to selection on major and minor mutations affecting a metrical trait. Heredity 50:47–65.

Lande, R., and S. J. Arnold. 1983. The measurement of selection on correlated characters. Evolution 37:1210–1226.

Liu, M. Y. 1982. Insecticide resistance in the diamond-back moth. J. Econ. Entomol. 75:153–155.

McKenzie, J. A., M. J. Whitten, and M. A. Adena. 1982. The effect of genetic background on the fitness of diazinon resistance genotypes of the Australian sheep blowfly, *Lucilia cuprina*. Heredity 49:1–9.

Mood, A. M., F. A. Graybill, and D. C. Boes. 1963. Introduction to Theory of Statistics, 3rd ed. New York: McGraw-Hill.

Ogita, Z. 1961a. An attempt to reduce and increase insecticide-resistance in *D. melanogaster* by selection pressure. Genetical and biochemical studies on negatively correlated cross-resistance in *Drosophila melanogaster*. I. Botyu-Kagaku 26:7–18.

Ogita, Z. 1961b. Genetical studies on actions of mixed insecticides with negatively correlated substances. III. Botyu-Kagaku 26:88–93.

Pathak, P. K., and E. A. Heinrichs. 1982. Selection of biotype populations 2 and 3 of *Nilaparvata lugens* by exposure to resistant rice varieties. Environ. Entomol. 11:85–90.

Pluthero, F. G., and S. F. H. Threlkeld. 1983. Mutations in *Drosophila melanogaster* affecting physiological and behavioral response to malathion. Can. Entomol. 116:411–418.

Roush, R. T. 1984. A populational perspective on the evolution of resistance. Paper presented at the Symp. Annu. Mtg. Entomol. Soc. Am., San Antonio, Tex., December 3–6, 1984.

Tabashnik, B. E., and B. A. Croft. 1982. Managing pesticide resistance in crop-arthropod complexes: Interactions between biological and operational factors. Environ. Entomol. 11:1137–1144.

Taylor, C. E., and G. P. Georghiou. 1982. Influence of pesticide persistence in evolution of resistance. Environ. Entomol. 11:746–750.

Thomas, V. 1966. Inheritance of DDT resistance in *Culex pipiens fatigans* Wiedemann. J. Econ. Entomol. 59:779–786.

Via, S. 1984a. The quantitative genetics of polyphagy in an insect herbivore. I. Genotype-environment interaction in larval performance on different host plant species. Evolution 38:881–895.

Via, S. 1984b. The quantitative genetics of polyphagy in an insect herbivore. II. Genetic correlations in larval performance within and among host plants. Evolution 38:896–905.

Via, S., and R. Lande. 1985. Genotype-environment interaction and the evolution of phenotypic plasticity. Evolution 39:505–522.

Wood, R. J., and J. A. Bishop. 1981. Insecticide resistance: Populations and evolution. Pp. 97–127 *in* Genetic Consequences of Man Made Change, J. A. Bishop and L. M. Cook, eds. New York: Academic Press.

Wright, S. 1968. Evolution and the Genetics of Populations. Genetic and Biometric Foundations, Vol. I. Chicago, Ill.: University of Chicago Press.

Pesticide Resistance: Strategies and Tactics for Management. 1986. National Academy Press, Washington, D.C.

Managing Resistance to Rodenticides

J. H. GREAVES

To manage rodenticide resistance, rodenticide susceptibility must be conserved and the frequency of resistant phenotypes must be reduced to an acceptable level and kept there. Several attempts to manage resistance to anticoagulant rodenticides in the Norway rat, Rattus norvegicus, *are reviewed, and the responses of users, suppliers of rodenticides, and official agencies to the problem of resistance are discussed.*

Although improvements in rodent-control techniques and further analysis of genetical-ecological aspects of the problem would be useful, the technical means for making long-term progress already exist. Certain short-term factors, however, seem to predispose the interested parties to act in ways that facilitate rather than retard or reverse the continued development of resistance.

INTRODUCTION

Resistance to warfarin and some other anticoagulant rodenticides was recorded first in the Norway rat, *Rattus norvegicus*, in Scotland in 1958 (Boyle, 1960) and has since been found in other countries and species. The subject has been reviewed most recently by Lund (1984) and Greaves (1985). Briefly, anticoagulant resistance in the Norway rat is generally due to a single major gene, of which there seem to be more than two alleles whose effects are subject to the action of modifiers and whose phenotypic expression is usually dominant. (For a detailed discussion on biochemistry of resistance, see the paper by MacNicoll in this volume.)

RESISTANCE MANAGEMENT

Resistance to a rodenticide becomes a problem when the proportion of resistant phenotypes in the targeted rodent population increases to where the rodenticide cannot effectively control infestation. To manage resistance we must conserve rodenticide susceptibility or reduce the phenotypic frequency of resistance to, and keep it at, an acceptable level, preferably close to the underlying mutation frequency. The only way to reach this objective is to place resistant individuals at a selective disadvantage. In theory this may be accomplished either by selecting against resistant individuals; within populations or by selecting against populations containing resistant individuals; doing so in practice is more complex. This general approach is usually reinforced by natural selection, since resistance alleles are usually deleterious in the absence of artificial selection with the pesticide.

The concept of resistance management involves (1) setting practical management objectives, (2) determining how to reach the objectives, (3) assigning resources commensurate with the size and nature of the task, and (4) identifying managers who will be accountable for reaching the objectives. That such resistance managers rarely, or more probably never, exist reflects the fact that the problem of resistance crosses the boundaries within which management functions normally are confined. This is why few, if any, of the theoretical approaches to resistance management (Georghiou, 1983) have been implemented successfully. Managing resistance requires a management structure comparable perhaps with those that have been successfully developed to control communicable diseases.

PRACTICAL ATTEMPTS TO MANAGE RESISTANCE IN BRITAIN

Nipping Resistance in the Bud

For several years Britain maintained official vigilance for new outbreaks of resistance using the procedures described by Drummond and Rennison (1973) and tried to exterminate the resistant rats with acute rodenticides. These operations normally involved joint action by the research and field advisory services of the Ministry of Agriculture and staff of the local municipal health departments, as well as official teams of pest-control operatives.

The method was used 11 times (Drummond, 1971). In seven cases no subsequent evidence of resistance was found. Thus, nipping resistance in the bud seems to have worked. The significance of these apparent successes, however, is difficult to assess since insufficient evidence is available on the genetic nature of the resistance. Therefore, it is not known whether the successes were due to the promptness and efficiency of the countermeasures or because the resistance was of a kind inherently unlikely to survive and spread. In the four unsuccessful cases the resistance was of the monogenic, dominant type.

In principle it should be possible to eradicate local populations of rats showing monogenic resistance by these means, except that a resistant infestation develops 12 to 18 months before it is discovered (Drummond, 1970, 1971), during which time it might spread a radial distance of 5 to 10 kilometers (km). Thus, prompt and sustained countermeasures probably should be conducted within a radius of 20 km to eliminate any new outbreak of monogenic resistance.

Eradicating Widely Established Resistant Populations

A pilot scheme to eradicate warfarin-resistant rats was conducted in a rural area of five square miles in Wales, using the acute rodenticides zinc phosphide, arsenious oxide, antu, and norbormide (Bentley and Drummond, 1965). It failed because of the limited efficacy of the available rodenticides and also probably because such a small experimental area is vulnerable to invasion by rats from the surrounding countryside. Further, the objective may have been defined inappropriately as the total eradication of rats, both resistant and susceptible, rather than eliminating primarily the resistant individuals. Resistance monitoring might have shown that switching from warfarin to other, nonselective rodenticides had brought the resistance under control.

The failure of this particular scheme, however, does not vitiate the concept of selective targeting of relatively large areas for managing resistance. Today a similar scheme would have a greatly increased chance of success, owing to improvements both in rodent control-technology and in our understanding of the problem.

Containment of Resistant Populations

A third approach adopted in Britain as a short-term expedient was to throw a kind of guarded perimeter strip 5 km wide around a resistance area that was about 60 km in diameter. A rat-control program was instituted on the perimeter "containment zone." All sites were inspected regularly and, if infested, treated with acute rodenticides (Drummond, 1966). Resistant rats, however, were found 8 km outside the perimeter within two years (Pamphilon, 1969), casting doubt on the efficacy of the scheme and indeed on whether the entire resistant population had been enclosed within the perimeter. Such considerations further emphasize the importance of resistance monitoring in any management scheme.

TABLE 1 Relative Fitness of Genotypes in Norway Rat Populations in the Presence and Absence of Anticoagulant Treatment

	Genotypes		
Conditions	RR	RS	SS
Anticoagulants Present[a]	0.37	1.00	0.68
Anticoagulants Absent[b]	0.46	0.77	1.00

SOURCE: [a]Greaves et al (1977); [b]Partridge (1979).

Natural Selection

Resistance to anticoagulants in Norway rats seems to be a pleiotropic effect of a defect in vitamin K metabolism such that the dietary requirement for the vitamin is increased (Hermodson et al., 1969). Two independent studies in Britain suggest that this physiological defect alone may eliminate resistance from natural populations when artificial selection with anticoagulant rodenticides is withheld.

In the first study, when acute rodenticides were substituted for anticoagulants in a sizable experimental area, the frequency of phenotypic resistance decreased steadily from 57 to 39 percent in two years. Simultaneously, in a control area where approximately one-half of the farmers were using anticoagulants, the resistance frequency remained stable at about 44 percent (Greaves et al., 1977). Analysis of the genotypic frequencies indicated that the stability of the resistance in the control area represented a balanced polymorphism in which selection favored heterozygotes (Table 1).

The second study concerned a single, somewhat isolated rat infestation on a farm. During the 18 months when no treatment was applied to the infestation, the frequency of phenotypic resistance decreased from approximately 80 to 33 percent. Evaluation of the phenotypic frequencies by an optimization procedure suggested that in the absence of selection with anticoagulants, heterozygotes as well as resistant homozygotes were at a substantial disadvantage compared with susceptibles (Table 1) (Partridge, 1979). No detailed analysis, however, has yet been made of the ecological-genetical processes that control the level of anticoagulant resistance in wild rodent populations.

NEW RODENTICIDES

Although the previous experiences suggest that substantial progress could be made in managing resistance (even with blunt instruments), the increasing prevalence of resistance to anticoagulants has given considerable impetus to research on new rodenticides. The most outstanding new products are three highly toxic, broad-spectrum anticoagulants: brodifacoum, bromadiolone, and difenacoum. Warfarin-resistant strains may show various, usually minor, degrees of cross-resistance to the new compounds, but they can usually eliminate resistant rats. These compounds are potentially extremely valuable tools for managing resistance.

Inadequate application methods, however, allow some rodents to survive treatment. Populations then tend to increase both the degree and the frequency of resistance. Thus, resistance to all three compounds is increasing (Lund, 1984). For example, difenacoum has suffered a very marked loss of efficacy against Norway rats in one area of England, where continual selection with the new anticoagulants seems to have raised the frequency of phenotypic resistance to warfarin to around 85 percent (Greaves et al., 1982a,b). The introduction of new products to control resistant rodents, therefore, probably has accelerated rather than retarded the evolution of resistance in this area.

Simply substituting new rodenticides for old ones to cope with resistance rests on one of two assumptions, which if not palpably false may be insecure: (1) resistance to new rodenticides will not evolve, or (2) the process of developing new rodenticides to counter new forms of resistance can be repeated indefinitely. The essential question to ask about any technique in the context of resistance management is not whether it can control resistant rats but whether it can control resistant rats selectively, because only then will it be possible to reverse the evolution of resistance or prevent it from proceeding at its natural pace.

THE CAUSE OF RESISTANCE

The origin of resistance may be a random event such as a mutation, but its development into a practical problem results solely from human activities. We must examine the behavior and attitudes of groups that are affected by rodenticide resistance to help us decide how to manage the problem.

Users

The main users of rodenticides—farmers, environmental health workers, and professional pest-control operators—often are unaware of the possibility of resistance until a control method fails. Alternatively, if the resistance has had any notoriety, they often blame all failures on resistance, although the failures may be due to faults in formulation or method of application. Such factors produce a confused picture of resistance. Users, therefore, should report control problems promptly and accept expert advice on how to deal with them.

If resistance is the problem an alternative rodenticide often gives acceptable results. The alternative rodenticide, however, may be more expensive, more hazardous, more difficult to use, or less effective than the original compound. Consequently, users often revert to the original product, taking advantage of any recession in resistance, until further control failures occur. This behavior maintains resistance, causes persistent control problems, and promotes the spread of the resistant strain. It also may promote coadaptation, the process by which a resistance gene may be integrated into the gene pool of the population. One of the main objectives, therefore, in managing resistance must be to prevent the use of compounds to which resistance has developed or in circumstances that make the further development of resistance likely.

Industry

For many years industry has joined with others in voicing concern about the strategic threat to crop protection posed by pesticide resistance, bearing in mind the high cost of introducing new products and that every new compound seems to be vulnerable to the development of resistance. When resistance is first encountered, however, firms tend to respond with caution, which is engendered by (1) confidence in the excellence of their products; (2) an awareness that many reports of resistance turn out to be spurious; (3) the knowledge that for a while the resistance, if real, is likely to be highly localized; and (4) trepidation that publicity about the resistance may adversely affect their competitive position in the market. This caution may militate against early action to control the resistance.

A practical and indispensable response by industry is to develop new rodenticides to control the resistant strains. The timing of this response tends to be governed by economics. Thus, it tends to occur late, when markets are being eroded significantly by the increasing prevalence of resistance, or when the expiry of exclusive commercial rights make an existing product less viable, or when a new concept for a competitive new product is invented.

Because rodenticides are specialized, minor-use compounds, investment in research on new compounds frequently is regarded as unprofitable. Consequently, little effective investment has been made in this area except when a special commercial interest has been at stake, or when there has been some form of official sponsorship or interest. Despite these difficulties several new rodenticides have reached the market, thus lessening the resistance problem.

When new rodenticides with a useful degree of toxicity to resistant strains are registered, normal marketing strategy dictates that they be promoted for their "anti-resistant" and other favorable properties. Such action may be counterproductive, in that the indiscriminate introduction of a new product may speed up the evolution of resistance. This dilemma, although it may not be perceived as such, is heightened when the first indications of resistance to a new product are recognized.

The problem of how rodenticides may best be deployed to manage resistance is complex, requiring some research and analysis. Since selective action (increased deployment of certain compounds and restraint on the use of others in particular localities) is required, effective regulation of the sale and use of certain compounds is essential. Industry may not be able to do this alone, chiefly because companies cannot control the use of their products once they are sold. It cannot be accomplished, however, without the consent and cooperation of industry.

Official Agencies

The primary role of many official agencies is to provide a source of impartial, expert advice to individual users of rodenticides and to undertake or sponsor the investigational work necessary for sound advice. Sometimes they may organize or conduct practical rodent-control operations. Official agencies are usually responsible for administering legislation concerned with the control of infestations and the use of rodenticides. They are in a powerful position to influence whatever action is taken to manage resistance in rodent populations.

Information on the extent to which such influence is actually exercised is limited. What has been done ranges (in different countries) from almost no action to fairly direct intervention. In Britain, for example, action by the Ministry of Agriculture has included field investigations of new outbreaks of resistance, development of diagnostic tests for resistance, research into its formal genetics, local programs to control or eliminate resistant populations, and collaboration with industry in research on new rodenticides. These efforts, in part, have prevented the situation from getting out of hand. Indeed many countries are benefiting from the work done in Britain, most notably from the introduction of new rodenticides to control resistant strains.

Nevertheless, the prevalence of resistance to rodenticides is not decreasing, and in some countries it is getting worse. In this sense the success of official intervention in resistance management has been limited. To the extent that they have continued to advocate the use of rodenticides that could be expected to further the development of resistance, the activities of these agencies, like those of users and suppliers, are counterproductive.

CONCLUSION

The foregoing outline of how the rodenticide resistance problem has been addressed points toward two general conclusions. First, the logical structure of the problem seems to be clear in its technical aspects: rodenticide resistance can be controlled by eliminating resistant populations faster than new ones can develop. Such control requires information about the location and characteristics of the resistant populations, prevents the use of rodenticides that accelerate the development and spread of resistance, and increases the use of nonselective or counter-selective control techniques against the populations concerned. Although further improvements in techniques for controlling resistant populations would be welcome, the existing technical means may be adequate. For practical implementation we need to understand more precisely the genetical-ecological processes that control the level of resistance in natural populations and, thus, how the available rodent-control techniques could be deployed advantageously. Adequate resistance monitoring also is necessary to steer and verify the progress of any practical scheme.

The human factors affecting the management of resistance are less easy to assess since they concern subjective judgments of value, most notably of how the certainty of short-term costs should be balanced against less-certain long-term gains. Since resistance is responsive to selection, the actions of users and suppliers of rodenticides and of advisory and regulatory agencies play a crucial role in its management. The exigencies of rodent control in the real world create pressures, however, that predispose the various participants to cooperate involuntarily in the continued evolution of resistance rather than to reverse or retard it.

Progress has been made in areas of technique, but rodenticide resistance continues to develop, probably because resistance, like communicable disease, cuts across the boundaries of most ordinary management structures. We need to improve coordination and above all to redirect the efforts of the interested parties. Such coordination may be possible through consensus and through vigorous promotion. The alternatives are either to increase official regulation in the field of rodent control or to allow resistance to continue to evolve at its own unregulated pace.

REFERENCES

Bentley, E. W., and D. C. Drummond. 1965. The resistance of rodents to warfarin in England and Wales. Pp. 58–76 *in* Report of the International Conference on Rodents and Rodenticides. Paris: European and Mediterranean Plant Protection Organization.

Boyle, C. M. 1960. Case of apparent resistance of *Rattus norvegicus* Berkenhout to anticoagulant poisons. Nature (London) 188:517.

Drummond, D. C. 1966. Rats resistant to warfarin. New Sci. 30:771–772.

Drummond, D. C. 1970. Variation in rodent populations in response to control measures. Symp. Zool. Soc. London 26:351–367.

Drummond, D. C. 1971. Warfarin-resistant rats—some practical aspects. Pestic. Abstr. News Sum. 17:5–8.

Drummond, D. C., and B. D. Rennison. 1973. The detection of rodent resistance to anticoagulants. Bull. W.H.O. 48:239–242.

Georghiou, G. P. 1983. Management of resistance in arthropods. Pp. 769–792 *in* Pest Resistance to Pesticides, G. P. Georghiou and T. Saito, eds. New York: Plenum.

Greaves, J. H. 1985. The present status of resistance to anticoagulants. Acta Zool. Fenn. 173:159–162.

Greaves, J. H., R. Redfern, P. B. Ayres, and J. E. Gill. 1977. Warfarin resistance: A balanced polymorphism in the Norway rat. Genet. Res. 30:257–263.

Greaves, J. H., D. S. Shepherd, and J. H. Gill. 1982a. An investigation of difenacoum resistance in Norway rat populations in Hampshire. Ann. Appl. Biol. 100:581–587.

Greaves, J. H., D. S. Shepherd, and R. Quy. 1982b. Field trials of second-generation anticoagulants against difenacoum-resistant Norway rat populations. J. Hyg. 89:295–301.

Hermodson, M. A., J. W. Suttie, and K. P. Link. 1969. Warfarin metabolism and vitamin K requirement in the warfarin resistant rat. Am. J. Physiol. 217:1316–1319.

Lund, M. 1984. Resistance to the second-generation anticoagulant rodenticides. Pp. 89–94 *in* Proc. 11th Vertebr. Pest Conf., D. O. Clarke, ed. Davis: University of California.

Pamphilon, D. A. 1969. Keeping the super-rats down. Munic. Eng. (London) 146:1327–1328.

Partridge, G. G. 1979. Relative fitness of genotypes in a population of *Rattus norvegicus* polymorphic for warfarin resistance. Heredity 43:239–246.

Pesticide Resistance: Strategies and Tactics for Management.
1986. National Academy Press, Washington, D.C.

Response of Plant Pathogens to Fungicides

M. S. WOLFE and J. A. BARRETT

Genetic variation for fungicide resistance must occur if a pathogen is to respond to fungicide use. The rate of pathogen response depends on a complex interaction between the exposure of the pathogen to the fungicide, the biology of the pathogen, and the environment. An example of this interaction is the response of the barley mildew pathogen Erysiphe graminis *f. sp.* hordei *to the widespread use of triazole fungicides in the United Kingdom, which also illustrates the interaction of fungicide resistance and host pathogenicity.*

The current strategies of fungicide use tend to exacerbate the problem of restraining pathogen response. Other strategies, based on different forms of diversification, may be helpful in practice, at least under western European conditions. Experiments were conducted with fungicide treatments of the seed of single components of mixtures of host varieties having different resistance genes. On the farm this system can give good disease control and predictably high yields at low cost. Durability is not predictable, except that it is likely to be better than with current strategies, with the additional benefit of restricting the response of the pathogen to resistant hosts.

INTRODUCTION

This paper is an amalgam of first principles and practical experience gleaned largely from research on the control of *Erysiphe graminis* f. sp. *hordei* on barley. The use of fungicides changes the environment of the pathogen, and to understand its response requires a knowledge of how such changes affect selective differences between different genotypes in the population. Only then can a way that is acceptable biologically and for practical crop production be developed to modify the response.

FUNGICIDE USE

The Attraction of Fungicides

Why are fungicides used? Broadly, there are three reasons. The first is to control disease during crop development. Among field crops the view is encouraged that a particular species or variety is susceptible and thus losing yield to a disease, that the plant breeders have failed to deal with the problem, and that fungicides will provide the answer. The perception of susceptibility in commercial production, however, is based on an assessment relative to complete absence of disease. Truly susceptible host lines are eliminated during the breeding process and are rarely seen in agriculture; those that are deemed susceptible but remain in cultivation often have yields of only 20 percent (or less) below their potential maximum. Fungicides are used extensively to remove this limitation so as to achieve the "ideal" of a disease-free crop.

Initially at least, fungicides remove these restraints consistently and reliably because the recommended dose rates are determined from field trials with adequate pathogen inoculum applied to the currently most susceptible commercial varieties. For the farmer the fungicide controls the disease perfectly because his varieties, on average, will be less susceptible than those used in manufacturers' trials, and his farm conditions will tend to be less favorable for disease development.

For these same reasons many fungicide applications expose the pathogen to a fungicide for no economic return, but the psychological impact of the clean crop more than offsets this hidden factor. A similar psychological problem arises from using fungicides to eliminate blemishes completely from produce for direct consumption. Perfect produce has become the norm for the marketplace even though it may not be essential, productivity is not improved, and exposure of pathogens to fungicides is maximized. The demands for clean crops and perfect produce mean that fungicides are used increasingly as prophylactic treatments—known to cereal farmers in eastern England as the sleep-easy factor—despite the consequences.

The second reason for the use of fungicides is to improve the storage of produce. Perfect control of storage diseases increases the size and duration of the market available for the product. Thus, the marketplace again encourages widespread use of fungicides, particularly since plant breeders do little or nothing directly to breed for resistance to storage diseases.

Third, with fungicides growers can increase production of a particular crop and reduce their dependence on conventional controls of crop rotation and sanitation. Moving away from the costs and constraints of conventional controls is double-edged: the fungicide usage per unit area is increased, as is the total area of the crop and the size of the potential medium for the target pathogen. The increased potential for the crop provided by the fungicide is often so dramatic initially that some manufacturers suggest that breeders need no longer breed for host resistance. Any decrease in attention to inherent host resistance, however, is almost certain to exacerbate and accelerate selection of fungicide resistance, simply because pathogen survival is made easier.

Fungicide Application and Type

The area treated with a fungicide contains the effective treated area, defined as the proportion of the crop at any one time in which the fungicide level is higher than the threshold of control of the common fungicide-sensitive genotypes of the pathogen. For example, if equal amounts of two different fungicides are applied to a crop but one is more systemic and persistent than the other, the effective treated area of the first will be greater. Disease control will be greater, but so will the advantage accruing to resistant genotypes of the pathogen.

Fungicides may be formulated for use as seed treatments, or as foliar sprays, or both. Seed treatments are potentially more effective because they may control the pathogen when the population is at its smallest and thus delay epidemic development, particularly if the compound is systemic and persistent. The corollary is that the pathogen population has a longer exposure to the treatment. If a fungicide is formulated both as a seed treatment and as a foliar spray and the compound is used widely and sequentially in the two forms, the effective treated area and the advantage to resistant genotypes are greatly increased.

Broad-spectrum fungicides, as opposed to selective fungicides, may compound the problem if they remove competitors or hyperparasites that would assist the activity of a selective fungicide. Thus, the greatest potential for fungicide resistance comes from the large-scale prophylactic use of a broad-spectrum, systemic, and persistent material applied to the seed and then to the foliage. The fungicide initially controls the disease dramatically, and it is easily sold to farmers who are mostly risk-averse. The alternative of a nonpersistent, selective foliar spray, applied only when the disease level passes a defined threshold, is risky and demands accurate monitoring, forecasting, and assessment of yield loss, but it reduces the time over which the pathogen is exposed to the fungicide and thus reduces the probability of resistance evolving.

PATHOGEN RESPONSE

A Priori Considerations

Any response to fungicide use depends, first, on whether genetic mechanisms exist to reduce or eliminate the effects of the fungicide. The mechanisms may occur at low frequencies before the fungicide is introduced, they may occur as mutations, or both. The rate at which the pathogen responds then depends on the interaction between the mechanisms available and their genetic control, the use of the fungicide, the biology of the organism, and the environment.

One major factor is whether the organism is diploid or haploid in the asexual stage. If haploid then any mutation to fungicide resistance is immediately expressed, and the frequency of the mutant will be influenced by its effect on fitness. With a diploid organism the situation is more complex; there may be a cryptically high frequency of resistance, depending on the fitness of the heterozygotes and resistant homozygotes relative to the wild type, in the presence and absence of the fungicide (Barrett, in press).

The rate of response of a pathogen also depends on its breeding system, principally on whether there is an obligate sexual or parasexual sequence in the life cycle. An effective sexual stage allows for more rapid formation of novel combinations of appropriate characters through recombination, which may increase the fitness of the resistant pathogen genotypes. With no sexual stage, linkage disequilibrium between resistance and other characters is likely to persist, which may limit or delay adaptation of the pathogen to the treated host population.

The spread of fungicide resistance depends on the distribution of propagules: populations of foliar pathogens with airborne spores will respond more rapidly than soil-borne pathogens. Finally, the ability of a pathogen to respond to fungicidal control depends on its ability to cope with other environmental stresses. An organism at the limits of its ability to survive in a particular environment will be less able to respond to an extra stress. For example, the greater the level of disease resistance and diversity in the host crop the less likely it will be for a pathogen to develop and spread resistance to a fungicide.

Dynamics

Wolfe (1982) summarized the interaction of selection for resistance and for other characters. Whether fungicide resistance increases in a population is determined by the size of the effective treated and untreated areas and the fitness of the forms of the pathogen with different sensitivities to the fungicide on each of these areas. There will tend to be large differences in fitness on the treated crop and smaller differences on the untreated. If the differences on the untreated crop area are small, then a small area of treated crop may allow resistant forms of the pathogen to predominate in the population as a whole. If the fitness differences on the untreated crop are large, then fungicide-resistant forms of the pathogen may not become apparent until there is a large treated area. The overall fitness of sensitive and resistant forms of the pathogen, therefore, depend on the area of fungicide treatment. Growth rate differences between isolates measured in the laboratory may have little relevance to the fate of those isolates in the field.

Monitoring the range of forms of a pathogen with reduced sensitivity to a fungicide is difficult. The phenotypes isolated first may not be the ones that eventually become common, because recombination and selection may change the expression of resistance during its spread. Indeed, if selection is maintained it is never possible to predict when the response will cease. In the example of barley mildew adapting to the use of ethirimol, Brent et al. (1982) noted a shift to an apparent equilibrium between sensitivity and resistance in the pathogen population. In this case, however, selection for resistance declined when ethirimol was replaced by other fungicides and more resistant varieties: the apparent equilibrium may have been a temporary peak associated with maximum use of the fungicide.

AN EXAMPLE

The worst case in terms of selection for resistance is where a systemic, persistent, and broad-spectrum fungicide is applied sequentially on the major part of the crop area to control a well-adapted foliar pathogen that is efficiently dispersed by airborne spores and has an effective sexual stage. Among field crops this combination of characters is exemplified by the use of triazole fungicides to control barley mildew in western Europe.

Shortly after introduction of these fungicides into commercial use in the United Kingdom, the first isolates with some resistance were identified in small populations surviving on treated crops (Fletcher and Wolfe, 1981). From 1981 the air spora was monitored continuously (Wolfe et al., 1984a) by means of a simple spore trap mounted on a car roof (Wolfe et al., 1981; Limpert and Schwarzbach, 1981). The numbers of colonies that incubated on seedlings with different doses of the fungicide increased annually relative to the numbers on untreated seedlings. The early surveys could not always detect isolates with fungicide resistance in the small populations on treated crops; by 1984, however, such isolates were detected easily on untreated crops.

The increase in frequency of the less-sensitive phenotypes showed two interesting characteristics. The first was that the rate of increase varied during the year. This variation was repeated between years, which suggested that during the spring, following seed treatment and early foliar sprays, there

TABLE 1 Mean Pathogenicity (Pathog.) on Six Differential Barley Hosts of Powdery Mildew Isolates with Different Levels of Sensitivity to Triadimenol Obtained from Untreated and Treated Seedlings in a Car Spore Trap in East Anglia, 1981–1983

Seedling Source	1981		1982		1983	
	ED_{50}	Pathog.	ED_{50}	Pathog.	ED_{50}	Pathog.
Untreated	0.028	32	0.060	40	0.080	35
0.025[a]	0.045	27	0.080	35	0.093	35
0.125[a]	0.085	7	0.093	25	0.108	35

[a]Grown from seed treated at 0.025 or 0.125 g a.i./kg.

SOURCE: Wolfe (in press [a]).

was rapid selection toward resistance. During the summer the response slackened or reversed, presumably following dissipation of the fungicide. At the beginning of autumn, however, frequency sharply increased, probably due partly to release of ascospores from cleistothecia formed at the time of relatively high frequencies of resistance at the end of spring and partly to the influence of emerging crops of treated winter barley. During autumn and winter the frequency of resistant forms again declined.

In pathogen populations on individual field crops of treated winter barley, the frequency of the most resistant forms was high on seedlings in the autumn because of the selection imposed by the high concentration of fungicide in the seedling leaf tissue (Wolfe et al., 1984a). As the plants grew and the concentration decreased, the frequency of these forms decreased and forms with intermediate resistance became predominant. On the untreated crops sensitive forms were initially predominant, but, again, forms with intermediate resistance eventually became more common, presumably due to spores migrating from other crops, most of which would have been treated at some stage.

The second major feature of interest was the relationship between resistance and pathogenicity. During the early stages of the overall increase in resistance, the more resistant forms of the pathogen were less pathogenic on the range of host varieties in common use at the time (Table 1). In subsequent seasons, however, pathogenicity of the sensitive fraction remained constant, but the resistant fraction gradually increased to the same level.

The increase in pathogenicity in the resistant part of the population occurred earlier for some characters than for others. For example, resistance increased first in Scotland and northern England in populations having a high frequency of pathogenicity for varieties with the Mla6 resistance gene. This created linkage disequilibrium, and isolates having these characters rapidly became common throughout the United Kingdom. The potential value of Mla6 was thus diminished in areas where it was not in current use. Simultaneous with these changes the resistant variety Triumph became extensively cultivated and increasingly susceptible. Triazole fungicides thus became widely used on Triumph; isolates resistant to triazoles are now commonly pathogenic on Mla6 or Triumph or both.

As fungicide resistance in the pathogen population increases, there may be loss of disease control and a reduction in the yield advantage expected from treatment. Initially such effects have a patchy distribution. Not all resistant isolates will be associated with poor fungicide performance and, conversely, not all poor fungicide performance will result from the occurrence of fungicide-resistant isolates. Inevitably, during the first seasons of using a new fungicide, there will be some instances of poor control due to incorrect application and other environmental problems. This small proportion will fluctuate from season to season; a real deterioration in fungicide performance will be signalled by a continuing increase in instances of poor control.

For example, with triazoles and the control of barley mildew, following the increase in frequency of resistant forms in eastern England, performance of triazoles both in disease control and in yield benefit rapidly declined (Table 2). The effect was most marked in varieties with the Mla12 resistance gene; the yield increase following treatment declined from 25 percent in 1982 ($P < 0.001$) to 3 percent in 1984 (not significant), during which time ethirimol—a different seed treatment that was less widely used—gave a consistent yield advantage of around 10 percent ($P < 0.05$). A similar yield advantage during this period was obtained with ethirimol applied to Carnival (Mla6), but there was no advantage with triazole treatment, probably because of the higher frequency of resistant isolates carrying pathogenicity for Mla6 compared with those pathogenic against Mla12. A more complex interaction with these fungicides was obtained with Triumph and Tasman because of the declining resistance of the varieties during this same period. Nevertheless, the performance of the triazoles declined relative to that of ethirimol.

CONTROLLING THE PATHOGEN RESPONSE

Reducing exposure of the pathogen to the fungicide is the most obvious way to deter resistance, and this can be helped by making disease forecasting more precise and educating growers to the problems. Commercial pressures against such actions, however, may be strong. Reducing the fungicide dose may or may not delay resistance development. If the dose is reduced to a level at which some sensitive genotypes survive, there may be some delay; however, the pathogen may cause unacceptable yield loss. On the other hand any delay caused by an increased dose is likely to be followed by emergence of highly resistant strains of the pathogen. Other changes in the formulation of the compound or inefficiency of application may also alter the fitness

TABLE 2 Yield (t/ha) of Spring Barley Varieties with Different Mildew Resistance Genes, Untreated or Treated with Ethirimol or Triadimenol, 1982–1984

Variety	Year	Untreated	Ethirimol-trt.	Triadimenol-trt.
Mla12				
Egmont	1982	5.01	5.49	6.25
	rel.	100	110	125
Patty	1983	3.51	3.90	4.12
	rel.	100	111	114
Patty	1984	6.90	7.46	7.13
	rel.	100	108	103
Mla6				
Carnival	1982	5.38	5.87	5.64
	rel.	100	109	105
Carnival	1983	3.83	4.11	3.84
	rel.	100	107	100
Carnival	1984	6.60	7.07	6.53
	rel.	100	107	99
Mlk/a7				
Triumph	1982	5.40	—	5.81
	rel.	100	—	108
Tasman	1983	3.57	3.85	3.70
	rel.	100	108	104
Tasman	1984	5.66	6.43	6.05
	rel.	100	114	107

NOTE: Standard error for 1982, ±0.11; 1983, ±0.23; 1984, ±0.14.

SOURCE: Wolfe (in press [a]).

differences between sensitive and resistant genotypes and make prediction difficult.

Reducing the use of a particular compound may need to be accompanied by other means of limiting pathogen increase, such as diversifying between fungicides with different modes of action known or thought to be matched by different pathogen mechanisms. For commercial and technical reasons, there are considerable constraints to the kinds of action that can be recommended. The current system is the use of mixtures, usually a tank mix of a systemic and a nonsystemic compound. The data to support this approach are inconclusive. Adding a nonsystemic material may only temporarily reduce the absolute population size of the pathogen, while the systemic material will be more persistent so that after the initial combined action of the fungicides, the pathogen population will be exposed uniformly to the systemic compound on all plants and thus selected for resistance.

A more effective system, analogous to the use of variety mixtures (Wolfe, 1981), may be to ensure that the compounds eliciting different responses are applied to adjacent plants. The pathogen must then either adapt to a single plant or become versatile between plants. Compared with a uniformly treated stand, there is a greater space between plants receiving the same treatment, so that increase of the population resistant to that treatment is delayed. Further, any genotypes with combined resistance to all of the fungicides used are likely to be less fit on any one plant than the genotype specifically adapted to the treatment on that plant.

Currently, this approach can be contemplated only for fungicides applied to seed. Even here treatment on one seed may spread to other seeds treated differently, and different treatments may vary in their effects on the flow rate of seed either in a mixing process or in a seed drill. Recent developments in film coating of seeds may eliminate such problems. Fungicides can be applied to seed in a carrier material, improving the precision of individual seed treatment. The material is fixed firmly to the seed, and the flow characteristics of the seed are similar to those of seed treated with other fungicides (M. D. Tebbit, Nickersons Seed Specialists Ltd., personal communication, 1984). Seeds can also be simultaneously color coded so that intimacy of mixing can be confirmed.

Future developments in application technology may allow a similar approach with foliar sprays. For example, ultra-low-volume equipment such as the electrostatic sprayer raises the possibility of using a square matrix of containers holding different fungicides, mounted on a frame with a system of rapid on-off switching so that a fine mosaic of different materials can be applied.

INTEGRATED DISEASE CONTROL

Unfortunately, much of the discussion on controlling pathogen response to fungicides makes no reference to the host crop. In the simplest case, with partially resistant host varieties, the number of treatments and the dose can be reduced, thereby reducing selection on the pathogen for resistance to the fungicide and indeed for pathogenicity to the host (Wolfe, 1981).

Sometimes it is more effective to use intimate mixtures of host varieties with different resistance genes (Jensen, 1952; Wolfe and Barrett, 1980; Wolfe, 1985). Particularly if diversity between mixtures is maintained in space and in time, disease control is more consistent and durable than if the components are used in monoculture. By changing the composition of mixtures as new varieties become available, both the yield potential and the diversity are maximized, which suits both the farmer and the plant pathologist.

From 1980 through 1984 four barley varieties with different resistance genes and the four mixtures of three varieties that can be made from them were grown in field trials at the Plant Breeding Institute, Cambridge, England (Wolfe et al., 1984b; Wolfe et al., 1985). Over the trial series the mixtures outyielded the pure stands by 7 percent ($P < 0.001$). The best strategy found

TABLE 3 Average Yields (t/ha) and Infection (total percent leaf cover) for 1983 and 1984 of the Three Spring Barley Varieties Carnival, Patty, and Tasman, Grown as Pure Stands or Mixtures, Untreated or Treated with a Triazole or Ethirimol at the Normal Field Rate

	Yield (t/ha)				Infection (total % leaf cover)			
	Pure	Rel.	Mixed	Rel.	Pure	Rel.	Mixed	Rel.
Untreated	5.19[2]	100	5.61[1]	108	25.7	100	19.3	75
Triazole								
1/3	5.25[3a]	101	5.62[2]	108	22.6[2]	88	15.2	59
N	5.37[2]	103	5.44[1]	105	16.4	64	13.6	53
Ethirimol								
1/3	5.35[3a]	103	5.68[2]	109	20.4[a]	79	10.2	40
N	5.67[2]	109	5.65[1]	109	9.7	38	6.3	25

NOTE: The 1/3 treatment of the mixtures is the mean of three mixtures in each, of which only one component is treated with triazole or ethirimol. The 1/3 treatments of pure stands are calculated values obtained from the sum of the three pure varieties treated, plus twice their sum untreated, divided by nine.

[a]Calculated values.
[1]SE = ±0.16.
[2]SE = ±0.09.
[3]SE = ±0.07.

SOURCE: Wolfe (in press [b]).

for the farmer, given the choice of only those four varieties each year, would have been to grow any one or more of the mixtures. Based on this research variety mixtures are now grown commercially in the United Kingdom and Denmark, with generally favorable reports from the farmers involved. A much larger scale of development is being undertaken in the German Democratic Republic, particularly because of the high cost of fungicides in eastern Europe.

Despite the obvious advantages of the variety mixtures, disease control is sometimes considered to be inadequate, and some mixtures are treated with fungicides even though the benefit may be uneconomic. For this reason and to provide long-term protection for the varieties and the fungicides, experiments have been conducted with fungicide-integrated mixtures (Wolfe, 1981; Wolfe and Riggs, 1983). The seed of one component of a three-variety mixture is treated with a fungicide and then mixed with the two untreated components. Data for two field experiments in 1983 and 1984 are summarized in Table 3. In these experiments Carnival (M1a6 resistance), Patty (M1a12), and Tasman or Triumph (both M1a7 plus M1Ab) were grown alone, untreated, or treated either with ethirimol or a triazole fungicide. They were also grown as a mixture and in plots where only one component was treated. All plots were surrounded by guards to reduce interplot interference.

Although it reduced infection, treating pure stands with triazole did not increase yields significantly, probably because fungicide resistance increased during the period. The effect of ethirimol treatment on yield, however, was highly significant ($P < 0.001$) and was associated with greater disease control. Mixing varieties without a fungicide treatment increased yield significantly ($P < 0.05$) and reduced infection, although fungicide treatments of the mixture had no significant effect.

An interesting but not significant result was that the highest absolute yields were obtained with the mixtures in which single components had been treated. For both fungicides the yields of these 1/3 treatments were significantly higher ($P < 0.01$) than the equivalent calculated treatment of pure stands. Moreover, there was considerably less infection on these mixtures than on untreated mixtures; they were only slightly more infected than the mixtures that received the conventional fungicide treatment. Comparing the 1/3 treatments of the mixture with the conventional treatment of the pure stands, the mixture yields were higher, significantly so for the triazole treatments, and infection levels were the same.

Thus, for the farmer, using the 1/3 treatment of a variety mixture would produce a yield as high and a crop as clean as from conventionally treated pure stands, but at a lower cost. Epidemiologically the fungicide seed treatment protects the crop at the beginning of the epidemic, when variety mixing is least effective. Later in the growth cycle the crop is protected more by the varietal heterogeneity, after the fungicide concentration has declined below the threshold for disease control. Biologically the pathogen is less able to overcome each variety and fungicide component, and less fungicide is delivered into the environment. We may also expect to maintain higher yields with the partly treated mixtures than with the conventionally treated pure varieties.

CONCLUSION

The response of a pathogen population to fungicide use depends on genetic variation for resistance being present in the population. When such variation is present and can be demonstrated, the rate and form of the response will depend on a complex interaction of the genetic and breeding system and general biology of the target organism, the range of host varieties in use, cultivation practices, and the physical environment. The example of powdery mildew of barley shows how responses can be manipulated using different forms of crop husbandry. The ability to modify the pathogen response requires at least an understanding of the genetics and population dynamics of the pathogen so that the consequences of changes in cultivation practices can be predicted. Without a reasonable understanding of the population biology of the pathogen and of the consequences of crop husbandry methods, it is not possible either to understand the responses or to suggest changes in agricultural practices that might modify the response. The only certain conclusion is that if variation for resistance exists, and the fungicide is used extensively and homogeneously, then its effectiveness will soon decline. Unfortunately, the pathogen may ultimately find a way around any strategy designed to control it.

ACKNOWLEDGMENT

We wish to acknowledge financial help from ICI Plant Protection Ltd. for part of the experimental work.

REFERENCES

Barrett, J. A. In press. *In* Populations of Plant Pathogens: Their Dynamics and Genetics, M. S. Wolfe and C. E. Caten, eds. Oxford: Blackwell.

Brent, K. J. 1982. Case study 4: Powdery mildews of barley and cucumber. Pp. 219–230 *in* Fungicide Resistance in Crop Protection, J. Dekker and S. G. Georgopoulos, eds. Wageningen, Netherlands: Centre for Agricultural Publishing and Documentation.

Fletcher, J. T., and M. S. Wolfe. 1981. Insensitivity of *Erysiphe graminis* f. sp. *hordei* to triadimefon, triadimenol and other fungicides. Pp. 633–640 *in* Proc. Br. Crop Prot. Conf. Fungic. Insectic. Vol. 2. Lavenham, Suffolk: Lavenham.

Jensen, N. E. 1952. Intra-varietal diversification in oat breeding. Agron. J. 44:30–34.

Limpert, E., and E. Schwarzbach. 1981. Virulence analysis of powdery mildew of barley in different European regions in 1979 and 1980. Pp. 458–465 *in* Proc. 4th Int. Barley Genet. Symp. Edinburgh: Edinburgh Univ. Press.

Wolfe, M. S. 1981. Integrated use of fungicides and host resistance for stable disease control. Philos. Trans. R. Soc. London, Ser. B 295:175–184.

Wolfe, M. S. 1982. Dynamics of the pathogen population in relation to fungicide resistance. Pp. 139–148 *in* Fungicide Resistance in Crop Protection, J. Dekker and S. G. Georgopoulos, eds. Wageningen, Netherlands: Centre for Agricultural Publishing and Documentation.

Wolfe, M. S. 1985. Current status and prospects of multiline cultivars and variety mixtures for disease resistance. Annu. Rev. Phytopathol. 23:251–253.

Wolfe, M. S. In press [a]. Dynamics of the response of barley mildew to the use of sterol synthesis inhibitors. EPPO Bull., Vol. 15.

Wolfe, M. S. In press [b]. Integration of host resistance and fungicide use. EPPO Bull., Vol. 15.

Wolfe, M. S., and J. A. Barrett. 1980. Can we lead the pathogen astray? Plant Dis. 64:148–155.

Wolfe, M. S., and T. J. Riggs. 1983. Fungicide integrated into host mixtures for disease control. P. 834 *in* Proc. 10th Int. Congr. Plant Prot. Brighton, 1983. Vol. 2.

Wolfe, M. S., P. N. Minchin, and S. E. Slater. 1981. Powdery mildew of barley. Annu. Rep. Plant Breed. Inst. 1980:88–92.

Wolfe, M. S., P. N. Minchin, and S. E. Slater. 1984a. Dynamics of triazole sensitivity in barley mildew, nationally and locally. Pp. 465–470 *in* Proc. 1984 Br. Crop Prot. Conf., Pests and Dis. Washington, D.C. College Park, Md.: Entomological Society of America.

Wolfe, M. S., P. N. Minchin, and S. E. Slater. 1984b. Annu. Rep. Plant Breed. Inst. 1983:87–91.

Wolfe, M. S., P. N. Minchin, and S. E. Slater. 1985. Powdery mildew of barley. Annu. Rep. Plant Breed. Inst. 1984:91–95.

Pesticide Resistance: Strategies and Tactics for Management.
1986. National Academy Press, Washington, D.C.

Experimental Population Genetics and Ecological Studies of Pesticide Resistance in Insects and Mites

RICHARD T. ROUSH and BRIAN A. CROFT

Current data on the population genetics and ecological aspects of pesticide resistance in insects and mites are reviewed. Very little is known about initial frequencies of resistance alleles. In some cases dominance depends on pesticide dose. In untreated habitats the fitnesses of resistant genotypes appear to be 50 to 100 percent of those susceptible genotypes. Up to about 20 percent of the individuals in treated populations escape exposure. Important parameters for further research include initial allele frequencies and immigration rates.

INTRODUCTION

One objective of population genetics is to describe evolutionary change. Even though pesticide resistance has long been recognized as evolutionary change (Dobzhansky, 1937), most detailed empirical population studies of insecticide and acaricide resistance have been conducted only during the last decade. Although more work is needed, these experiments complement experiences of field entomologists and provide new insights into management of resistance.

The rate of allelic substitution in a closed population is a function of allele frequency, dominance, and relative fitnesses of genotypes. Arthropod populations, however, are rarely completely closed. Gene flow (''migration'' in the genetic sense) between populations varies tremendously, depending on species and ecological factors affecting insect and mite dispersal. Thus, the evolution of resistance can be described only by considering both genetic and ecological factors.

Most population genetics theory assumes that the traits under consideration are controlled by only one or two loci (for contrast see paper by Via, this volume). Many studies have shown that resistance of practical significance in the field is almost always controlled by one or two loci (Plapp, 1976; Brown, 1967). Although polygenic resistance does occur in nature (Liu et al., 1981), it is much more common in the laboratory (Whitten and McKenzie, 1982; Roush, 1983). Therefore, it is not unreasonable to assume that the toxicology of resistance is due to a single allelic variant at one locus (for additional discussion see papers by May and Dobson, Uyenoyama, and Via, this volume).

INITIAL ALLELE FREQUENCY

Little is known about allelic frequencies prior to pesticide selection, although they may range from 10^{-2} (Georghiou and Taylor, 1977) to 10^{-13} (Whitten and McKenzie, 1982). These frequencies should be measurable, but this has been accurately done only with dieldrin resistance in *Anopheles gambiae*, where the frequency is unusually high (Wood and Bishop, 1981). Initial allele frequency is a function of selection against the resistant genotypes and mutation rate (Crow and Kimura, 1970). Although some data exist on selective disadvantages, mutation rates are only estimates. Based on mutation rates for other traits in organisms such as *Drosophila* (Dobzhansky et al., 1977), these rates may vary from 10^{-4} to 10^{-8} or may be as low as 10^{-16} if resistance requires a change at two or more sites in the gene (Whitten and McKenzie, 1982).

Measuring initial resistance gene frequencies directly is difficult. The phenotype of a resistance gene and an efficient means to detect it can be known only when resistance develops in the field. By that time most populations have been exposed to the pesticide. One alternative, laboratory selection, often produces artifacts such as polygenic resistance (Whitten and McKenzie, 1982; Roush, 1983). Laboratory-susceptible strains collected before pesticide use commonly suffer population bottlenecks (LaChance, 1979) that distort rare allele frequencies (Nei et al., 1975).

Despite these difficulties initial resistance allele frequencies could and should be measured. Some resistance management strategies depend on allele frequency. For example, high pesticide doses may delay resistance, but only if allele frequency is very low and other conditions are met (Tabashnik and Croft, 1982). Such frequencies could be measured in field populations by screening for resistance before using a new insecticide at a dose that kills more than 99 percent of susceptible individuals. Survivors would have to be held for testing for major resistance alleles. A more efficient approach would be to develop a sophisticated detection test (e.g., electrophoretic) for a cosmopolitan resistant pest (e.g., *Musca domestica* L., *Tetranychus urticae*

Koch, *Myzus persicae* [Sulz.], and *Heliothis armigera* [Hübner]) in one country and to take the test to another country where the pesticide has never been used. With international cooperation it would then be possible to take advantage of differing pesticide-use patterns to estimate initial allele frequencies.

DOMINANCE

Dominance refers to the resemblance of heterozygotes (usually F_1 offspring) to one of their parents. If heterozygotes (RS) more closely resemble the toxicological phenotypes of the resistant homozygous (RR) parents, resistance is dominant. If the heterozygotes show little or no resemblance, resistance is recessive. For many genetic traits, particularly visible mutants, dominance and fitness can be defined independently. For example, "stubby wing" of the house fly can be defined as recessive to the wild type by morphology, even though there may be recessive effects on reproductive fitness. In the field, however, dominance for survival of pesticides may also mean higher relative fitness compared to the susceptible genotypes.

In the field, dominance of the toxicological phenotype may depend on dose (Curtis et al., 1978). A dose that would kill RS heterozygotes but not resistant (RR) homozygotes means that the heterozygotes resemble the susceptible homozygotes (SS), and resistance is effectively recessive. Conversely, a dose that would kill susceptible homozygotes but not the heterozygotes makes resistance functionally dominant, since heterozygotes and RR homozygotes are phenotypically similar. This concept of adjusting the dose is often called alteration of dominance, but could be called alteration of relative fitness. The ultimate reduction in relative fitness results from doses so high that even RR genotypes are killed, which is generally not feasible. At least two research groups have reported on toxicological dominance in the field. Interestingly, the results have not always been consistent with laboratory data.

Resistance to lindane and cyclodienes, including dieldrin, ordinarily shows clear discrimination between all three genotypes in laboratory assays (Brown, 1967). Therefore, some pesticide doses in the field should kill all susceptibles and heterozygotes but not all resistant homozygotes. This occurs in anopheline mosquitoes (Rawlings et al., 1981): SS, RS, and RR adults marked with fluorescent dusts were released into lindane-sprayed village huts in Pakistan. The higher treatments killed all three genotypes at first, but eventually allowed some RR individuals to survive as residues decayed. Thus, resistance was rendered effectively recessive.

Similarly, McKenzie and Whitten (1982) implanted eggs of RR, RS, and SS sheep blow flies (*Lucilia cuprina* [Wiedemann]) into artificial wounds in dieldrin-treated sheep. Larvae were later collected from the wounds, reared to adulthood, and tested with a discriminatory dose to determine genotype. The RS individuals had a relative viability intermediate between the RR and SS larvae, even as the dose decayed. Although higher doses might have made resistance recessive, these results differed from those of Rawlings et al. (1981) despite a similar form of resistance (Whitten et al., 1980).

McKenzie and Whitten (1982) also studied relative viabilities of diazinon-resistant genotypes. Diazinon resistance was incompletely dominant in laboratory assays of sheep blow fly larvae (Arnold and Whitten, 1976). Therefore, the RR and RS genotypes should have similar relative viabilities in the field, that is, resistance should be dominant. To the contrary the RS genotypes actually showed relative viabilities very similar to the SS genotypes (i.e., resistance was effectively recessive under field conditions), even as the diazinon residues decayed to allow considerable survival of the SS homozygotes. The reason for the contrasting results for dieldrin and diazinon is unclear, but dominance in the field and in the laboratory should not be assumed to be similar.

Dominance is important not only in relation to pesticide pressure but also in the absence of pesticide pressure. The important phenotype in this case is relative fitness, which is even more difficult to measure than toxicological dominance in the field. The phenotypic dominance of fitness is most easily discussed in the context of relative fitnesses in untreated habitats.

RELATIVE FITNESSES

Untreated Habitats

Resistant genotypes must be at a reproductive disadvantage in the absence of pesticides. If not, resistance alleles would be more common prior to selection (Crow, 1957). Small selection intensities, however, can maintain very low allele frequencies over evolutionary time (Crow and Kimura, 1970). For resistance management the selective differences between resistant and susceptible genotypes must be accurately quantified.

Resistant and susceptible strains of arthropods often are reported to differ in developmental time, fecundity, and fertility. Mating competitiveness might also differ, but of the reports found on this, neither detected differences (Gilotra, 1965; Roush and Hoy, 1981). Table 1 compares R and S strains from some commonly cited studies where reproductive factors were well quantified and where the R strains could be classified as field- or laboratory-selected. In a field-selected strain resistance was diagnosed or suspected before the strain was brought into the laboratory. A laboratory strain was produced by selection from an initially susceptible colony. Whenever possible all data relevant to fecundity (i.e., egg and larval survival) or developmental time (egg, larval, pupal, or mean generation time) were combined. (For

TABLE 1 Fitness Components of Resistant (R) Compared with Susceptible (S) Strains

Species	Insecticide	Fecundity (R/S)	Developmental Time (S/R)	References
		Field Selected Resistant Strains		
Musca domestica	DDT	—	0.99 (NS)	March and Lewallen (1950)
M. domestica[1]	DDT	1.07	0.71[a]	Pimentel et al. (1951)
M. domestica	DDT	0.83 (NS)	0.99 (NS)	Babers et al (1953)
M. domestica	DDT (probably KDR)	—	1.05	Bogglid and Keiding (1958)
Anopheles albimanus	Dieldrin	1.02	—	Gilotra (1965)
Blatella germanica	Chlordane	0.88[a]	1.03	Grayson (1954); Perkins and Grayson (1961)
Anthonomous grandis	Endrin	0.96 (NS)	1.01	Bielarski et al (1957)
Tribolium castaneum[2]	Malathion	0.19[a]	—	Brower (1974)
		Laboratory Selected Resistant Strains		
M. domestica	DDT	0.67[a]	0.88[a]	Babers et al (1953)
B. germanica	DDT	0.67[a]	0.94[a]	Grayson (1953); Perkins and Grayson (1961)
T. castaneum	DDT	0.36[a]	0.86[a]	Bhatia and Pradhan (1968)
Dermestes maculatus	Lindane	0.12[a]	0.85[a]	Shaw and Lloyd (1969)
A. grandis	Endrin	0.78[a]	0.98[a]	Thomas and Brazzel (1961)

[a]R strain statistically less fit than S strain ($p < .05$).
[1]Selected for five generations in laboratory.
[2]Selected for 10 generations in laboratory.

SOURCE: See references column.

simplicity throughout this paper the SS genotype will have a relative fitness of 1.00 compared to the RS and RR genotypes.)

Two conclusions are apparent from Table 1. First, reproductive disadvantages are not always associated with resistance. Second, laboratory-selected strains suffer more reproductive disadvantages than resistant strains colonized from the field. Even two of the field strains that showed disadvantages had been selected for 5 to 10 generations in the laboratory. The differences between the laboratory and field strains are consistent with the conclusions of Whitten and McKenzie (1982) and Roush (1983) that laboratory and field selection often produce different kinds of resistance. Although the genetic basis of resistance in most of these strains is unknown, resistance in the laboratory-selected DDT-resistant *Blatella germanica* was polygenic (Cochran et al., 1952).

Studies of the type shown in Table 1 are interesting, but they cannot provide accurate data for resistance management. Strains often differ in fitness, independent of resistance (Babers et al., 1953; Bogglid and Keiding, 1958; Perkins and Grayson, 1961; Birch et al., 1963; Varzandeh et al., 1954; Roush and Plapp, 1982). Even when RR and SS genotypes differ, the more important question is whether there are differences between RS and SS genotypes. During the early stages of selection for resistance and the later stages of resistance reversion, most R alleles in large, randomly mating populations will be carried by heterozygotes. Assuming that selection is not intense, the genotypic frequencies are likely to approximate Hardy-Weinberg proportions (p^2:2pq:q^2). Thus, at resistance allele frequencies of 20 percent, for example, 32 percent of the population will carry RS, and only 4 percent will carry RR. Clearly resistance management will be best served by comparisons of RR, RS, and SS genotypes in similar genetic backgrounds.

Methods

There are two basic methods available for making genotype comparisons. One is to analyze fecundity and developmental-time differences for all three genotypes (Ferrari and Georghiou, 1981; Roush and Plapp, 1982). The other is to monitor changes in genotypic or phenotypic frequencies in untreated populations where the resistance alleles are initially at some intermediate frequency (often 50 percent). These experiments can be conducted and analyzed by iteratively fitting curves for fitness estimates to the observed data (White and White, 1981). Although not always conducted in cages, the studies will be referred to as "population cage" studies because of their clear analogies to the cage studies long conducted by *Drosophila* geneticists. The resistance population-cage data available only as LD_{50}s or resistance ratios are not included here, because such data give only a qualitative appraisal of genotypic fitnesses.

Although both approaches have advantages and disadvantages, the population-cage approach is probably better for most purposes. Fecundity and developmental-time estimates can be measured accurately and fairly quickly. Although these are the only fitness components reported to differ between resistant and susceptible strains, other aspects of fitness could also differ. A population-cage experiment increases the prospect that such differences will be detected.

Another problem for component studies is data analysis. Both cage- and fitness-component studies have generally been conducted on discrete rather than overlapping generations, which is somewhat unrealistic for the field but creates a dilemma in the analysis of fitness-component data. In discrete generations a strain that produces half as many offspring may be only half as fit. For continuous generations population growth rates are more important, as represented by the intrinsic rate of increase, *r* (Ferrari and Georghiou, 1981).

Population growth rates can be more affected by small developmental-time differences than by similar differences in fecundity, as seen in the expression for intrinsic growth rate, $r = \log_e R_0/T$, where R_0 is the net replacement rate (number of daughters per female) and T = mean generation time (Roush and Plapp, 1982). A 50 percent reduction in fecundity (R_0) may affect *r* by much less than 50 percent if R_0 is large and mean generation time remains unchanged. For example, if $R_0 = 100$ for susceptible females and $R_0 = 50$ for resistant females in the laboratory, the difference in *r* is only 15 percent. On the other hand realistic values of R_0 in the field may be only about 5 (Georghiou and Taylor, 1977), where a 50 percent reduction in R_0 (5 to 2.5) means a 43 percent reduction in *r*. Thus, quantifying fitness with *r* or similar terms (Roush and Plapp, 1982) depends on an implicit assumption about R_0. For logistical reasons population cages must be maintained at a relatively constant density, so R_0 is about 1, which is probably closer to field conditions than if R_0 is around 50. In addition cages can be maintained in continuous generations, if appropriate to the species.

A third advantage of the population-cage approach is that all three genotypes can be compared against a homogenized genetic background. Crossing unrelated R and S strains often results in heterotic F_1 heterozygotes, giving biased or ambiguous estimates of fitness specific to the RS genotypes (Roush and Plapp, 1982). The easiest way to establish a population cage in an unbiased way is with F_1 heterozygotes. When fitness differences have been implicated by population cages, the fitness-component approach may be useful for identifying the factors that differ.

Fitness estimates should be obtained in the field whenever possible. It is rare, however, that one can monitor populations known to be isolated from R or S immigration and where an allele has been raised to moderately high frequency by pesticide pressure that has ceased. It is generally more feasible to maintain population-cage experiments in the laboratory. Curtis et al. (1978) compared estimates of fitness from both field and laboratory data and obtained similar results. Therefore, laboratory results may be realistic if the laboratory conditions simulate the field as much as possible. The most realistic studies of this kind may be conducted on species whose behavior and ecology are not too disrupted by laboratory or greenhouse settings, including *Musca domestica*, *Tetranychus urticae*, *Blatella germanica*, and *Tribolium castaneum*. It may be particularly important to conduct these studies under different temperatures.

Data

In a seminal study Curtis et al. (1978) estimated relative fitnesses from field data on changes in frequencies of resistant and susceptible phenotypes of two species of *Anopheles* mosquitoes during several generations after treatment was discontinued. Although there were some uncertainties about the estimates (Curtis et al., 1978; Wood and Bishop, 1981; Roush and Plapp, 1982), the DDT- and dieldrin-resistant phenotypes in *An. culicifacies* had relative fitnesses of about 0.44 to 0.97. One important assumption was that susceptibles were not immigrating into the sites, thus causing fitnesses to be underestimated. Some immigration is likely for *An. culicifacies*, but immigration is less likely for *An. stephensi* (Wood and Bishop, 1981). In this species DDT-resistant phenotypes had estimated fitnesses of 0.91 from field data and 0.96 from a field-selected population held in the laboratory.

Muggleton (1983) used methods similar to those of Curtis et al. (1978) in a laboratory study of the fitnesses of malathion-resistant phenotypes of the stored products pest *Oryzaephilus surinamensis*. Relative fitnesses were about 0.63 to 0.76 compared with the S phenotypes when the populations were held at 25°C, but the R phenotypes may have had an advantage at temperatures over 30°C.

Only a few studies report data on the fitnesses of RS heterozygotes. In all of these, the fitness disadvantages suffered by the heterozygotes were not more than half of those for resistant homozygotes. In two studies the heterozygotes suffered no reproductive disadvantage (White and White, 1981; Roush and Plapp, 1982), that is, the reproductive effects of resistance were recessive.

Three studies used a fitness-component approach. Ferrari and Georghiou (1981) studied intrinsic growth rate, *r*, in RR, RS, and SS genotypes of *Culex quinquefasciatus*. The RR strain had an *r* of 0.79, but F_1 heterozygotes had an *r* of 0.95. Emeka-Ejiofor et al. (1983) compared the developmental times of dieldrin-resistant, DDT-resistant, and susceptible strains, and F_1 crosses of *An. gambiae*. The differences were small in all comparisons. Roush and Plapp (1982) found that diazinon-resistant (RR) house flies had about

57 to 89 percent of the reproductive potential of an SS strain, but RS flies had 100 percent of that potential.

White and White (1981) reported on a population-cage study of diazinon resistance in sheep blow flies. The frequency of resistance phenotypes declined quickly from an initial frequency of about 90 percent, then slowed dramatically (White and White, 1981), as is typical for selection against a recessive allele (Crow and Kimura, 1970). Approximately 10 percent of the population was still resistant at generation 38 (White and White, 1981). The fitness estimates for generations 13 to 38 were 0.61 for RR and 1.0 for RS and SS.

Coadaptation

The above studies were conducted on long-established R strains and may underestimate the fitness disadvantages suffered by RR and RS genotypes during the early stages of a resistance episode if "resistance coadaptation" is common. According to this theory the fitnesses of resistant genotypes are improved by "coadaptive" modifying genes that change the genetic background (Whitehead et al., 1985). Coadaptation of fitness and resistance may, however, be rare (Roush, 1983). The only reliable approach to evaluating whether coadaptation has occurred in a strain is to use repeated backcrossing to a susceptible strain (Crow, 1957) to isolate the major resistance gene in a susceptible genetic background.

Perhaps the first researcher to use repeated backcrossing and to report on fitness was Helle (1965). The Leverkusen-S strain of *Tetranychus urticae* was selected for more than 30 generations to produce an R strain. This strain was inferior to the S strain in fitness, and resistance reverted after relaxing selection. Contrary to what would be expected if coadaptation was occurring, fitness of the R strain was improved, not worsened, by repeated backcrossing.

More recently a backcrossing study on sheep blow fly has demonstrated that resistance coadaptation can occur. McKenzie et al. (1982) found that diazinon resistance was not deleterious in population cages established from F_1 and BC_3 RS flies, but was significantly deleterious in cages established from BC_6 and BC_9 RS flies. The decline in the frequency of the R allele in the BC_9 cages can be approximated by fitnesses of 0.5 for RR and 0.75 for RS. The major resistance modifier(s) were on a different chromosome than the major resistance locus (McKenzie and Purvis, 1984).

In contrast fitness coadaptation was not found in diazinon resistant house flies collected in Mississippi (Whitehead et al., 1985). Even after six generations of backcrossing to a laboratory-susceptible strain, there were no significant differences in developmental time or fecundity. There are major differences, however, between house flies in Mississippi and sheep blow flies in Australia. Fitness modifiers can only be at an advantage when in the presence of the resistance allele. Thus, selection for fitness modification must be fairly weak until the resistance allele reaches high frequency. Charlesworth (1979) gives a similar argument. The frequency of the diazinon-resistance allele appears to be about 0.27 in Mississippi house flies (Whitehead et al., 1985). The resistance allele in the sheep blow fly was maintained in very high frequency by continuous diazinon use against the insect for more than 10 years, which is rather unusual (McKenzie et al., 1982). Thus, it is reasonable that modification occurred in the sheep blow fly but not in the house fly.

In sum, fitness modification has been observed in only one of three cases. More such studies are needed. The available data show that fitnesses of RR range from 0.5 to 1.0; fitnesses of RS range from 0.75 to 1.0. At least in laboratory studies, organophosphorous (OP) insecticide-resistant genotypes generally seem to suffer larger reproductive disadvantages than DDT- or cyclodiene-resistant genotypes, consistent with a suggestion by Zilbermints (1975).

Treated Habitats

How do fitnesses in treated habitats compare with those in untreated habitats? Data on increases in frequencies of DDT- and dieldrin-resistant phenotypes of *Anopheles* spp. in the field show that resistant genotypes may have fitnesses of 1.3 to 6.1 (Curtis et al., 1978; Wood and Cook, 1983). Such fitnesses are a complex function of genotypic mortality (which depends on treatment intensity) and reproductive potential, refugia, and immigration (Georghiou and Taylor, 1977). In some circumstances the overall fitnesses of R phenotypes are probably much higher than 6.1.

ECOLOGICAL STUDIES

Although selection for resistance can proceed very quickly in closed populations where each individual is exposed, such intense treatment is uncommon in resistance episodes. Usually, some portion of the controllable individuals escapes significant exposure in protected or overlooked spots or "refugia" within the treated area. Also, some individuals, usually adults, will disperse into the treated areas from outside after pesticide residues have decayed. Both concepts are interrelated and emphasize the maintenance of susceptible individuals in the population.

Refuges

The importance of refugia is clear in models (Georghiou and Taylor, 1977) and can be readily noted in field experience. In spider mites, for example, resistance generally appears first in greenhouses, where all host plants are likely to be thoroughly treated, and later in orchard and field crops, where treatment is less intense or complete (Dittrich, 1975). Few estimates have been made of the portion of populations that ordinarily escape treatment. Such data could be gathered from mark-recapture or population sampling data. For example, population sampling data show that about 20 percent of *Heliothis* larvae in cotton fields escape lethal exposure (Wolfenbarger et al., 1984). The portion of 12 apple pests escaping in refugia ranges from 0.2 percent (apple maggot, *Rhagoletis pomonella*) to 17 percent (San Jose scale, *Quadraspidiotus perniciosus*), depending on species (Tabashnik and Croft, 1985). From practical considerations 20 percent may be an upper limit for the portion in refugia. Failure to obtain at least 80 percent control from insecticide or acaricide applications is probably unsatisfactory for almost any pest and would lead to changes in treatment practices until higher levels of control were achieved.

Immigration

A recent experimental laboratory study on house flies has demonstrated the importance of both susceptible immigration and the influence of pesticide persistence on such immigration (Taylor et al., 1983; Uyenoyama, Via, this volume). Yet immigration is difficult to quantify in terms that relate to resistance development. Rates of immigration for a species depend not only on distances to the source of the untreated population and its size but also on weather and the quality and distribution of host plant species (Stinner, 1979; Follett et al., 1985; Whalon and Croft, 1986).

A better understanding of dispersal is a key component of many emerging pest-management tactics, but resistance management has some rather special needs. It is not enough to conduct mark-recapture studies on adults. Knowing where the individuals mate and oviposit is also necessary for understanding the impact they have on the susceptibility of a population. Genetic markers, including pesticide resistance and allozymes, may be particularly useful in such studies.

Based on a survey of orchard entomologists, ratios of migrants to the resident population among 12 apple pests range from 0.1 to 10^{-5}, depending on species (Tabashnik and Croft, 1985). As was true for initial R allele frequencies, and in contrast to factors like refugia, current estimates of immigration rates vary over several orders of magnitude. This emphasizes not only the need to tailor resistance management programs to individual species but also the need to improve estimates of immigration.

RESEARCH NEEDS

Most resistance models are based on fairly reasonable genetic assumptions (Tabashnik, this volume). Most resistance seems to be associated with single-locus changes. Fitness disadvantages clearly occur, although they may be "slight" to "moderate" rather than "severe," as defined in some modeling studies (Georghiou and Taylor, 1977; Tabashnik and Croft, 1982). Nonetheless, more studies must be conducted on the fitnesses of resistant genotypes, with emphasis on coadaptation, to determine if the studies reviewed here are representative across a range of species. More important, however, better estimates must be obtained for R allele frequencies in untreated populations, since current estimates vary over several orders of magnitude.

Although migration and refugia are important, they are poorly understood compared with their potential impact. The quantification of immigration, in particular, requires continued improvement in understanding the basic ecology of pest species. Presumably, such understanding will also allow better control of these species without pesticides and will further deter resistance development, which is at the heart of modern pest management.

ACKNOWLEDGMENTS

We thank J. C. Schneider, M. J. Whitten, and B. E. Tabashnik for discussion. Paper approved as No. 5985 by Director, Mississippi Agricultural and Forestry Experiment Station.

REFERENCES

Arnold, J. T. A., and M. J. Whitten. 1976. The genetic basis for organophosphorous resistance in the Australian sheep blowfly, *Lucilia cuprina* (Wiedemann) (Diptera: Calliphoridae). Bull. Entomol. Res. 66:561–568.

Babers, F. H., J. J. Pratt, Jr., and M. Williams. 1953. Some biological variations between strains of resistant and susceptible house flies. J. Econ. Entomol. 46:914–915.

Bhatia, S. K., and S. Pradhan. 1968. Studies on resistance to insecticides in *Tribolium castaneum* Herbst 1. Selection of a strain resistant to p, p′ DDT and its biological characteristics. Indian J. Entomol. 30:13–32.

Bielarski, R. V., J. S. Roussel, and D. F. Clower. 1957. Biological studies of boll weevils differing in susceptibility to chlorinated hydrocarbon insecticides. J. Econ. Entomol. 50:481–482.

Birch, L. C., T. Dobzhansky, P. O. Elliott, and R. C. Lewontin. 1963. Relative fitness of geographic races of *Drosophila serata*. Evolution 17:72–83.

Bogglid, O., and J. Keiding. 1958. Competition in house fly larvae. Oikos 9:1–25.

Brower, J. H. 1974. Radio-sensitivity of an insecticide-resistant strain of *Tribolium castaneum* (Herbst). J. Stored Prod. Res. 10:129–131.

Brown, A. W. A. 1967. Genetics of insecticide resistance in insect vectors. Pp. 505–552 *in* Genetics of Insect Vectors of Disease, J. W. Wright and R. Pal, eds. New York: Elsevier.

Charlesworth, B. 1979. Evidence against Fisher's theory of dominance. Nature (London) 278:848–849.

Cochran, D. G., J. M. Grayson, and M. Levitan. 1952. Chromosomal and cytoplasmic factors in transmission of DDT resistance in the German cockroach. J. Econ. Entomol. 45:997–1001.

Crow, J. F. 1957. Genetics of insect resistance to chemicals. Annu. Rev. Entomol. 2:227–246.

Crow, J. F., and M. Kimura. 1970. An Introduction to Population Genetics Theory. New York: Harper and Row.

Curtis, C. F., L. M. Cook, and R. J. Wood. 1978. Selection for and against insecticide resistance and possible methods of inhibiting the evolution of resistance in mosquitoes. Ecol. Entomol. 3:273–287.

Dittrich, V. 1975. Acaricide resistance in mites. Z. Angew. Entomol. 78:28–45.

Dobzhansky, T. 1937. Genetics and the Origin of Species. New York: Columbia University Press.

Dobzhansky, T., F. J. Ayala, G. L. Stebbins, and J. W. Valentine. 1977. Evolution. San Francisco, Calif.: Freeman.

Emeka-Ejiofor, S. A. I., C. F. Curtis, and G. Davidson. 1983. Tests for effects of insecticide resistance genes in *Anopheles gambiae* on fitness in the absence of insecticides. Entomol. Exp. Appl. 34:163–168.

Ferrari, J. A., and G. P. Georghiou. 1981. Effects of insecticidal selection and treatment on reproductive potential of resistant, susceptible, and heterozygous strains of the southern house mosquito. J. Econ. Entomol. 74:323–327.

Follett, P. A., B. A. Croft, and P. H. Westigard. 1985. Regional resistance to insecticides in *Psylla pyricola* from pear orchards in Oregon. Can. Entomol. 117:565–573.

Georghiou, G. P., and C. E. Taylor. 1977. Genetic and biological influences in the evolution of insecticide resistance. J. Econ. Entomol. 70:319–323.

Gilotra, S. K. 1965. Reproductive potentials of dieldrin-resistant and susceptible populations of *Anopheles albimanus* Wiedemann. Am. J. Trop. Med. Hyg. 14:165–169.

Grayson, J. M. 1953. Effects on German cockroaches of twelve generations of selection for survival to treatments with DDT and benzene hexachloride. J. Econ. Entomol. 45:124–127.

Grayson, J. M. 1954. Differences between a resistant and a nonresistant strain of the German cockroach. J. Econ. Entomol. 47:253–256.

Helle, W. 1965. Resistance in the Acarina: Mites. Pp. 71–93 *in* Recent Advances in Acarology, Vol. II, J. D. Rodriguez, ed. New York: Academic Press.

LaChance, L. E. 1979. Genetic strategies affecting the success and economy of the sterile insect release methods. Pp. 8–18 *in* Genetics in Relation to Insect Management, M. A. Hoy and J. J. McKelvey, Jr., eds. New York: The Rockefeller Foundation.

Liu, M-Y., Y-J. Tzeng, and C-N. Sun. 1981. Diamondback moth resistance to several synthetic pyrethroids. J. Econ. Entomol. 74:393–396.

March, R. B., and L. L. Lewallen. 1950. A comparison of DDT-resistant and nonresistant house flies. J. Econ. Entomol. 43:721–722.

McKenzie, J. A., and A. Purvis. 1984. Chromsomal localisation of fitness modifiers of diazinon resistance genotypes of *Lucilia cuprina*. Heredity 53:625–634.

McKenzie, J. A., and M. J. Whitten. 1982. Selection for insecticide resistance in the Australian sheep blowfly, *Lucilia cuprina*. Experientia 38:84–85.

McKenzie, J. A., M. J. Whitten, and M. A. Adena. 1982. The effect of genetic background on the fitness of diazinon resistance genotypes of the Australian sheep blowfly, *Lucilia cuprina*. Heredity 49:1–9.

Muggleton, J. 1983. Relative fitness of malathion-resistant phenotypes of *Oryzaephilus surinamensis* L. (Coleoptera: Silvanidae). J. Appl. Ecol. 20:245–254.

Nei, M., T. Maruyama, and R. Chakraborty. 1975. The bottleneck effect and genetic variability in populations. Evolution 29:1–10.

Perkins, B. D., Jr., and J. M. Grayson. 1961. Some biological comparisons of resistant and nonresistant strains of the German cockroach, *Blattella germanica*. J. Econ. Entomol. 54:747–750.

Pimentel, D., J. E. Dewey, and H. H. Schwardt. 1951. An increase in the duration of the life cycle of DDT-resistant strains of the house fly. J. Econ. Entomol. 44:477–481.

Plapp, F. W., Jr. 1976. Biochemical genetics of insecticide resistance. Annu. Rev. Entomol. 21:179–197.

Rawlings, P., G. Davidson, R. K. Sakai, H. R. Rathor, K. M. Aslamkhan, and C. F. Curtis. 1981.

Field measurement of the effective dominance of an insecticide resistance in anopheline mosquitoes. Bull. W.H.O. 49:631–640.

Roush, R. T. 1983. A populational perspective on the evolution of resistance. Speech presented at Nat. Meet. Entomol. Soc. Am. Detroit, Mich., November 27–December 1, 1983.

Roush, R. T., and M. A. Hoy. 1981. Laboratory, glasshouse and field studies of artificially selected carbaryl resistance in *Metaseiulus occidentalis*. J. Econ. Entomol. 74:142–147.

Roush, R. T., and F. W. Plapp, Jr. 1982. Effects of insecticide resistance on biotic potential of the house fly. (Diptera: Muscidae). J. Econ. Entomol. 75:708–713.

Shaw, D. D., and C. J. Lloyd. 1969. Selection for lindane resistance in *Dermestes maculatus* de Geer (Coleoptera: Dermestidae). J. Stored Prod. Res. 5:69–72.

Stinner, R. E. 1979. Biological monitoring essentials in studying wide area moth movement. Pp. 199–208 *in* Movement of Highly Mobile Insects, R. L. Rabb and G. G. Kennedy, eds. Raleigh: North Carolina State University.

Tabashnik, B. E., and B. A. Croft. 1982. Managing pesticide resistance in crop-arthropod complexes: Interactions between biological and operational factors. Environ. Entomol. 11:1137–1144.

Tabashnik, B. E., and B. A. Croft. 1985. Evolution of pesticide resistance in apple pests and their natural enemies. Entomophaga 30:37–49.

Taylor, C. E., F. Quaglia, and G. P. Georghiou. 1983. Evolution of resistance to insecticides: A cage study on the influence of migration and insecticide decay rates. J. Econ. Entomol. 76:704–707.

Thomas, J. G., and J. R. Brazzel. 1961. A comparative study of certain biological phenomena of a resistant and a susceptible strain of the boll weevil, *Anthonomus grandis*. J. Econ. Entomol. 54:417–420.

Varzandeh, M., W. N. Bruce, and G. C. Decker. 1954. Resistance to insecticides as a factor influencing the biotic potential of the house fly. J. Econ. Entomol. 47:129–134.

Whalon, M. E., and B. A. Croft. 1986. Dispersal of apple pests and natural enemies in Michigan. Michigan State University, Agricultural Experiment Station: Research Report No. 467.

White, R. J., and R. M. White. 1981. Some numerical methods for the study of genetic changes. Pp. 295–342 *in* Genetic Consequences of Man Made Change, J. A. Bishop and L. M. Cook, eds. New York: Academic Press.

Whitehead, J. R., R. T. Roush, and B. R. Norment. 1985. Resistance stability and coadaptation in diazinon-resistant house flies (Diptera: Muscidae). J. Econ. Entomol. 78:25–29.

Whitten, M. J., and J. A. McKenzie. 1982. The genetic basis for pesticide resistance. Pp. 1–16 *in* Proc. 3rd Australas. Conf. Grassl. Invert. Ecol., K. E. Lee, ed. Adelaide, Australia: S.A. Government Printer.

Whitten, M. J., J. M. Dearn, and J. A. McKenzie. 1980. Field studies on insecticide resistance in the Australian sheep blowfly, *Lucilia cuprina*. Aust. J. Biol. Sci. 33:725–735.

Wolfenbarger, D. A., J. A. Harding, and S. H. Robinson. 1984. Tobacco budworm (Lepidoptera: Noctuidae): Variations in response to methyl parathion and permethrin in the subtropics. J. Econ. Entomol. 77:701–705.

Wood, R. J., and J. A. Bishop. 1981. Insecticide resistance: Populations and evolution. Pp. 97–127 *in* Genetic Consequences of Man Made Change, J. A. Bishop and L. M. Cook, eds. New York: Academic Press.

Wood, R. J., and L. M. Cook. 1983. A note on estimating selection pressures on insecticide resistance genes. Bull. W.H.O. 61:129–134.

Zilbermints, I. V. 1975. Genetic change in the development and loss of resistance to pesticides. Pp. 85–91 *in* Proc. 8th Int. Congr. Plant Prot., Vol 2.

4

Detection, Monitoring, and Risk Assessment

RESISTANCE DETECTION MEANS IDENTIFYING a significant change in the susceptibility of a pest population to pesticides, ideally very soon after the emergence of resistance. *Resistance monitoring* attempts to measure changes in the frequency or degree of resistance in time and space. Resistance monitoring is most useful when undertaken early in a resistance episode. Monitoring can also be used to evaluate the effectiveness of alternative tactics that are employed to overcome, delay, or prevent the development of resistance.

In contrast to detection and monitoring of resistance in the field after the fact, *resistance risk assessment* is predicting the probability of resistance emerging as a result of use of a pesticide in a given use environment. A risk assessment is subject to a varying margin of error and should, in any event, be applied with care. Resistance risk assessments can be made for certain plant pathogens with some precision when the toxicological, epidemiological, and population considerations of the pathogen are well known from previous resistance episodes (Staub and Sozzi, 1984). In such cases, resistance management actions may be taken to prevent resistance before it occurs and is detected in the field. Likewise, there are extensive historical data bases on resistance trends for some insects that make it possible to carry out resistance risk assessments, thereby making it possible to manage resistance by restricting the use of certain pesticides, or by managing their application in some specific fashion (Keiding, this volume). More often than not, though, the data base on the resistance potential of a given pest and pesticide combination is too limited to allow for resistance risk predictions that are reliable enough for use in devising strategies to manage resistance.

Detection, monitoring, and assessment of resistance risk are interrelated. They are generally used during different, sometimes overlapping periods in a resistance episode, and each has a distinctly different objective. A resistance risk assessment may be made when a new compound is proposed for use on a new target pest, or in a new crop or region. A resistance detection program should be initiated when a resistance risk assessment—or common experience—suggests a likelihood of resistance developing. With pesticides involving new chemistry and modes of action, the resistance risk potential will rarely be known. The resistance potential of known products, or of their chemical analogues, often can be assessed with reasonable precision. Once resistance is detected, the ideal program shifts into a monitoring phase. During this phase the spread and degree of resistance are periodically determined.

Specific, well-known objectives of these interrelated activities include

- Provide an early assessment of the risk for resistance before a pesticide is widely used.
- Determine whether ineffective control following applications of a pesticide are due to resistance.
- Provide an early warning system so that alternative pest-control tactics can be implemented.
- Delineate the geographic extent and movement of the resistant species over time.
- Validate the effectiveness of resistance management tactics introduced at a specific time and place.
- Provide effective crop protection.

METHODS AVAILABLE FOR RESISTANCE DETECTION, MONITORING, AND RISK ASSESSMENT

Resistance detection and monitoring methods for pest species have in the past been based on classical bioassay techniques (see examples in Keiding and in Brent, this volume; FAO, 1982; Georgopolous, 1982). With these methods, test organisms are exposed to a gradient of pesticide doses or concentrations, and features of mortality, growth, or population abundance are evaluated. More recently, biochemical tests for identifying unique detoxification enzymes associated with resistant pests have been refined for use in survey of both resistant individuals and populations (Miyata, 1983). Even more recent are immunological tests for resistance based on identification of detoxification enzymes using monoclonal antibodies (e.g., Devonshire and Moores, 1984). One expected benefit from biotechnology research is DNA probes, which may be used to identify specific genetic sequences such as alleles conferring resistance in a pest species. It appears likely that a much greater degree of resolution and more specific identification of resistance alleles in pest individuals and populations will be available in the near future. These tools should enable monitoring of resistance much earlier than is currently possible.

RESEARCH ON RESISTANCE DETECTION AND MONITORING

Research is needed at several levels to determine the speed and degree of resistance that may develop in a given pesticide-use environment (see Chapters 2 and 3).

At the molecular level, experimental assays in vitro and in vivo may be used to compare responses to proposed new compounds with currently used compounds eliciting known (or unknown) resistance mechanisms. Generally, it is assumed that a biochemical mechanism that is genetically conferred is the cause of resistance in most species.

At the organismal level, tests with large and diverse populations may be helpful to determine the degree and speed with which resistance may develop in a species. The impacts of a variety of factors on the speed of resistance developing can be studied, including the resistance mechanism, allele dominance and frequency, immigration of susceptible types into the system, the competitiveness of resistant types, etc.

At the population level, the probability of resistance developing under varying ecological conditions and field-use practices may be examined through field tests using the methods employed by pest-control personnel or in trial runs made in conjunction with pest-management operations. In this type of test, problems are often encountered with experimental design, making it difficult to control treatments on highly mobile pests.

RECOMMENDATION 1. **The following research is needed to evaluate the biological and practical feasibility of resistance detection and monitoring in key pests.**

- **Develop new and improved standard methods to detect and monitor resistance for key pests, where needed.** Extensive work in this area has been done by industry and by the World Health Organization (WHO), the Food and Agricultural Organization (FAO), the European Plant Protection Organization (EPPO), the Entomological Society of America (ESA), and other similar organizations. Continued and expanded cooperation is needed. Detection and monitoring methods should be as simple, rapid, accurate, precise, field-adaptable, and inexpensive as possible. Major differences in methods exist among pest types, i.e., insects, weeds, microorganisms, and these differences properly (and sometimes improperly) can influence how data are interpreted. Monitoring systems need to consider the unique attributes of each pest group and differences among and within species in different geographic areas.

- **Determine the relationship between detection and/or frequency of resistance as measured by laboratory bioassay tests, and the likelihood and severity of failure of a pesticide under field conditions.** Data from resistance monitoring, coupled with field observations, can then be the basis for rational decision making.
- **Collect and compile baseline susceptibility data for pesticides effective against key pests.** An important use of these data will be to estimate doses that kill essentially all susceptible individuals (for example, twice the LD_{99}). Such doses could then be used for sampling efforts that can quickly detect resistance. The nature of the data needed for different species may vary seasonally over time, geographically, and according to when various pesticides were first introduced commercially.
- **Develop specialized evaluation methods and statistical procedures for early detection of resistance at low levels, when required.** Such methods may differ considerably from routine monitoring methods, and may involve specialized genetic screening tests.
- **Evaluate new and developing immunological, biochemical, and biotechnological methods for monitoring resistance in the field.** Resistance tests for most pests should be directed at the population level; however, assessments of individuals also is possible based on new biochemical and immunological methods that are becoming available. These assessments may prove important for some pests, although many of the currently used bioassays to monitor plant pathogens evaluate individuals (i.e., isolates) rather than populations.
- **Research on each of the above methods should consider accuracy and precision, cost of collecting samples, previous pesticide histories, environmental conditions, and other sources of experimental variation that may affect pest susceptibility.** To determine the appropriate size and frequency of a resistance monitoring program, the following should all be considered: statistical levels of accuracy required for detection, time delays involved in monitoring, and time required to set resistance management into action.

IMPLEMENTATION

Where feasible, a resistance monitoring system should be based partly on an areawide, regular survey scheme and should respond to local reports of control failures for key pests throughout their potential range of infestation and economic impact. Once resistance is detected, the scope and extent of the monitoring should be expanded to determine the size, type, and spread of resistance. Ideally, monitoring results will become available on a timely basis—certainly within a production season—to allow for development and implementation of appropriate management tactics. Levels of resistance that can be reliably detected in the field may vary greatly depending on pest species and the environment in which pesticide is used. Thus, to ensure economical crop protection, it may also be important to take into account the variable periods of time required for a pest to develop resistance, and for resistance to reach a level at which crop production efforts may fail without a change in control strategy and/or chemicals.

Examples of pests for which a resistance monitoring program might be appropriate and feasible include the insects *Heliothis* sp., *Spodoptera* sp., boll weevil, Colorado potato beetle, and aphids; mites; the fungal plant pathogens *Penicillium* sp., *Cercospora* sp., *Botrytis*, *Monilinia*; downy mildews; and certain other pest groups, including selected grass weeds, rodents, etc.

Monitoring technologies must be developed to evaluate management strategies, validate tactics (Chapter 5), accurately determine critical frequencies for pests under different conditions (i.e., crop, climate, economics), and guide implementation of optimum tactics. At present, some theoretical concepts that have been inadequately tested in the field are being advocated for use in resistance management planning. This practice can be dangerous and emphasizes the need to address deficiencies in knowledge through comprehensive research efforts of applied biologists, population biologists, toxicologists, and modelers.

Efforts should be made to identify and exploit more systematically the expertise of industry, academia, and public-sector agencies for conducting research and monitoring pesticide resistance. Both the extension service and industry have access to data on geographical extent and degree of resistance development in particular regions. A critical issue that will always need attention is confirming the validity of resistance reports. Industry can assist in eliminating false reports of resistance by rapidly sharing any data that suggest a change in resistance in a given pest population. A major commitment on the part of pesticide companies to resistance detection and monitoring and to the communication of their findings will be extremely helpful to any public information and recommendation system. The committee commends those companies that have already demonstrated both a willingness and commitment to these goals.

RECOMMENDATION 2. **Working groups involving both private and public sectors should continuously identify the priority of pests for resistance monitoring, based on estimates of economic, environmental, and social costs and benefits. Such working groups should be convened by state agricultural experiment stations, working in conjunction with extension, industry, and university scientists. The involvement and input of grower groups should also be encouraged.**

RESISTANCE-RISK ASSESSMENT

Resistance-risk assessment is carried out intuitively by a wide variety of personnel associated with pesticide discovery, development, or use. Relatively few, however, have attempted to present more formal or structured methods for organizing or implementing assessment systems. Exceptions exist including the WHO program for health-related pest insects (Chapter 6), house flies in Danish farms (Keiding, this volume), and with certain highly specific fungicides applied for plant disease control (Staub and Sozzi, 1984).

RECOMMENDATION 3. **Research methods and data bases needed to carry out resistance risk assessments need to be developed more fully and systematically. Components such as historical data bases, detection and monitoring data, resistance models, laboratory selection tests for resistance, and use data could be incorporated into overall systems that can be used to aid in risk-assessment decisions with a higher degree of benefit.**

IMPLEMENTATION OF RESISTANCE-RISK ASSESSMENT

The results of resistance-risk assessments should serve as aids to decision-makers and should not be considered conclusive forecasts of the outcome of a resistance episode. The designers of resistance-risk assessment programs must ensure that the results of these programs are balanced scientifically and consider species and local differences.

Greater communication is needed among all personnel associated with the development, use, regulation, and research on pesticides and pesticide resistance. Information systems to monitor resistance currently are maintained by a variety of international, national, and local institutions (e.g., WHO, FAO, USDA, EPA, U.S. Department of Defense, university laboratories, mosquito control districts, pest-management areas). Additional data bases will certainly be developed in the future. There is need to coordinate and share information from these systems to the entire pesticide user community to be used in resistance-risk assessment.

RECOMMENDATION 4. **Appropriate international, federal, state, and local agencies should establish and maintain data bases both to support monitoring and detection systems and to serve as a repository and clearing house for data on monitoring resistance. The data bases should contain information on pest species, chemical-use profiles, local conditions, resistance mechanisms, levels of resistance, test methods, and cross-resistances. Studies are needed on ways to coordinate the diverse resistance data base activities better among these groups and institutions.**

Both public agencies and pesticide companies should play an expanded role in financing activities to monitor resistance and ultimately resistance-risk assessment. Industry should concentrate on supporting research and monitoring related to its individual products, while publicly funded institutions should emphasize activities such as basic research on monitoring methods and disseminating monitoring information on resistance. Moreover, it is critical for the activities and investments of the public and private sectors to be coordinated more systematically and integrated so that the best possible informational data base emerges from a given level of combined resources.

Results of resistance-risk assessment programs should be available to the entire pesticide development/user community for evaluation, confirmation, and improvement over time.

RECOMMENDATION 5. **Programs should be developed to help decision-makers use information from resistance-risk assessment in pesticide related activities such as pesticide design, regulatory programs, use directions, and resistance management. Methods and means are needed to share results of resistance-risk assessment programs among all users involved in pesticide production, regulation, and use.**

REFERENCES

Devonshire, A. L., and G. D. Moores. 1984. Immunoassay of carboxylesterase activity for identifying insecticide-resistant *Myzus persicae*. Pestic. Biochem. Physiol. 18:235–239.

FAO (Food and Agriculture Organization). 1982. Recommended methods for the detection and measurement of resistance of agricultural pests to pesticides. Plant Protection Bull. 30:36–71 and 141–143.

Georgopoulos, S. G. 1982. Detection and measurement of fungicide resistance. Pp. 24–31 *in* Fungicide Resistance in Crop Protection, J. Dekker and S. G. Georgopoulos, eds. Wageningen, Netherlands: Centre for Agricultural Publishing and Documentation.

Miyata, T. 1983. Detection and monitoring methods for resistance in arthropods. Pp. 99–116 *in* Pest Resistance to Pesticides, G. P. Georghiou and T. Saito, eds. New York: Plenum.

Staub, T., and D. Sozzi. 1984. Fungicide resistance: A continuing challenge. Plant Dis. 68:1026–1031.

WORKSHOP PARTICIPANTS

Detection, Monitoring, and Risk Assessment

BRIAN A. CROFT (*Leader*), Oregon State University
KEITH J. BRENT, Long Ashton Research Station
WILLIAM BROGDON, Centers for Disease Control
THOMAS M. BROWN, Clemson University
C. F. CURTIS, London School of Hygiene and Tropical Medicine
WILLIAM FRY, Cornell University
MARJORIE A. HOY, University of California, Berkeley

ALAN JONES, Michigan State University
JOHANNES KEIDING, Danish Pest Infestation Laboratory
JOSEPH M. OGAWA, University of California, Davis
STEVEN RADOSEVICH, Oregon State University
CHARLES STAETZ, FMC Corp.
T. STAUB, Ciba-Geigy, Ltd., Switzerland
ROBERT TONN, World Health Organization, Switzerland
MARK WHALON, Michigan State University

Pesticide Resistance: Strategies and Tactics for Management.
1986. National Academy Press, Washington, D.C.

Prediction or Resistance Risk Assessment

JOHANNES KEIDING

Resistance risk, or the potential for development of field resistance to pesticides, depends on genetic and biological factors characteristic of the pest species and the local population and of operational factors, that is, the way pest control is carried out and the history of pesticide use. Thus, for resistance risk assessment (RRA) these factors must be considered and investigated. As an example the RRA in house fly populations on Danish farms from 1948 to 1983 is discussed. Farmers, pesticide producers, and scientists closely cooperated in this work. As a result many new insecticides and types of applications have been rejected owing to high resistance risk, while others have been recommended. Reference is made to RRA for insecticides and acaricides in selected national and international programs to control important veterinary and agricultural pests. For RRA in insecticides the following general points are discussed: (1) the use of laboratory versus field selection, (2) geographical differences, and (3) the fitness of resistant genotypes and phenotypes. RRA for fungicides, herbicides, rodenticides, and veterinary nematicides is discussed briefly. The paper concludes with lists of elements of RRA and research needs and discussions of the organization, interpretation, and use of RRA.

INTRODUCTION

Before a new pesticide is introduced for wide-scale field use, it is important to estimate the potential for significant ''field resistance'' (Davies, 1984) in the pests for which it is intended. Resistance risk assessment (RRA) concerns

TABLE 1 Genetic, Biological, and Operational Factors Influencing Resistance Risk

General Factor	Specific Factors
Genetic	Existence of genetic resistance characters (R-genes, R-alleles)
	Frequency of occurrence of resistance characters
	Number of genes needed to cause resistance
	Interaction of genes
	Dominance of genes
	Penetrance of genes
	Past selection by other chemicals
	Fitness of the R-geno- and phenotypes in the presence or absence of the insecticide
Biological	Reproduction (generations, offspring, etc.)
	Climatic and other ecological conditions
	Behavior
	Isolation, migration, and refugia
Operational	History of insecticide applications
	Persistence of insecticide
	Method of insecticide application (frequency, coverage, life stage(s) exposed, residual effect, etc.)

SOURCE: Modified from Georghiou and Taylor (1977).

the present occurrence of resistance and its potential development, including the rate and extent of development. An RRA should refer to a specific pest species, geographical area, ecological situation, history of pesticide use, and type of formulation/application. Estimating the potential for developing resistance to a pesticide can be very difficult, yet an assessment can make the introduction and use of new pesticides more intelligent and thus avoid big problems. In this paper I will discuss (1) how to estimate resistance risk: methods, factors, conditions, difficulties, and research needs; (2) how to organize and coordinate the investigations; and (3) how to interpret and use the results. As I am most familiar with resistance to insecticides, I shall start by discussing RRA for chemical control of insects, ticks, and mites and then deal with special problems concerning other pesticides, fungicides, herbicides, rodenticides, and compounds to control parasitic nematodes.

INSECTICIDE AND ACARICIDE RESISTANCE

Resistance risk depends on genetic, biological, and operational factors, and these must be included in any resistance risk assessment. As shown in Table 1 a resistance risk cannot be assigned to a given insecticide or a given pest species—it must relate to the local pest population, with its characteristics and conditions, and the way the insecticide is applied. (For a more detailed discussion of factors influencing the development of resistance, see Georghiou and Taylor, 1977 and Georghiou, this volume.)

HOUSE FLY RESISTANCE

The Danish Experience

As an example of how to estimate resistance risk in practice, I shall briefly describe the work of the Danish Pest Infestation Laboratory (DAPIL) on house fly resistance to insecticides in Denmark and elsewhere (Keiding, 1977). In Denmark the house fly, *Musca domestica*, is primarily a pest on farms with pigs and calves, and in recent years poultry. Chemical fly control is carried out in animal houses using residual sprays, space sprays, and spot treatments with impregnated strips or bait paints or with larvicides. Development of resistance has been favored by (1) the organized and extensive use of insecticides and (2) the relatively low migration of flies between the farms.

Since 1945 DAPIL[1] has received good cooperation from the farmers' associations and many farms, the pesticide industry, and the research laboratories overseas doing basic research on insecticide resistance in our and other house fly strains. The cooperation with the farmers gave DAPIL the essential current information on the effect of various insecticides, formulations, and applications that enabled us to follow the development of resistance and to detect and study early cases. Such cooperation is also necessary for the organization of field trials. The use of insecticides for fly control and the development of resistance from 1945 to 1983 are shown in Figure 1.

The main elements we found to be important in conducting our RRA were as follows:

Surveillance of Resistance Occurrence

- Obtain information, complaints, inquiries, and so forth, from farmers, pest control operators, and others
- Determine resistance by standard methods in the laboratory and the field
- Conduct systematic surveys to determine the distribution and level of various types of resistance in the state

Research on and Surveillance of Cross-resistance and Type of Resistance

- Conduct cross-resistance tests
- Determine resistance mechanisms and their diagnoses (e.g., by use of a synergist)

---

[1]DAPIL combines an advisory service, evaluation of new insectides, formulations and applications and research and development on pest control, biology, and resistance.

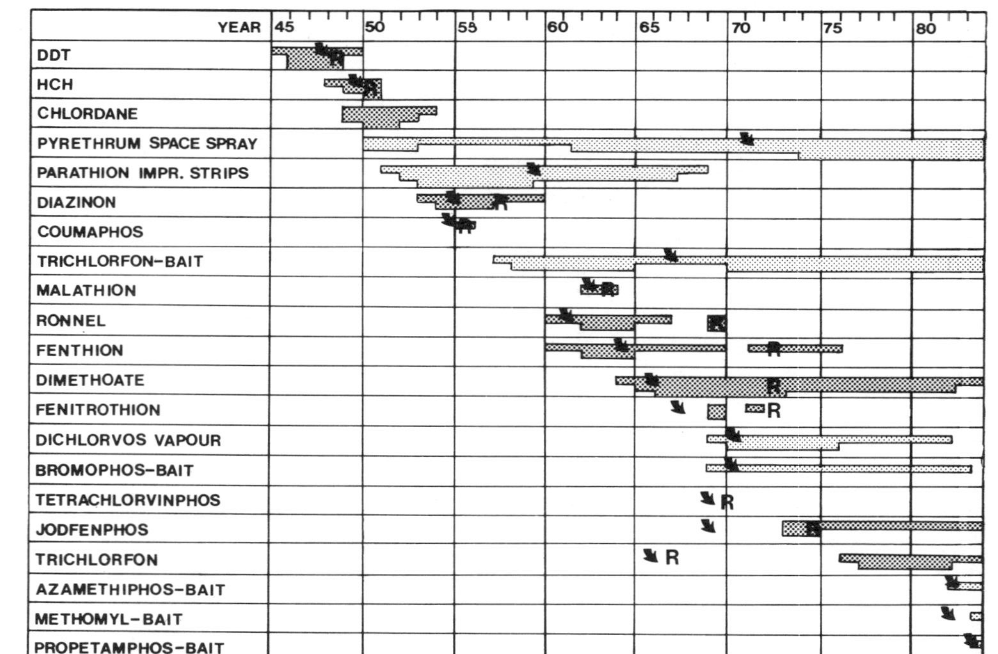

FIGURE 1 Countrywide use of insecticides for house fly control on Danish farms 1945–1983 and development of resistance. Treatments: The insecticides were used as residual sprays except where other applications (impregnated strips, space sprays, paint-on baits, or vapor generators) are indicated. The width of each band indicates the extent to which the insecticide concerned was used, from relatively few, many, to the majority of Danish farms. Occurrence of resistance: Arrows indicate the first confirmed case(s) of resistance of practical importance, and **R** indicates that resistance causing control failures occurs on most farms.

- Determine the genetics of the resistance (genes involved, dominance, fitness of genotypes)
- Survey resistance types and frequency of phenotypes
- Establish a collection of strains representing the important resistant types and their combinations

Studies on the Dynamics of Resistance Development: Operational and Ecological Factors

- Conduct studies under field conditions, rather than laboratory experiments
- Follow development of resistance through small-scale field trials
- Monitor, over several years, the development of resistance and cross-resistance to widespread use of insecticides, formulations, application, effect of alternating treatments, and the like
- Collect information on the time of development and the stability of resistance
- Study the basic population dynamics and behavior of the pest under field conditions and under different ecological conditions

The cooperation between the pesticide industry and DAPIL on the house fly problem has played an important role in the possibility of assessing the resistance risk of new compounds and of using this assessment to: (1) conduct cross-resistance tests using a suitable range of our collection of resistant strains; (2) monitor field populations for resistance to the new compound; (3) conduct small-scale field trials with the new compound, possibly in two or more formulations/applications, to see if resistance may develop rapidly; (4) use the information from (1), (2), and (3) to decide whether and how to introduce the new compound for fly control and how to use it; (5) follow the development of resistance to the new compound when it is widely used and adjust the yearly recommendations for fly control accordingly; and (6) make available to industry our general and specific knowledge of the resistance situation and the factors involved, for example, by annual reports. The cooperation with scientists in other countries resulted in much useful, timely information on mechanisms and genetics of resistance (Keiding, 1977; Sawicki and Keiding, 1981), which could be used for our RRA.

First, DAPIL used the RRA to explain to, convince, or persuade companies that certain insecticides or applications with high resistance and cross-resistance risks should not be introduced, or that it might be advantageous to make available an insecticide application with a low resistance risk. For example, in 1948 DAPIL found that high DDT resistance extended to available DDT analogues—these were not introduced. In the mid-1950s DAPIL persuaded industry not to sell any organochlorine insecticides for fly control. Owing to rapid development of resistance in small-scale field trials, the following insecticides were not introduced for fly control on Danish farms:

the organophosphorus compounds coumaphos (1955), coumithoate (1957), formothion (1965), phosmet (1968), tetrachlorvinphos (1969), and azamethiphos residual spray (1981); the carbamate mobam (1967).

In other cases DAPIL found that high resistance to new compounds was already present, due to cross-resistance. This happened for most OP compounds and carbamates in the 1970s, when dimethoate had been commonly used for five to seven years and high resistance had become widespread. Researchers in England (Sawicki, 1974, 1975; Sawicki and Keiding, 1981) studied the resistance (R) mechanisms, their genetics, and interaction of this resistance and showed that insensitive target (cholinesterase) and several detoxification processes were involved. This research explained the cross-resistance and demonstrated the importance of the sequence of use of insecticides (Sawicki, 1975; Keiding, 1977; Sawicki and Keiding, 1981).

The most striking example of RRA was that of pyrethroid resistance. Investigations from 1970 to 1973 showed that house flies on Danish farms had a common potential for developing high resistance to pyrethroids when the selection pressure with pyrethroids was strong, for example, by frequent use of pyrethroid aerosols (Keiding, 1976). If aerosols were used less frequently, however, once a week or less, allowing some unexposed flies to reproduce, the resistance might remain low and the aerosols would remain effective. Knowing that treatments with residual sprays give a strong selection pressure, DAPIL advised the companies and the authorities not to introduce residual sprays with pyrethroids for fly control on farms. The advice was followed, even though there was no proper legal basis for banning residual pyrethroids for fly control until 1980.[2] In the meantime DAPIL received further support for this decision.

In 1977 and 1978 DAPIL found that heterogeneous resistance to candidate residual pyrethroids was widespread on Danish farms, and the resistance factor *kdr*, which causes resistance to DDT and pyrethroids (in connection with other factors), occurred in practically all fly populations investigated (Keiding, 1978, 1979, 1980; Keiding and Skovmand, 1984). The predicted rapid development of general pyrethroid resistance when residual pyrethroids were used was confirmed in Switzerland (Keiding, 1980), in Germany (Skovmand and Keiding, 1980; Künast, 1979), and England (Chapman and Lloyd, 1981). In Denmark we continue to avoid the residual pyrethroids for fly control. The aerosols with pyrethrum, and the like, are still effective, and pyrethroid resistance is low or moderate.

---

[2]The Danish "Act on Chemical Compounds and Products," passed in 1980, empowers the Danish Ministry of Environment to require, before registration, experimental data on cross-resistance and the potential for developing resistance. If the data indicate that resistance will quickly make the product ineffective and/or its use will result in resistance to useful products, the registration may be refused (Sawicki, 1981). Registrations also may be withdrawn if general development of resistance is found after a period of use.

The experience with fly control illustrates another principle, already mentioned in this volume (see section on Genetics, Biochemical, and Physiological Mechanisms of Resistance to Pesticides), that development of resistance depends on the type of formulation and application used, owing to difference in selection pressure. Thus, for the flies, bait applications promote less resistance than residual sprays, and knock-down sprays less than residual pyrethroids.

Several results from DAPIL's lengthy studies have been positive, resulting in recommendations of formulations. Three OP compounds are registered as bait formulations, but not as residual sprays, because of the resistance risk. OP compounds were effective on flies resistant to organochlorines in the early 1950s, and various compounds and applications were recommended (Figure 1). Fenthion, and especially dimethoate, were effective and were recommended when other OPs failed. Tests with a variety of resistant fly strains, including high multiresistance, showed susceptibility to the development inhibitors diflubenzuron and cyromazine, used as larvicides, without significant resistance development after selection pressure (Keiding and El-Khodary, 1983). In addition the extensive data collected on the development of resistance in house fly populations on farms since 1948 are being put into a data base, which should provide greater possibilities for analyzing resistance risks under various conditions (Keiding et al., 1983).

Resistance in Other Regions

Sequential development of resistance in field populations of house flies also has been studied in Czechoslovakia (Rupes et al., 1983), California (Georghiou and Hawley, 1971), and Japan (Yasutomi and Shudo, 1978) and has been used as a guide for choosing new insecticides. In addition house fly samples from many parts of the world have been tested for resistance by Keiding, Hayashi, Kano, and others (Keiding, 1977; Taylor 1982). These surveys have provided information on the global occurrence of various types of resistance and the resistance risks. An important finding was that DDT resistance occurs everywhere, but only in some areas in northern Europe is the *kdr* factor for DDT resistance common (Keiding, 1977; Keiding and Skovmand, 1984). Since *kdr* is also an important factor for pyrethroid resistance, the risk for development of high pyrethroid resistance is still lower in all the areas where *kdr* is rare or absent.

China recently surveyed for resistance in more than 400 field samples of house flies. As no sign of pyrethroid-R or the *kdr* factor was found, China recommended the use of residual pyrethroids for fly control (Gao Jin-ya, Institute of Zoology, Acad. Sinika, Beijing, personal communication, 1983). In Japan where *kdr* is rare, high pyrethroid resistance was not found until after six years of fly control with a residual pyrethroid (Motoyama, 1984);

in areas of Europe where *kdr* is common, high pyrethroid resistance developed in a few months.

Examples from Other Insect and Mite Species

National Programs The following are examples of some of the many systematic, long-term programs on development, types, and risk of resistance. The development of multiple resistance in the cattle tick *Boophilus microplus* in Australia has been investigated for about 30 years by the Commonwealth Scientific and Industrial Research Organization (CSIRO) tick laboratory in Queensland (Nolan and Roulston, 1979; Roulston et al., 1981). This work includes all the factors of importance to RRA: (1) close cooperation with farmers to obtain early detection of resistance, (2) investigation of resistance mechanisms and their genetics to define resistant types and cross-resistance spectra, (3) surveys of the distribution of resistant types, (4) studies and modeling of the dynamics of resistance development, (5) cooperation with industry to test new acaricides against tick strains representing the resistant strains, (6) field trials with promising acaricides and types of application, and (7) advice to farmers on control methods. Investigations of resistance in the sheep blow fly, *Lucilia cuprina*, in Australia, begun about 25 years ago, contain the same elements as mentioned for the cattle tick, including surveys of resistance gene frequency and fitness in field populations (Hughes, 1981, 1982, 1983; Hughes and Devonshire, 1982; McKenzie et al., 1980; Whitten and McKenzie, 1982).

Among agricultural pests are the following examples. Comprehensive investigations were begun more than 20 years ago on leaf- and planthoppers attacking rice in Japan. These include extensive resistance surveys, studies of resistance mechanisms and genetics, and trials of many new insecticides, especially the effect of using mixtures or alternating treatments (Saito and Miyata, 1982; Hama, 1975, 1980). Surveys, resistance mechanisms, and genetic research have been conducted on the aphids *Myzus persicae* in Britain (Sawicki et al., 1978) and *Phorodon humuli* in Czechoslovakia (Hrdý, 1975, 1979; Sula et al., 1981). National RRA programs have been conducted in Egypt on cotton pests, especially the leafworm, *Spodoptera*; in Australia on spider mites (Dittrich, 1979) and *Heliothis* ssp. (Davies, 1984); and in the United States on spider mites and *Heliothis* (Sparks, 1981; Bull, 1981). Spider mites in several countries also have been investigated (Dittrich, 1975).

International Programs The World Health Organization (WHO) has organized global programs for detecting and monitoring resistance in vectors and pests of medical importance, especially vector mosquitoes; WHO also has supported many studies on resistance genetics and resistance types occurring in vectors (WHO, 1980), as well as trials on the dynamics of resistance development (Curtis, 1981). Moreover, WHO has organized a Pesticide Evaluation Scheme including tests of new insecticides from industry with some resistant strains of mosquitoes, flies, and the like. The United Nations' Food and Agriculture Organization (FAO) has organized a global survey of pesticide susceptibility of stored-grain pests (Champ and Dyte, 1976), including some typing of OP resistance.

Laboratory Versus Field Selection

Experience and theoretical considerations have shown that the predictive value of investigating resistance risk through laboratory selection is limited. If resistance develops when an insect population is exposed to selection pressure with an insecticide through a number of generations, the ability of resistance exists, but the level, type, and rate at which it develops may be quite different from what happens under field conditions. If a laboratory selection is negative and no resistance develops, one may conclude very little. There is no guarantee that resistance will not develop in the field (Pal and Brown, 1971). For example, in the 1950s, laboratory strains of house flies were selected at the Riverside Laboratory in California for 19 to 149 generations with various OP compounds; only a slow and moderate increase of tolerance was obtained, compared to what later developed in the field (Pal and Brown, 1971).

There may be several reasons for the differences between laboratory and field selection: for example, (1) because of the smaller gene pool in laboratory selection, rare resistance genes and ancillary genes may be missing; (2) a difference in insecticide pressure often results in lower mortality in the laboratory than in the field; laboratory selection may exploit polygenic variation while field selection tends to act on alleles of single resistance genes (Whitten and McKenzie, 1982); (3) a difference in the fitness of resistance genotypes; and (4) a difference in natural selection. Therefore, if laboratory selection is used for RRA: (1) the gene pool should be as big as possible and should be taken from natural populations initially; and (2) insecticide pressure, ecological conditions, and natural selection should simulate that occurring in the field. Small-scale control trials on isolated or semi-isolated field populations with monitoring of resistance often may be better than laboratory trials, provided such field selection is feasible, for example, on farms with house flies in Denmark, pests in greenhouses (Helle and van de Vrie, 1974), isolated fields, and so forth. If small-scale field trials on resistance are not feasible, the first practical applications must be monitored for resistance development. This activity should be organized in collaboration with farmers, state institutions, research laboratories, and industry, and the results should be made available to all interested parties so that the first experiences can be used for RRA in other areas.

Geographical Differences

The biological and operational factors influencing the development of resistance in a pest species may vary greatly depending on climate, farming practice, use of insecticides, and the like, and the resistance risk will vary accordingly. The genetic factors also may differ, not only in frequency of resistance genes but also which genes and mechanisms cause resistance locally, as has been found for DDT and pyrethroid resistance in house flies. These possible regional and local differences must be considered for any RRA in a given area and for the use of resistant strains to test for cross-resistance of new compounds.

Fitness of Resistant Geno- and Phenotypes

The relative fitness of the resistant geno- and phenotypes under field conditions may be difficult to estimate, but the stability or reversion of resistance in the field when the insecticide pressure is relaxed is important. Estimating fitness under laboratory conditions has a limited value (Keiding, 1967), not only because conditions differ from the field but because strains with different periods of adaptation to laboratory conditions may be compared. Relatively little is known about the importance of the fitness factor for insecticide resistance. Fitness of resistant types, however, is not constant. With time the resistance genome may be integrated with fitness factors by natural selection, a process called coadaptation (Keiding, 1967).

Mathematical Models

A number of simulation models (Taylor, 1983; Section III in this proceedings) have contributed significantly to our general understanding of resistance dynamics and are being used for developing strategies to reduce the development of resistance. Their usefulness, however, depends on whether the assumptions and the parameters are realistic. For example, in RRA we need information about factors such as local frequency and number of resistance genes, fitness factors, selection pressure, population dynamics, and migration. Such information must be gathered in the field, and assumptions must be tested in the field where possible (Davies, 1984; Denholm, 1981).

RESISTANCE OF PLANT PATHOGENS

Resistance risk assessment in fungicides has been well discussed in several recent reviews (Dekker, 1981, 1982a,b; Wade, 1982; Staub and Sossi, 1983).

TABLE 2 Specific and Systemic Action of Some Fungicides to Emergence of Fungicide Resistance

Fungicide or Fungicide Group	Mode of Action		Occurrence of Resistant Strains		Risks[a] for Failure of Disease Control
	Specific	Systemic	In Vitro	On Plants	
Copper compounds	–	–	–	–	very low
Dithiocarbamates	–	–	–	–	very low
Chlorothalonil	–		–	–	very low
Phthalimides	–	–	–	–	very low
Organic Hg compounds	–	–	+	+	low
Aromatic hydrocarbons	+	–[b]	+	+	high
sec Butylamine	+	–	+	+	high
Dicarboximides	+	–[b]	+	+	moderate to high
Dodine	+	–	+	+	moderate
Organic tin compounds	+	–	+	+	moderate
Acylalanines	+	+	+	+	high
Benzimidazoles	+	+	+	+	high
Dimethirimol	+	+	0[c]	+	high
Ethirimol	+	+	0[c]	+	moderate
Organic P compounds	+	+	+	+	moderate
Carboxanilides	+	+	+	+	moderate to low
Fenarimol, nuarimol	+	+	+	–	low
Imidazoles	+	+	+	–	low
Morpholines	+	+	+	–[d]	low
Triazoles	+	+	+	–[d]	low
Triforine	+	+	+	–	very low

+ with the property; – without the property.

[a]The risk for failure of disease control is a rough estimation, since it also depends on other factors (type of disease, strategy of fungicide application, etc.).

[b]Chloroneb and procymidone have systemic properties.

[c]Concerns obligate parasites.

[d]Occurrence of strains with decreased sensitivity to some of these compounds has been reported.

SOURCE: Dekker (1981).

As with insecticide resistance the RRA is influenced by inherent genetic and biological factors in the pest fungus, including reproduction rate, spore mobility, and host range. Moreover, climate and weather play a role, and the operational factors determining the selection pressure (i.e., area treated, coverage and frequency of treatments, duration of exposure, and persistence of the fungicide) are highly important for the development of field resistance. More specifically than in insecticides, resistance risk in fungicides is connected with the biochemical mode of action of the fungicide. The resistance risk, therefore, can be classified according to type of fungicide (Table 2).

Within a certain mode of action, for example, the benzimidazoles, a high degree of cross-resistance is found. Fungi, unlike insects, produce so many spores that resistant mutants can be detected even at a very low frequency. Therefore, the genetic ability for resistance can be demonstrated easily in the laboratory for most pathogens that can be grown on artificial media. Thus, a standard method for RRA in fungicides is to grow spores of pathogens on a medium containing an amount of fungicide just above the minimum inhibitory concentration in which only resistant cells survive.

Using laboratory tests resistant mutants have been found for all the specific-site fungicides (Table 2), but resistance may not be a problem in the field. Fitness of the resistant mutants in fungal pathogens is generally a decisive factor for development of field resistance. Therefore, assessments of fitness are important for RRA. These assessments may be done (1) by determining the relative growth of resistant and wild-type strains in vitro, (2) by testing the pathogenicity of strains on plants in the greenhouse, and (3) by infecting plants with a sensitive and a resistant strain and observing the result of competition in the absence of the fungicide over a number of pathogen generations. As with insects, however, laboratory and greenhouse tests may not realistically estimate fitness under field conditions. Therefore, field trials may be necessary for the full answer (Dekker, 1982a).

Although laboratory and greenhouse tests can provide much information on resistance risk, negative results cannot exclude the possibility of resistance developing in the field if the selection pressure in area and time is large enough. Moreover, the rate and extent of development of resistance depends mainly on biological, environmental, and operational field factors, as previously mentioned. Field experiments and monitoring of resistance in pathogens in areas subjected to various schemes of fungicide treatments are therefore essential for RRA of fungicides as well as of insecticides. Cooperation and rapid exchange of information between producers and users of fungicides, advisers, and research and regulatory institutes are necessary to cope with the rapidly developing problems of fungicide resistance. The international association of agrochemical industry associations (GIFAP—Groupement International des Associations Nationals de Fabricants de Produits Agronomiques) recognized this need in 1981 when it formed the Fungicide Resistance Action Committee (FRAC). The equivalent for insecticide resistance, the Insecticide Resistance Action Committee (IRAC), was formed in 1984.

HERBICIDE RESISTANCE

Assessing resistance risk to herbicides is simplified because resistance is confined mainly to the s-triazine herbicides, usually with general cross-resistance to all s-triazines and related degrees of resistance or tolerance to the asymmetrical triazinones, ureas, and many other nitrogen-containing photosynthetic inhibitors, but as a rule no cross-resistance or negative cross-resistance to herbicides with other modes of action (LeBaron and Gressel, 1982). A careful monitoring and verification of resistance to s-triazines in the field is important for RRA. As for most other pests the rate of resistance development is influenced by the selection pressure, which is a product of the persistence of the herbicide effect after treatment, the dose, the number of years the herbicide has been used alone in an area, and the proportion of the weed population that is exposed. S-triazines have a very specific action and a high persistence. Resistance problems may be expected in new herbicides having these characteristics. Using selection experiments for RRA, however, is difficult because of the time required for sufficient generations to be exposed and because the experiments have to be done in field areas of a sufficient size. (The relation between weed ecology and resistance risk is discussed by Slife in this volume.) The fitness of resistant strains does not seem to be of great importance for the development of herbicide resistance.

RODENTICIDE RESISTANCE

Rodenticide resistance is mainly a problem with one group of rodenticides, the anticoagulants. For practical reasons it is difficult to investigate resistance risk by meaningful selection experiments in the laboratory or in other confined colonies of rats, mice, and other rodents. The best method for RRA is a systematic monitoring of control failures and rodenticide resistance in connection with rodent-control campaigns using a given rodenticide. Good collaboration is therefore essential between the people organizing, conducting, and supervising the control campaigns and a laboratory that can carry out the standard resistance tests on trapped rodents and that can interpret the results. Thus, it is very important to have as complete information as possible on the history of rodenticide use in the area. When resistance has been found a central laboratory should, if possible, keep a colony of each type of resistant strain for use in toxicity tests with new rodenticides to gather information on cross-resistance. Studies on resistance mechanisms and genetics are also important for RRA, as discussed under insecticide resistance. (For more detailed discussions on rodenticide resistance see papers by MacNicoll, Greaves, and Jackson in this volume.)

NEMATODE RESISTANCE

Nematicide resistance of parasitic nematodes in domestic animals has been found and investigated mainly in sheep, but it may also occur in goats and horses (Prichard et al., 1980; Bjørn, 1983). Resistance has developed mostly to the benzimidazole compounds having a general cross-resistance within this group, but no cross-resistance to other types of nematicides. Surveys of nematode resistance are hampered because critical tests to determine the effect of the compound require that a large number of treated host animals be killed. Indications of resistance, however, may be obtained by fecal egg counts after treatments or may be confirmed by in vitro tests on egg hatch for the benzimidazoles.

Laboratory selection is fairly simple; colonies of nematodes are exposed to treated hosts for a number of generations. The conditions for resistance development, however, are different in the field, for example, as to natural selection and selection pressure by the nematicide. In the field a high proportion of the nematode population may be unexposed, since it is outside the host (Le Jambre, 1978).

CONCLUSION

Elements of RRA

The important elements of RRA as discussed above are listed in Table 3 (the succession of the elements are not necessarily chronological nor in order of importance).

Any RRA program must establish good coordination, collaboration, and exchange of information between (1) the producers (the agrochemical industry), (2) the advisers and organizers of pesticide use, (3) the users of pesticides, and (4) the research institutes. An RRA program may be organized by an international body, for example, FAO or WHO, or it may be a national or state institution. International collaboration and rapid exchange of information are essential, however, by informal reports, correspondence, conferences, and visits. The traveling pesticide experts from industry may play a special role for rapid information dissemination to national institutions that may serve as a link between users, scientists, and industry. In this way resistance problems may be realized early, such that suitable monitoring and research can be organized, for example, supported by industry. Examples of such collaboration are the work on the cattle tick in Australia, the house fly in Denmark, and rice pests in Japan. Other examples and a discussion of the interagency cooperation are given by Davies (1984).

The WHO and the FAO have organized data bases on the occurrence of pesticide resistance (Georghiou and Mellon, 1983). The results of unpublished investigations, including those in industry, would be useful. One means of providing such information about new findings would be a newsletter on pesticide resistance; WHO had one for several years, but it was discontinued in 1976.

Interpretation and Use of the Assessments

Two types of interpretation can come from these assessments: scientific-technical interpretation and economic interpretation. For example, a scien-

TABLE 3 Elements of Resistance Risk Assessment

A. Consider the pesticide: mode of action, chemistry, and stability.

B. Evaluate the pest species: genetic diversity, resistance potential.
1. Conduct laboratory selection experiments.[a]
2. Conduct field selection experiments.[b]
3. Survey for the occurrence and development of resistance in field populations of the pest to pesticide use.[c]
4. Determine the cross-resistance spectrum.[d]
5. Determine the resistant type (mechanism, genetics).[e]
6. Determine the fitness of resistant biotypes.[f]
7. Monitor for local and regional distribution of resistant types.[g]
8. Investigate factors influencing the development of resistance: genetic, biological, and operational.[h]
9. Develop mathematical models on the dynamics of resistance development.[h]
10. Conduct computer simulations of resistance development.[i]
11. Check and improve simulation models by field experiments.
12. Investigate the effect of sequential use of pesticides for resistance development.

[a]These experiments have a limited predictive value owing to restricted gene pool, difference of conditions, exposure to pesticides, natural selection, and so forth, and in some pests (e.g., weeds and rodents) they are difficult to perform.

[b]These experiments, especially in isolated or semi-isolated localities, may be more informative, but also more difficult to arrange. The risk of spreading resistant strains is a limitation.

[c]Surveying is very important and should be a regular activity for applications of new pesticides. Information on the history of pesticide use influencing the previous selection of resistance factors is essential (see item 12). If resistance has reverted in a field population, it usually develops quickly when the pesticide is reintroduced.

[d]Determine cross-resistance when resistance to a pesticide is detected. Patterns of cross-resistance are often known or should be investigated.

[e]This activity is important for predicting and understanding cross-resistance, including the components of resistance and their genetics. It is also important to know whether resistance depends on one or more resistance factors.

[f]Fitness of resistant biotypes under field conditions is of general importance for resistance development, particularly to fungicides.

[g]Such occurrence may vary locally and regionally.

[h]Knowledge of the dynamics of resistance development and of the parameters in the field is essential for constructing realistic models and for predicting the rate and extent of resistance development.

[i]Computer simulations are important to evaluate the effects of various genetic, biological, and operational factors and to develop strategies for delaying or avoiding resistance.

SOURCE: Keiding (unpublished).

tific-technical interpretation may estimate the probability of resistance developing in a pest in an area, the rate and extent of resistance, and the factors influencing it, while an economic interpretation would estimate its economic consequences. Use of the assessments are valuable for regulatory authorities and industry in reaching agreement on formulations and recommended applications for the pesticide. If industry is more interested in getting a quick profit or recouping investments, however, which could lead to applications and recommendations that conflict with the long-term interest of the users and perhaps of the company, the regulatory authorities and their advisers may want to regulate the use of the pesticide to comply with the strategy of pest control recommended and the hazards of rapid development of resistance.

Additionally, the assessments may find that resistance found in the laboratory may not apply to the field; resistance found in one area may not occur or develop in the same way in another; and certain pesticides may be useful even if some resistance has developed because they are so much cheaper than the substitutes, as is true with DDT for controlling some malaria vectors.

Research Needs

The research needed to improve the ability to assess resistance risk may be related to the elements of RRA. The following is a brief list of some general research fields for RRA, with reference to the ''element numbers'' in Table 3. The need and importance of the research may vary between the groups of pests and pesticides.

- Develop and improve methods for detecting and monitoring types of resistance, especially at low frequencies (see Brent in this volume) (3,7)
- Research resistance mechanisms, cross-resistance (4,5)
- Study the genetics of resistance (5, 6, 8)
- Determine the fitness of resistant biotypes (6, 8)
- Conduct field investigations of the biology, ecology, and population dynamics of the pest (8, 9, 10, 11)
- Conduct field investigations on selection pressure by various applications of pesticides and control schemes (8, 9, 10, 11)
- Develop and use more realistic models on the dynamics of resistance development (9, 10, 11)
- Investigate the effect of sequential use of pesticides for resistance development (8, 12)

REFERENCES

Bjørn, H. 1983. On aspects of anthelmintic resistance of parasitic nematodes in domestic animals. A review. Pp. 1–116 *in* Report from Institute of Hygiene and Microbiology. Copenhagen: Royal Veterinary and Agriculture University.

Bull, D. L. 1981. Factors that influence tobacco budworm resistance to organo-phosphorous insecticides. Bull. Entomol. Soc. Am. 27:193–197.

Champ, B. R., and C. E. Dyte. 1976. Pp. 1–297 *in* Pesticide Susceptibility of Stored Grain Pests. Rome: FAO.

Chapman, P. A., and C. J. Lloyd. 1981. The spread of resistance among houseflies from farms in the United Kingdom. Pp. 625–631 *in* Proc. Br. Crop Prot. Conf. Lavenham, Suffolk: Lavenham, 1981.

Curtis, C. F. 1981. Possible methods of inhibiting or reversing the evolution of insecticide resistance in mosquitoes. Pestic. Sci. 12:557–564.

Davies, R. A. H. 1984. Insecticide resistance: an industry viewpoint. Pp. 593–600 *in* Proc. Br. Crop Prot. Conf. Lavenham, Suffolk: Lavenham, 1984.

Dekker, J. 1981. Impact of fungicide resistance on disease control. Pp. 857–863 *in* Proc. Br. Crop Prot. Conf. Lavenham, Suffolk: Lavenham, 1981.

Dekker, J. 1982a. Can we estimate the fungicide resistance hazard in the field from laboratory and greenhouse tests? Pp. 128–138 *in* Fungicide Resistance in Crop Protection, J. Dekker and S. G. Georgopoulos, eds. Wageningen, Netherlands: Centre for Agricultural Publishing and Documentation.

Dekker, J. 1982b. Countermeasures for avoiding fungicide resistance. Pp. 177–186 *in* Fungicide Resistance in Crop Protection, J. Dekker and S. G. Georgopoulos, eds. Wageningen, Netherlands: Centre for Agricultural Publishing and Documentation.

Denholm, I. 1981. Present trends and future needs in modeling for the management of insecticide resistance. Pp. 847–855 *in* Proc. Br. Crop Prot. Conf., Vol. 3. Lavenham, Suffolk: Lavenham, 1981.

Dittrich, V. 1975. Acaracide resistance in mites. Z. Angew. Entomol. 78:28–45.

Dittrich, V. 1979. The role of industry in coping with insecticide resistance. Pp. 249–253 *in* Proc. Symp. 9th Int. Congr. Plant Prot., Vol. 1, T. Kommédahl, ed. Minneapolis, Minn.: Burgess.

Georghiou, G. P., and M. K. Hawley. 1971. Insecticide resistance resulting from sequential selection of house flies in the field by organophosphorus compounds. Bull. W.H.O. 45:43–51.

Georghiou, G. P., and R. B. Mellon. 1983. Pesticide resistance in time and space. Pp. 1–46 *in* Pest Resistance to Pesticides, G. P. Georghiou and T. Saito, eds. New York: Plenum.

Georghiou, G. P., and C. E. Taylor. 1976. Pesticide resistance as an evolutionary phenomenon. Pp. 759–785 *in* Proc. 15th Int. Congr. Entomol., Washington, D.C. College Park, Md.: Entomological Society of America.

Hama, H. 1975. Resistance to insecticides in the green rice leafhopper. Jpn. Pestic. Inf. 23:9–12.

Hama, H. 1980. Mechanism of insecticide resistance in green rice leafhopper and small brown planthopper. Rev. Plant. Prot. Res. (Japan) 13:54–73.

Helle, W., and M. van de Vrie. 1974. Problems with spider mites. Outlook Agric. 8:119–125.

Hrdý, I. 1975. Insecticide resistance in aphids. Pp. 739–749 *in* Proc. Br. Insectic. Fungic. Conf. Lavenham, Suffolk: Lavenham, 1975.

Hrdý, I. 1979. Insecticide resistance in aphids. Pp. 228–231 *in* Proc. Symp. 9th Int. Congr. Plant Prot., Vol. 1, Washington, D.C., Minneapolis: Burgess.

Hughes, P. B. 1981. Spectrum of cross-resistance to insecticides in field samples of the primary sheep blowfly, *Lucilia cuprina*. Int. J. Parasitol. II:475–479.

Hughes, P. B. 1982. Organophosphorus resistance in the sheep blowfly, *Lucilia cuprina* (Wiedemann) (Diptera: Calliphoridae): A genetic study incorporating synergists. Bull. Entomol. Res. 72:573–582.

Hughes, P. B. 1983. Biochemical and genetic studies on resistance to organophosphorus insecticides in *Lucilia cuprina*. Ph.D. dissertation. Macquarie University, Australia.

Hughes, P. B., and A. L. Devonshire. 1982. The biochemical basis of resistance to organophosphorus insecticides in the sheep blowfly, *Lucilia cuprina*. Pestic. Biochem. Physiol. 18:289–297.

Keiding, J. 1967. Persistence of resistant populations after the relaxation of the selection pressure. World Rev. Pest Control 6:115–130.

Keiding, J. 1976. Development of resistance to pyrethroids in field populations of Danish houseflies. Pestic. Sci. 7:283–291.

Keiding, J. 1977. Resistance in the housefly in Denmark and elsewhere. Pp. 261–302 *in* Pesticide Management and Insecticide Resistance, D. A. Watson and A. W. A. Brown, eds. New York: Academic Press.

Keiding, J. 1978. Insecticide resistance in houseflies. Danish Pest Infest. Lab. Annu. Rep. 1977:37–54.

Keiding, J. 1979. Insecticide resistance in houseflies. Danish Pest Infest. Lab. Annu. Rep. 1978:43–55.

Keiding, J. 1980. Insecticide resistance in houseflies. Danish Pest Infest. Lab. Annu. Rep. 1979:40–53.

Keiding, J., and A. El-Khodary. 1983. Investigations of cross-resistance involving laboratory strains (of houseflies). Danish Pest Infest. Lab. Annu. Rep. 1982:58–59.

Keiding, J., and O. Skovmand. 1984. The occurrence of the *kdr*-factor in houseflies and the potential resistance to pyrethroids in Denmark and elsewhere. P. 737 *in* 17th Int. Congr. Entomol., Hamburg, 1984. (Abstr.)

Keiding, J., O. Skovmand, and H. O. H. Nielsen. 1983. Establishing a data base for data on insecticide resistance in houseflies at DPIL 1948–82. Danish Pest Infest. Lab. Annu. Rep. 1982.

Künast, C. 1979. Die Entwicklung der Permethrinresistenz bei der Stubenfliege (*Musca domestica*) im süddeutschen. Z. angew. Zool. 66:385–390.

LeBaron, H. M., and J. Gressel, eds. 1982. Herbicide Resistance in Plants. New York: John Wiley and Sons.

LeJambre, L. F. 1978. Anthelmintic resistance in gastrointestinal nematodes of sheep. Pp. 109–120 *in* The Epidemiology and Control of Gastrointestinal Parasites of Sheep in Australia, A. D. Donald, W. H. Southcott, and J. K. Dineen, eds. Australia: CSIRO.

McKenzie, J. A., J. T. Dearn, and M. J. Whitten. 1980. Genetic basis of resistance to diazinon in Victorian populations of the Australian sheep blowfly, *Lucilia cuprina*. Aust. J. Biol. Sci. 33:85–95.

Motoyama, N. 1984. Pyrethroid resistance in a Japanese colony of the housefly. J. Pestic. Sci. (Japan) 9:523–526.

Nolan, J., and W. J. Roulston. 1979. Acaricide resistance as a factor in the management of Acari of medical and veterinary importance. Pp. 3–13 *in* Recent Advances in Acarology, Vol. II, J. D. Rodriquez, ed. New York: Academic Press.

Pal, R., and A. W. A. Brown. 1971. Insecticide Resistance in Arthropods. Geneva: World Health Organization.

Prichard, R. K., C. A. Hall, J. D. Kelly, I. C. A. Martin, and A. D. Donald. 1980. The problem of anthelmintic resistance in nematodes. Aust. Vet. J. 56:239–251.

Roulston, W. J., R. H. Wharton, J. Nolan, J. D. Kerr, J. T. Wilson, P. G. Thompson, and M. Sehotz. 1981. A survey for resistance in cattle ticks to acaricides. Aust. Vet. J. 57:362–371.

Rupes, V., J. Pinterova, J. Ledvinka, J. Chmeia, J. Placky, M. Homolac, and V. Pospisil. 1983. Insecticide resistance in houseflies, *Musca domestica* in Czechoslovakia 1976–80. Int. Pest Control 25:106–108.

Saito, T., and T. Miyata. 1982. Studies on insecticide resistance in *Nephotettix cincticeps*. Pp. 377–382 *in* Proc. Int. Conf. Plant Prot. in Tropics, K. L. Heong, B. S. Lee, T. M. Lim, C. H. Teoh, and Yusof Ibrahim, eds. Kuala Lumpur: Malaysian Plant Protection Society.

Sawicki, R. M. 1974. Genetics of resistance of a dimethoate-selected strain of houseflies to several insecticides and methylenedioxyphenyl synergists. J. Agric. Food Chem. 22:344–349.

Sawicki, R. M. 1975. Effect of sequential resistance on pesticide management. Pp. 799–811 *in* Proc. 8th Br. Insectic. Fungic. Conf., Lavenham, Suffolk: Lavenham, 1975.

Sawicki, R. M. 1981. Problems in countering resistance. Philos. Trans. R. Soc. London, Ser. B 295:143–151.

Sawicki, R. M., and J. Keiding. 1981. Factors affecting the sequential acquisition by Danish houseflies (*Musca domestica*) of resistance to organophosphorus insecticides. Pestic. Sci. 12:587–591.

Sawicki, R. M., A. L. Devonshire, A. D. Rice, G. D. Moores, S. M. Petzing, and A. Cameron. 1978. The detection and distribution of organophosphorus and carbamate insecticide-resistant *Myzus persicae* (Sulz.) in Britain in 1976. Pest. Sci. 9:189–201.

Skovmand, O., and J. Keiding. 1980. Chemical fly control and resistance on farms in North Germany. Danish Pest Infest. Lab. Annu. Rep. 1979:39–40.

Sparks, T. C. 1981. Development of insecticide resistance in *Heliothis zea* and *Heliothis virescens* in North America. Bull. Entomol. Soc. Am. 27:186–192.

Staub, T., and D. Sossi. 1983. Recent practical experiences with fungicide resistance. Pp. 591–598 *in* 10th Int. Congr. Plant. Prot., Vol. 2. Lavenham, Suffolk: Lavenham, 1983.

Sula, J., J. Kuldová, and I. Hrdý. 1981. Insecticide-resistance spectrum in the hop aphid (*Phorodon humuli*) populations from different regions: Notes on resistance mechanisms. IOBC/WPRS Bull. IV. 3:46–54.

Taylor, C. E. 1983. Evolution of resistance to insecticides: The role of mathematical models and computer simulations. Pp. 163–173 *in* Pest Resistance to Pesticides, G. P. Georghiou and T. Saito, eds. New York: Plenum.

Taylor, R. N. 1982. Insecticide resistance in houseflies from the Middle East and North Africa with notes on the use of various bioassay techniques. Pestic. Sci. 13:415–425.

Wade, M. 1982. Resistance to fungicides. Span. 25:8–10.

Whitten, M. J., and J. A. McKenzie. 1982. The genetic basis for pesticide resistance. Pp. 1–16 *in* Proc. 3rd Australas. Conf. Grassl. Invert. Ecol., K. E. Lee, ed. Adelaide: S. A. Government Printer.

World Health Organization. 1980. Resistance of vectors of disease to pesticides. 5th Rep. WHO Exp. Comm. Vector Biol. Contr. Tech. Rep. Ser. 655.

Yasutomi, K., and C. Shudo. 1978. Insecticide resistance in the houseflies of the third Yumenoshima, a new dumping-island of Tokyo (II). Jpn. J. San. Zool. 29:205–208.

Pesticide Resistance: Strategies and Tactics for Management.
1986. National Academy Press, Washington, D.C.

Detection and Monitoring of Resistant Forms: An Overview

K. J. BRENT

Detection and monitoring are major components of pesticide resistance management, for several reasons. The different steps that should be taken in any detection and monitoring program, as well as examples of successful programs, are described. It is important to monitor for sensitivity and to establish a resistance management strategy early in the life of a new product. The need to distinguish clearly between detecting less-sensitive forms and concluding that practical resistance problems have arisen is also stressed. The most effective programs can be developed and carried out only with the collaboration of private and public organizations.

INTRODUCTION

What precisely is meant by "the detection and monitoring of resistance"? This basic question must be considered at the outset of any discussion on this topic, because much vagueness and misunderstanding exist about the terms involved and their meanings.

"Detection" indicates simply the obtaining of initial evidence for the presence of resistant forms in one or more field populations of the target organism. Consideration of the degree of resistance, the proportion of resistant variants in a population, or the effect on practical field performance of the pesticide is not involved.

"Monitoring" needs more consideration. To many people it denotes a routine, continuous, and random "watch dog" program, analogous to the official monitoring for levels of pesticide residues in foodstuffs. Such year-in, year-out surveillance aims to detect and then follow the spread of any markedly abnormal forms should they arise, or with sufficiently sensitive and quantitative methods, to reveal any gradual erosion of response, as has occurred with certain plant pathogens. Campaigns of this kind can be protracted and unrewarding, although sometimes they may be justified for certain very important pesticide uses when the risk of resistance is already known to be considerable. More specific, shorter term investigations are also (less aptly) referred to as monitoring. These are done either to gain initial or "baseline" sensitivity data before the widespread commercial use of a new pesticide or, more commonly, to examine individual cases of suspected resistance indicated by obvious loss of field efficacy of the product. Thus, monitoring can be used to indicate either continuous surveillance or ad hoc testing programs; this double use is acceptable, providing the meaning of the term is made clear in any particular context.

"Resistance" and "resistant" have many different shades of meaning. For precision either a particular usage must be specified as the correct one or resistance must be defined clearly whenever it is used. The first of these options is unattractive, because new, narrow definitions of commonly used and fairly general terms are seldom adopted universally or even remembered, and they force us to define a whole range of other narrow terms. Hence, "resistance," "tolerance," "insensitivity," and "adaptation" should not, as some suggest, be given separate, precise meanings. The second option, however, is both feasible and sensible and should be encouraged. Resistance can be used in a general way and interchangeably with the other terms to mean any heritable decrease in sensitivity to a chemical within a pest population. This can be slight, marked, or complete and may be homogeneous, patchy, or rare within a population. It can cause complete loss of action of an agrochemical or may have little practical significance. Thus, resistance and similar terms must, like monitoring, be defined carefully within each particular context.

In reports on monitoring, the absolute use of resistance (as in "the population was resistant") causes more problems of misinterpretation than relative use ("population A was more resistant than B"), and a quantitative definition of how resistance was categorized and measured should always be given. A "resistance index" or "resistance factor" (the ratio of the doses, commonly ED_{50}, required to act against resistant and sensitive forms, respectively) is often used, but the basis of its calculation needs careful consideration. The choice of sensitive reference strains (sometimes merely a single one is used) and any shift in their response with time can affect greatly the value of the index and inferences made, at least with regard to fungicide resistance. If a reference strain has been kept away from all chemical treatments for years in a laboratory culture, it may be abnormally sensitive.

The Fungicide Resistance Action Committee (FRAC) has recommended that the term "laboratory resistance" should be used to indicate strains of fungi with significantly reduced sensitivity as demonstrated by laboratory studies, whereas "field resistance" should be used to indicate a causal relationship between the presence of pathogenic strains with reduced sensitivity and a significant loss in disease control. The intention is to avoid false alarms such as have occurred when certain authors, having found some specimens to be more resistant than others in laboratory cultures or field samples, implied without evidence that these variants were causing or were about to cause problems in practical pest control. The use of the above terms as suggested by FRAC, however, can also be misleading: resistant forms found in the field in low numbers or with a low degree of resistance or fitness are certainly field and not laboratory resistant, yet such forms may not be affecting practical control. Whatever terms are selected there is no substitute for defining clearly the implications and limits of their use in all publications.

THE AIMS OF DETECTION AND MONITORING

There are at least seven distinct motives for resistance, detection, and monitoring, and whichever of them predominates will affect the scope and design of the surveys that are done. The aims, which are discussed in turn below, are as follows:

- Check for the presence and frequency of occurrence of the basic genetic potential for resistance (expressed resistance genes) in target organism populations.
- Gain early warning that the frequency of resistance is rising and/or that practical resistance problems are starting to develop.
- Determine the effectiveness of management strategies introduced to avoid or delay resistance problems.
- Diagnose whether rumored or observed fluctuations or losses in the field efficacy of an agrochemical are associated with resistance rather than with other factors.
- If resistance has been confirmed, determine subsequent changes in its incidence, distribution, and severity.
- Give practical guidance on pesticide selection in local areas.
- Gain scientific knowledge of the behavior of resistant forms in the field relation to genetic, epidemiological, and management factors.

Potential for Resistance

To obtain an initial indication of possible sources of future loss of effectiveness, we would need to be able to isolate and characterize rare mutants at, say, 1 in 10^{10} frequency. This is not feasible, however, without vast expense and effort. Resistant forms can be detected only after reaching much higher frequencies of 1 in 100 or perhaps 1 in 1,000 units (individual disease lesions, spores, pests, weeds), depending on the number of samples taken and the degree of statistical significance required. For example, if 1 in 100 units is resistant, 298 samples must be examined to achieve 95 percent probability of detection of 1 resistant unit; 2,994 samples must be checked if the frequency is 1 in 1,000. If a particular pesticide application normally allows 10 percent survivors (i.e., pest control is 90 percent effective), such detectable frequencies will occur only one or two applications prior to serious and obvious loss of practical control. With some pests and diseases this may be too late to allow any avoidance action to be introduced in the area concerned.

The relatively late first indication of the occurrence of resistance forms, however, can still give a valuable alert for certain purposes or situations. For example, it can indicate to other regions or countries that the potential for resistance exists. Or there may be time to introduce or modify avoidance strategies in cases where the rate of reproduction of target organisms is low (one or two generations per year), where lack of fitness in resistant mutants leads to an interrupted or fluctuating buildup (as with resistance of *Botrytis cinerea* to dicarboximide fungicides), or where a range of variants with different degrees of resistance arise and resistance tends to build up in a stepwise manner (as in the resistance of powdery mildews to 2–amino-pyrimidine and triazole fungicides). In such situations loss of efficacy is still a gradual process, even after relatively high frequency levels are first detected.

Shifts in Frequency or Severity of Resistance

After initial detection systematic monitoring can reveal subsequent changes (if any) in the frequency and degree of resistance and in its geographic distribution. For this reason repeated surveys have been done by public-sector organizations such as the Food and Agriculture Organization of the United Nations (FAO), the World Health Organization (WHO), and national agricultural and health research authorities. Surveys are also increasingly done by agrochemical companies, sometimes in cooperation with Resistance Action Committees. Examples are considered in the later section on achievements in resistance monitoring. Shifts in resistance can be very rapid. Sensitive populations have been known to be replaced completely by resistant ones over large areas within a year of first detection, particularly when the variants are highly resistant and retain normal or near normal fecundity and the ability to invade a host crop or animal. Shifts may be much more gradual, however, as mentioned above. It is essential to obtain information at each sampling site on the efficacy of field performance of the chemical following the latest and earlier applications, on the numbers and types of chemical treatments applied, and on management factors (e.g., cultivar grown, method of cultivation), in order to permit assessment of the practical impact of resistant forms at different stages of their buildup and to aid identification of factors that encourage or suppress resistance.

Checking Resistance Management Strategies

It is sometimes said that monitoring for resistance is a waste of time and money, because if positive results are obtained it is then too late to take effective action. This point of view may be valid under circumstances where the first variants detected are sufficiently resistant to cause loss of control and sufficiently fecund and competitive to accumulate rapidly and persist and where selection pressures are sufficiently heavy and widespread to induce large-scale shifts. Such has been the case with certain combinations of fungicides and plant pathogens, for example, the use of dimethirimol against cucumber powdery mildew (*Sphaerotheca fuliginea*) in Holland (Brent, 1982) or of benomyl against sugar beet leaf spot (*Cercospora beticola*) in Greece (Georgopoulos, 1982b). Insecticide resistance commonly arises in this way (Keiding, this volume). There is now, however, an increasing and very welcome trend toward establishing, in the light of risk assessments, some kind of strategy of resistance management at the very outset of the commercial life of a new chemical. Monitoring then is done not to warn of the need to initiate action but with the much better aim of checking whether an established strategy is working adequately or needs to be modified or intensified. This type of approach is indicated in Table 1.

Investigation of Suspected Resistance Problems

When observed losses of field efficacy are reported, they may be so dramatic that testing a few samples under controlled conditions against high doses of the chemical is sufficient to confirm resistance as the cause. The situation is sometimes less clear-cut: farmers may be using higher and higher rates of a chemical to achieve the same degree of control, or the period of persistence of protection may be gradually shortening. In such situations studies that are more extensive in area and time can reveal a great deal about the cause of these problems, and if there are correlations of reduced sensitivity of the target organism with loss of field performance, then the need for a change in the strategy of chemical use is indicated.

Subsequent Changes in Resistance

Later surveys, following a demonstration that resistant populations exist, can indicate whether shifts toward resistance are spreading or contracting in

TABLE 1 Phases of Monitoring and Resistance Management for a New Pesticide

Timing	Resistance Monitoring Activities	Other Management Activities
1–2 years before start of sales	Establish sampling and testing methods Survey for initial sensitivity data (include treated trial plots)	Assess risk Decide strategy of use
During years of use	Monitor randomly in treated areas for resistance, only if justified by risk assessment or special importance	Work the decided use strategy Watch practical performance closely
As soon as signs of resistance are seen visually or through monitoring	Monitor to determine extent and practical significance of resistance	If resistance problem is confirmed, review strategies and modify Study cross-resistance, fitness of variants and other factors affecting impact of resistance
Subsequently	Check rate of spread or decline of resistance	Watch performance, review strategies

SOURCE: Brent (unpublished).

geographic distribution, whether they are increasing or decreasing in frequency or severity, or whether an equilibrium is reached. Attempts should be made to correlate any such changes in resistance with either initial or modified strategies of chemical use or crop management.

Guidance in Pesticide Selection

Immediate practical guidance to individual growers, based on resistance monitoring on the farm, may be feasible in some situations. The only example known to the author is in the control of Sigatoka disease of bananas (caused by *Mycosphaerella* spp.) in Central America, where the United Fruit Company and du Pont have recommended that growers use a simple agar-plate test every month and postpone the use of benomyl if they find that the proportion of resistant ascospores exceeds 5 percent (du Pont, 1982).

Scientific Knowledge

The use of monitoring to aid our understanding of the nature of the resistance phenomenon is important because of our present limited state of knowledge of the population dynamics of resistant forms in relation to biological, agronomic, and environmental factors. For example, are different races of target organisms or cultivars of host plants more prone to resistance problems than others? There is evidence of this in the resistance of barley powdery mildew to fungicides (Wolfe et al., 1984). How far are theoretical models borne out in practice? Surprisingly few attempts have been made to validate the various proposed mathematical models of the progress of resistance in insects, plant pathogens, and weeds. How do factors such as dose applied, spray coverage, and timing affect the rate and severity of resistance development? The few studies that have been made for fungicides (Skylakakis, 1984; Hunter et al., 1984) have depended greatly on the development of precise and reproducible detection and monitoring procedures.

TIMING AND PLANNING OF SURVEYS

A new pesticide should work well initially on the target organisms against which it is recommended. If not, it would have failed in the large number of field trials that generally are done before marketing. Surveys should be started early, however, by testing field samples of each major target pest for degrees of sensitivity under controlled conditions before the chemical is used extensively (Table 1). Such testing provides valuable initial sensitivity (or baseline) data against which the results of any subsequent tests or surveys can be compared. These data could indicate the initial incidence of forms with resistance genes if their frequency and the number of samples tested were sufficiently high. Normally, however, testing will reveal the range of initial sensitivities of different populations of the pest; it also will provide an early opportunity to gain experience with and to check the precision of test methods that may be required at short notice if problems arise later. Some degree of variation in the results of initial sensitivity tests will occur, and it is necessary by replication or repetition of tests to separate experimental variation from real differences in response between populations. As part of the baseline exercise, it is very useful to check the sensitivity of surviving target populations shortly after successful use of the chemical in field trials: the less-sensitive elements of heterogeneous populations tend to predominate after treatment. Although these might persist and create problems later, often they lack fitness or are unstable and decline as the effects of the chemical wear off (Shephard et al., 1975).

Once initial data are obtained a decision must be made as to whether further surveys are needed. Unless there is a special reason—such as the critical importance of the particular target-chemical combination, an indication of high risk from a risk-assessment exercise, considerable variation between samples in the initial survey, or evidence from other regions for resistance phenomena—the effort and expense of further sampling will not be justified until signs of practical loss or erosion of efficacy are seen. A close watch should always be maintained, however, on the efficacy of treatment in practical use ("performance monitoring"), in comparison with initial field trial results and with the performance of other kinds of chemicals. If either an obvious major loss of effect or a gradual decline of performance are observed, all possible alternative causes of the difficulty (e.g., poor application, misidentification of target organism, increased pest or disease pressure) should be investigated, in addition to resistance. If possible, resistance sampling should be done at sites of poor and good control and at sites where the particular chemical has and has not been used. Positive correlations of degree of resistance with practical performance and with amount of use at the sampling sites must be sought. Sometimes highly resistant strains of fungi or insects have been detected readily at sites where the effectiveness of the product has been retained (Carter et al., 1982; Denholm et al., 1984).

If tests indicate an appreciable shift in sensitivity from the baseline position, then further monitoring, preferably at the same sites, may well be justified to reveal whether resistance is spreading, worsening, declining, fluctuating, or showing little change and how far it is associated with losses of control.

METHODS OF SAMPLING AND TESTING

In an extensive survey many sites (e.g., farms, fields, or glasshouses) containing the target organism throughout a region or country are examined, and one or a few representative samples of the population are taken at each site. At the extreme, area populations of insects or spores can be trapped by using suction traps for aerial populations of insects or by mounting test plants on a car top and driving through a cropping area to sample the powdery mildew spore population (Fletcher and Wolfe, 1981). In an intensive survey one or a few sites are visited, and many smaller samples—perhaps comprising single disease lesions or even spores, single insects, or single weed seeds—are collected on several occasions. Often, it is best that an extensive survey be done first, followed by a more detailed study if necessary. These two approaches are complementary, however, and it may be advantageous to use both concurrently or to adopt an intermediate method.

Information gathered at each sampling site should include the types, timing, and effectiveness of past chemical treatments and the amounts of target pests, disease, or weeds present. Differences in these factors should be compared with differences in sensitivity.

Sample size should relate to the circumstances. If searching for first signs of resistance in a largely sensitive population, a large bulk sample is more likely to find the "needle in a haystack." To determine the proportion of resistant forms in a population or the differences in degree of resistance, a number of small, specific samples should be tested.

Samples should be as fresh as possible, and repeated culture—in the absence or presence of chemical—should be avoided or minimized. One way to achieve this, which is particularly useful for obligate parasitic fungi, is to place treated test plants in pots in the field crop, allow them to collect inoculum, and then remove them for incubation in a controlled-environment facility or glasshouse to determine response. Conversely, it is valuable to retest samples after repeated subculture in vivo or in vitro to check for genetic stability of response.

For increased accuracy and to check degree of resistance, it is generally best to use a range of concentrations during initial testing rather than a single, arbitrary, discriminating dose. The response can be scored in various ways. The ED_{50} value is often used; it is a good "general purpose" value that is widely understood and can be measured relatively accurately, compared with an ED_{95} value. For large-scale surveys, however, and particularly where responses of sensitive and resistant forms are well separated (as with some fungicide and herbicide resistance and most insecticide and rodent resistance), the use of a single discriminating dose permits quick and adequate testing.

When resistance is clear-cut, different methods tend to reveal similar trends; only in marginal cases does the method of testing or scoring affect the picture. It is advantageous where possible, however, for one agreed method to be used by different workers nationally or internationally. The WHO standard tests for insecticide resistance in a range of insects of public health importance (WHO, 1970, 1980) have been used internationally since the first test, on anopheline mosquitoes, was introduced about 27 years ago. Test kits, based on diagnostic test dosages for susceptible, fully resistant, and sometimes intermediate populations, are available at cost for about a dozen pest species, including rodents. FAO-recommended methods to measure pest resistance in crop and livestock production and in crop storage have also been adopted widely: Busvine (1980) has drawn together details of tests against 20 important pests, published at intervals since 1969 in the FAO *Plant Protection Bulletin*; more recent issues of the bulletin contain new or updated procedures. Recommended methods for testing fungicide resistance in crop pathogens have also been published by FAO (1982), and general reviews of procedures are given by Georgopoulos (1982a) and Ogawa et al. (1983).

During testing it is important to investigate differences in pathogenicity, growth rate, reproductive rate, and other properties that contribute to the fitness of an organism. Often the more highly resistant forms are less fit or competitive than normal forms in the absence of chemical treatment, and knowledge of this can help to explain and predict their behavior.

Biochemical methods for detecting and monitoring resistant forms have been developed for insecticides and are increasingly used in surveys (Miyata, 1983; Devonshire and Moores, 1984). In some situations they can detect resistance at lower frequencies than do bioassays. They can also be more convenient and permit the degree of resistance to be measured quantitatively without the need to test several samples at different doses. Inhibition of photosystem II, as revealed by loss of chlorophyll fluorescence of herbicide-treated leaves, leaf discs, or isolated chloroplasts irradiated with short wave-length light, has proved a convenient method for monitoring atrazine-resistant weeds (Gasquez and Barralis, 1978, 1979). Another rapid method for testing response to photosynthesis inhibitors is the sinking-leaf disc technique. The buoyancy of discs floated on surfactant solutions appears to depend on the O_2/CO_2 ratio in the air spaces, which is decreased by the action of herbicides (Hensley, 1981). Biochemical monitoring is not yet used for fungicide resistance because mechanisms of resistance for field isolates are not well characterized and appear to involve changes at biosynthetic or genetic sites that are not easily detected. More research on this aspect seems justified. Specific diagnostic agents, such as cDNA probes or monoclonal antibodies, may offer new possibilities for future biochemical tests for all types of target organisms (Hardy, this volume). As pointed out by Truelove and Hensley (1982), however, biochemical methods should be used with caution, since resistance that depends on alternative mechanisms to the method under test could be missed; in this respect, bioassay tests on whole organisms remain the most reliable indicators of resistance.

ACHIEVEMENTS IN RESISTANCE MONITORING

Only a few examples of the many monitoring projects done in different countries and on different target organisms can possibly be considered here. Since the first case of insecticide resistance was reported by Melander in 1914 (Melander, 1914), response to insecticides has been monitored extensively in many countries (Georghiou and Mellon, 1983). Global programs have been organized by WHO to survey insecticide resistance in anopheline mosquitoes (WHO, 1976, 1980) and by FAO to survey insecticide resistance in pests of stored grain (Champ and Dyte, 1976) and acaricide resistance in ticks (FAO, 1979). These very large projects have provided valuable information on the geographic distribution and intensity of resistance, on its relationships to the successful use of chemicals, and to failures in control. The coordination and interpretation of results have benefited greatly from the general use of recommended methods of testing and reporting mentioned earlier.

Many national surveys have been conducted. An outstanding example is the study of resistance in house flies on farms in Denmark, discussed in this volume by Keiding, which has been sustained since 1948 and has shown clearly the large-scale shifts in response to successive introductions of different types of insecticide (organochlorines, organophosphorus compounds, and pyrethroids). Other notable programs have included studies of rice leafhoppers and planthoppers in Japan (Hama, 1980), cotton leaf worm in Egypt (El-Guindy et al., 1975), and the aphid *Myzus persicae* in the United Kingdom (Sawicki et al., 1978). In the last study biochemical (esterase-4) tests as well as bioassays were used; both approaches gave rapid and satisfactory results and to some extent were complementary in distinguishing different types of resistance.

International surveys comparable with those undertaken with pests have not been done for fungi. Although some recommended methods have been published by FAO, in practice a variety of test methods have been used by different workers. National or regional programs have included surveys of resistance of cucumber powdery mildew to dimethirimol in glasshouses in Holland (Bent et al., 1971) and later to other systemic fungicides (Schepers, 1984), the response of barley powdery mildew to ethirimol in the United Kingdom (Shephard et al., 1975; Heaney et al., 1984) and to triazole fungicides (Fletcher and Wolfe, 1981; Heaney et al, 1984; Wolfe et al., 1984), of metalaxyl resistance in *Phytophthora infestans* on potatoes in Holland (Davidse et al., 1981) and in the United Kingdom (Carter et al., 1982), benomyl resistance in sugar beet leaf spot in Greece (Georgopoulos, 1982b), and dicarboximide resistance in Botrytis on grape vines in West Germany (Lorenz et al., 1981). Each of these studies, as well as others not mentioned here, to some extent tells an individual story. Two main patterns can perhaps be distinguished: a rapid, widespread, and persistent upsurge of resistance and loss of disease control (as with dimethirimol and cucumber powdery mildew, metalaxyl and *P. infestans* in Holland, benomyl and sugar beet leaf spot) and a slower, fluctuating increase in resistance, with either partial or undetected loss of disease control (as in the cases of ethirimol or triazoles and barley powdery mildew, metalaxyl and *P. infestans* in the United Kingdom, and dicarboximides and Botrytis). The intensity and exclusivity of fungicide use and the degrees of resistance and fitness of the resistant forms are important factors in determining these patterns. In the former cases monitoring tended to follow reports of loss of control and results were obtained too late to permit any management strategy other than withdrawal of the product, but in the latter, where monitoring preceded any major breakdown in performance, avoidance strategies either were already operating or were introduced following the results of monitoring.

Since the early 1970s the incidence of triazine-resistant biotypes of various weeds in different crops has been monitored extensively in different parts of the United States, mainly by collecting seeds and growing progeny for glasshouse tests. The initial indications of resistance, obtained after 10 years of widespread use of these herbicides, came from farmer observations of obvious lack of control; the monitoring has served primarily to confirm resistance and to follow the problem in time and space (Bandeen et al., 1982). Atrazine resistance has also been observed in monitoring studies in several countries of continental Europe (Gressel et al., 1982). The rate of development of resistance appears to have varied between different parts of the United States and has been relatively slow in the United Kingdom (Putwain et al., 1982). Forms resistant to other herbicides, for example, phenoxy compounds and bipyridyls, have been detected in different countries, but their incidence has been sporadic, their resistance less marked, and little monitoring has been done.

COOPERATION AND COMMUNICATION

Detection of and monitoring for resistance call for close cooperation between scientists as individuals and as representatives of industrial and public-sector organizations. Although coordination does take place, such as in the work of the Fungicide Resistance Action Committee (FRAC) and Insecticide Resistance Action Committee (IRAC), much of the research is still too fragmented and haphazard. Industry has felt it has been excluded from some collaborative schemes and planning meetings organized by the public sector, but, equally, the RAC system does not fully involve the public sector, since it is primarily an intercompany concern. There is much that scientists in industry and the public sector can do to increase contact, review progress and priorities, and plan collaborative research. Such collaboration would be best focused on particular resistance problems and should be in work groups rather than in conferences, with one person or organization as the focal point for each topic. At this time of retrenchment of national research expenditures in many countries, the selection of priorities in resistance monitoring—which despite its importance is an expensive and essentially defensive area of research—is especially important.

The results of monitoring programs should be reported in the open scientific literature, not retained in confidential reports or computer files. The storage of information from many sources in a data bank from which it can be retrieved and disseminated readily is valuable, however; the data bank for insecticide resistance at the University of California (Riverside) is a good example (Georghiou, 1981).

Education in resistance monitoring is improving. Conferences are helpful, but the international courses on fungicide resistance—organized by Professor Dekker and colleagues and held at Wageningen and more recently in Malaysia—have proved particularly useful, since they included laboratory sessions and a tactical exercise in addition to lectures and group discussions. Perhaps similar courses could be organized on insecticide and herbicide resistance.

CONCLUSION

Detection and monitoring form an integral part of pesticide resistance management. To avoid misunderstanding and waste of effort, very careful definition, planning, and interpretation of these activities are required. Monitoring denotes different operations, ranging from global surveillance programs to much smaller investigations of cases of suspected resistance. Distinction must be made between detecting resistant forms and establishing that resistance has reached levels of severity and frequency sufficient to cause practical loss of pesticide performance. Criteria for defining resistance and sensitivity have differed greatly, especially when several different degrees of resistance occurred, and must always be made clear.

Test methods should be developed and initial sensitivity data sought before new compounds are brought into widespread use; avoidance strategies should also be established prior to widespread use, since monitoring cannot be relied on to give sufficient early warning of the need for such strategies.

Subsequent monitoring should be done if risks are considered high, if the particular pest-control system is especially important, or when visible signs of resistance problems arise. Selection of test procedures will depend on the nature of the pest and of the pesticide treatment, but the adoption of internationally recommended methods aids the comparison and coordination of results. Biochemical methods have already proved useful and have a promising future. Further collaboration between and within the industrial and public sectors in planning and conducting monitoring programs must be fostered.

ACKNOWLEDGMENTS

The author is grateful to a number of persons for providing information, and especially to Drs. A. Devonshire, G. P. Georghiou, H. LeBaron, and L. R. Wardlow.

REFERENCES

Bandeen, J. D., G. R. Stephenson, and E. R. Coweet. 1982. Discovery and distribution of herbicide resistant weeds in North America. Pp. 9–19 *in* Herbicide Resistance in Plants, H. M. LeBaron and J. Gressel, eds. New York: John Wiley and Sons.

Bent, K. J., A. M. Cole, J. A. W. Turner, and M. Woolner. 1971. Resistance of cucumber powdery mildew to dimethirimol. Pp. 274–282 *in* Proc. 6th Br. Insectic. Fungic. Conf., Vol. 1, Brighton, England, 1971.

Brent, K. J. 1982. Case study 4: powdery mildews of barley and cucumber. Pp. 219–230 *in* Fungicide Resistance in Crop Protection, J. Dekker and S. G. Georgopoulos, eds. Wageningen, Netherlands: Centre for Agricultural Publishing and Documentation.

Busvine, J. R. 1980. Recommended methods for measurement of pest resistance to pesticides. FAO Plant Prod. and Prot. Paper No. 21.

Carter, G. A., R. M. Smith, and K. J. Brent. 1982. Sensitivity to metalaxyl of *Phytophthora infestans* populations in potato crops in southwest England in 1980 and 1981. Ann. Appl. Biol. 100:433–441.

Champ, B. R., and C. E. Dyte. 1976. Report of the FAO global survey of pesticide susceptibility of stored grain pests. FAO Plant Production and Protection Paper No. 5.

Davidse, L. C., D. Looigen, L. J. Turkensteen, and D. van der Wal. 1981. Occurrence of metalaxyl-resistant strains of *Phytophthora infestans* in Dutch potato fields. Neth. J. Plant Pathol. 87:65–68.

Denholm, I., R. M. Sawicki, and A. W. Farnham. 1984. The relationship between insecticide resistance and control failure. Pp. 527–534 *in* Proc. Br. Crop Prot. Conf. Pests and Dis., Vol. 2, Croydon, England: British Crop Protection Council.

Devonshire, A. L., and G. D. Moores. 1984. Immunoassay and carboxylesterase activity for identifying insecticide resistant *Myzus persicae*. Pp. 515–520 *in* Proc. Br. Crop Prot. Conf. Pests and Dis., Vol. 2, Croydon, England: British Crop Protection Council.

du Pont. 1982. Black and Yellow Sigatoka, Improved Identification and Management Techniques. Coral Gables, Fla.: du Pont Latin America.

El-Guindy, M. A., G. N. El-Sayed, and S. M. Madi. 1975. Distribution of insecticide resistant strains of the cotton leafworm *Spodoptera littoralis* in two governorates of Egypt. Bull. Entomol. Soc. Egypt 9:191–199.

Fletcher, J. T., and M. S. Wolfe. 1981. Insensitivity of *Erysiphe graminis* f. sp. *hordei* to triadimefon, triadimenol and other fungicides. Pp. 633–640 *in* Proc. Br. Crop Prot. Conf. Pests and Diseases, Vol. 2, Croydon, England: British Crop Protection Council.

Food and Agriculture Organization. 1979. Pest resistance to pesticides and crop loss assessment. FAO Plant Production and Protection Paper No. 6/2.

Food and Agriculture Organization. 1982. Recommended methods for the detection and measurement of resistance of agricultural pests to pesticides. Plant Prot. Bull. 30:36–71, 141–143.

Gasquez, J., and G. Barralis. 1978. Observation et selection chez *Chenopodium album* L. d'individus resistants aux triazines. Chemosphere 11:911–916.

Gasquez, J., and G. Barralis. 1979. Mise en evidence de la resistance aux triazines chez *Solanum nigrum* L. et *Polygonum lapathifolium* L. par observation de la fluorescence de feuilles isolees. C. R. Acad. Sci. (Paris) Ser. D 288:1391–1396.

Georghiou, G. P. 1981. The occurrence of resistance to pesticides in arthropods: An index of cases reported through 1980. Rome: FAO.

Georghiou, G. P., and R. B. Mellon. 1983. Pesticide resistance in time and space. Pp. 1–46 *in* Pest Resistance to Pesticides, G. P. Georghiou and T. Saito, eds. New York: Plenum.

Georgopoulos, S. G. 1982a. Detection and measurement of fungicide resistance. Pp. 24–31 *in* Fungicide Resistance in Crop Protection, J. Dekker and S. G. Georgopoulos, eds. Wageningen, Netherlands: Centre for Agricultural Publishing and Documentation.

Georgopoulos, S. G. 1982b. Case study I: *Cercospora beticola* of sugar beet. Pp. 187–194 *in* Fungicide Resistance in Crop Protection, J. Dekker and S. G. Georgopoulos, eds. Wageningen, Netherlands: Centre for Agricultural Publishing and Documentation.

Gressel, J., H. U. Ammon, H. Fogelfors, J. Gasquez, Q. O. N. Kay, and H. Kees. 1982. Discovery and distribution of herbicide-resistant weeds outside North America. Pp. 32–55 *in* Herbicide Resistance in Plants, H. M. LeBaron and J. Gressel, eds. New York: John Wiley and Sons.

Hama, H. 1980. Mechanism of insecticide resistance in green rice leafhopper and small brown planthopper. Rev. Plant Prot. Res. (Japan) 13:54–73.

Heaney, S. P., G. J. Humphreys, R. Hutt, P. Montiel, and P. M. F. E. Jegerings. 1984. Sensitivity of barley powdery mildew to systemic fungicides in the UK. Pp. 459–464 *in* Proc. Br. Crop Prot. Conf. Pests and Diseases, Vol. 2, Croydon, England: British Crop Protection Council.

Hensley, J. R. 1981. A method for identification of triazine resistant and susceptible biotypes of several weeds. Weed Sci. 29:70–78.

Hunter, T., K. J. Brent, and G. A. Carter. 1984. Effects of fungicide regimes on sensitivity and control of barley mildew. Pp. 471–476 *in* Proc. Br. Crop Prot. Conf. Pests and Diseases, Vol. 2, Croydon, England: British Crop Protection Council.

Lorenz, G., K. J. Beetz, and R. Heimes. 1981. Resistenzentwicklung von *Botrytis cinerea* gegenuber Fungiziden auf Dicarboximid-Basis. Mitt. Biol. Bundesanst. Land- Forstwirtsch., Berlin-Dahlem 203:278–285.

Melander, A. L. 1914. Can insects become resistant to sprays? J. Econ. Entomol. 7:167–174.

Miyata, T. 1983. Detection and monitoring methods for resistance in arthropods based on biochemical characteristics. Pp. 99–116 *in* Pest Resistance to Pesticides, G. P. Georghiou and T. Saito, eds. New York: Plenum.

Ogawa, J. M., B. T. Manji, C. R. Heaton, J. Petrie, and R. M. Sonada. 1983. Methods for detecting and monitoring the resistance of plant pathogens to chemicals. Pp. 117–162 *in* Pest Resistance to Pesticides, G. P. Georghiou and T. Saito, eds. New York: Plenum.

Putwain, P. D., K. R. Scott, and R. J. Holliday. 1982. The nature of the resistance to triazine herbicides: Case histories of phenology and population studies. Pp. 99–116 *in* Herbicide Resistance in Plants, H. M. LeBaron and J. Gressel, eds. New York: John Wiley and Sons.

Sawicki, R. M., A. L. Devonshire, A. D. Rice, G. D. Moores, S. M. Petzing, and A. Cameron. 1978. The detection and distribution of organophosphorus and carbamate insecticide-resistant *Myzus persicae* (Sulz.) in Britain in 1976. Pestic. Sci. 9:189–201.

Schepers, H. T. A. M. 1984. Resistance to inhibitors of sterol biosynthesis in cucumber powdery mildew. Pp. 495–496 *in* Proc. Br. Crop Prot. Conf. Pests and Diseases, Vol. 2, Croydon, England: British Crop Protection Council.

Shephard, M. C., K. J. Brent, M. Woolner, and A. M. Cole. 1975. Sensitivity to ethirimol of powdery mildew from UK barley crops. Pp. 59–66 *in* Proc. 8th Br. Insectic. Fungic. Conf., Brighton, 1975.

Skylakakis, G. 1984. Quantitative evaluation of strategies to delay fungicide resistance. Pp. 565–572 *in* Proc. Br. Crop Prot. Conf. Pests and Diseases, Vol. 2, Croydon, England: British Crop Protection Council.

Truelove, B., and J. R. Hensley. 1982. Methods of testing for herbicide resistance. Pp. 117–131 *in* Herbicide Resistance in Plants, H. M. LeBaron and J. Gressel, eds. New York: John Wiley and Sons.

World Health Organization. 1970. Insecticide resistance and vector control. 17th Rep. WHO Exp. Comm. on Insectic. WHO Tech. Rep. Ser. No. 443.

World Health Organization. 1976. Resistance of vectors and reservoirs of disease to pesticides. 22nd Rep. WHO Exp. Comm. on Insectic. WHO Tech. Rep. Ser. No. 585.

World Health Organization. 1980. Resistance of vectors of disease to pesticides. 5th Rep. WHO Exp. Comm. on Vector Biol. Contr. WHO Tech. Rep. Ser. No. 655.

Wolfe, M. S., P. M. Minchin, and S. E. Slater. 1984. Dynamics of triazole sensitivity in barley mildew nationally and locally. Pp. 465–470 *in* Proc. Br. Crop Prot. Conf. Pests and Diseases, Vol. 2, Croydon, England: British Crop Protection Council.

5

Tactics for Prevention and Management

THE FREQUENCY OF RESISTANCE in a pest population is in large part a result of selection pressure from pesticide use. Strategies to manage resistance aim to reduce this pressure to the minimum, using tactics designed to increase the useful life of a pesticide and to decrease the interval of time required for a pest to become susceptible to a given pesticide again (Chapter 3). *Strategy* is used here in the sense of an overall plan or methods exercised to combat pests, whereas *tactic* is used to mean a more detailed, specific device for accomplishing an end within an overall strategy. This chapter will focus on promising strategies and tactics.

Judicious use of pesticides reduces the selection pressure on pest populations for developing resistance. Use of pesticides only as needed not only avoids or delays resistance but tends to protect nontarget beneficial species. These practices are an essential part of Integrated Pest Management (IPM), which implies the optimum long-term use of all pest-control resources available. Excessive use or abuse of pesticides for short-term gains (e.g., minor yield increase) may be the worst possible practice long-term because it may lead to the permanent loss of valuable, efficient, and often irreplaceable pesticides. Such practices represent a serious issue affecting all segments of society. Catastrophic events, such as the failure of an entire pesticide class against a target species, have in the past, and may again in the future, force dramatic changes in our crop production and pest-control practices.

Genetic, biological, ecological, and operational factors influence development of resistance. Operational factors, including pesticide chemicals and how they are used, obviously can be controlled (Georghiou and Taylor, 1977;

Georghiou, this volume). The biological factors are considered beyond our control, but current studies in biotechnology and behavior have shown that components of genetic, reproductive, behavioral, and ecological factors may be manipulated and have potential for use in management (Leeper, this volume).

While the basic principles of resistance management apply to all major classes of pests (insects, pathogens, rodents, and weeds), there are some important differences among these classes that influence the applicability of management strategies and tactics. Tactics are site and species specific. For example, many insects and plant pathogens have considerable mobility, whereas rodents and weeds generally have less. The usefulness of maintaining refuges can vary substantially among pest classes. Weed seeds, egg sacs of some nematodes, and the resting structures of some plant pathogenic fungi may remain dormant in soils for many years, thus preserving susceptible germ plasm. This does not occur for other classes of pests. Rates of reproduction, population pressure, and movement of susceptible individuals from refuges into a treated area are often very high with plant pathogens, moderate to high with insects, and comparatively low for weeds and rodents (Greaves, this volume). The residual nature or persistence of pesticides varies greatly, which will affect the success of various tactics to manage resistance. Generally, the greater the persistence, the greater the probability of resistance. The number of target species being controlled with a given pesticide varies with the class of pest. Biological control agents are critical for many insect pests but have not yet become as important in control of pests in other classes. Other differences exist, but their strategic significance is poorly understood.

Some of the most important issues that impinge on the development and selection of management tactics are: differences among classes of pests and pesticides; dynamics of resistance (differences between high- and low-risk pesticides, and variations in the rate of resistance development within species and geographic areas); complexes of pests on crops or locations requiring multiple pesticides for control; and lack of supporting data and validation in the field. Pesticides considered to be at high risk for resistance generally have a single site of toxic action and, in fungicides, are usually systemic, while low-risk compounds have multiple sites of action. Our current insecticides and most of our new systemic fungicides tend to have single sites and would, therefore, fall within the high-risk category. On the other hand, few plants have evolved resistance to herbicides, which also tend to have single sites of action. Although experience with inorganic insecticides (i.e., lead arsenate) shows that resistance can also develop to multisite compounds, such resistance is rare.

The rate at which pesticide resistance develops is extremely variable among species as well as among different field populations of the same species.

Rate of reproduction, pest movement, relative fitness of resistant members of a population, mechanism(s) of resistance, etc., all contribute to the dynamics of resistance and determine the severity of its effect on economic efficacy and the viability of continued use of a given compound. Therefore, the applicability of specific management tactics must be established on the basis of specific cases and locations.

Although resistance poses a most serious threat to a pesticide's economic life and has resulted in total loss of previously valuable chemicals from some major pest-control programs, no pesticide has been lost from the marketplace solely because of resistance. Resistance is not absolute throughout a pest's range, and susceptible populations of some pests continue to exist. Furthermore, in an area where resistance has occurred, a pesticide's continued use may be required to control other pests that are still susceptible. This may confound management attempts, but documented cases of resistance do not necessarily warrant removal of a pesticide.

On the other hand, industry has a responsibility to adjust marketing plans (and perhaps propose label changes) to reflect a product's efficacy or inefficacy, leaving the marketplace to determine its actual value and life. In addition, public-sector research, extension, and regulatory programs have a key role to play in ensuring that growers are completely informed of resistance situations that are identified, so that rational decisions can be made among pest-control alternatives.

Several major deficiencies in scientific understanding currently frustrate efforts to develop and implement tactics to manage resistance. Resistant strains of pests selected in the laboratory may differ from field strains in some ways, including fitness and number of alleles conferring resistance. Therefore, tactics should be validated for a wide range of pests under field as well as laboratory conditions. Monitoring technologies must be developed to evaluate the strategies, validate the tactics, accurately determine critical resistance frequencies for pests under different conditions, and guide the implementation of optimum tactics (Chapter 4).

TACTICS FOR RESISTANCE MANAGEMENT

Several concepts discussed below have been proposed as tactics for managing specific cases of resistance. Most of these tactics have been used, often inadvertently or without confirming data, in pest-control practices. Owing to lack of rigorous field and laboratory evaluations, our inability to establish and detect critical frequencies of resistance, and the limitations of space, no attempt is made here to detail the strengths and weaknesses of the tactics. Sweeping generalizations about the applicability or feasibility of specific tactics are not justified. These caveats must be kept in mind in interpreting the data presented in Table 1. The ratings are usually only valid within the

TABLE 1 Tactics for Management of Resistance to Pesticides and Their Suitability for Classes of Pests

Tactics	Insecticides	Fungicides	Herbicides	Rodenticides
Variation in dose or rate	+ +(T)	+ +(T)	+ +(T)	+ +(T)
Frequency of applications	+ +(T)	+ + +	+	+
Local rather than areawide applications	+ + +	+	0	+ +
Treatments only to economic threshold	+ +	+ +	0	0
Less persistent pesticides	+ +	+ +(T)	+ + +	0
Life stages of pest	+ +(T)	+ +(T)	0	0
Pesticide mixtures	+ +(T)	+ + +(T)	+ + +(T)	0
Alternations, rotations, or sequences of pesticide applications	+ + +	+ + +	+ + +	+ + +
Pesticide formulation technology	+(T)	+(T)	+(T)	+ + +
Synergists	+ +	+	+(T)	0
Exploiting unstable resistance	+ +(T)	+ +(T)	+ +	+
Pesticide selectivity	+ +(T)	+(T)	0(−)	0(−)
New toxophores with alternate sites of action	+ + +	+ + +	+ + +	+ + +
Protection and use of natural enemies with pesticide tolerance	+ +	+	0	0
Reintroduction of susceptible pests	+(T)	0	0	0(−)

Code for Suitability Ratings:

+ + + Very useful, generally supported by laboratory data and/or field experience.
+ + Moderately useful.
+ Minor use, in exceptional cases only, or supported by few data.
0 Not applicable or assumed to be of no value.
(T) Supported by theoretical assumptions only. No data or experience.
(−) May actually be detrimental to managing resistant populations.

The suitability ratings presented in this table are very tenuous, may be theoretical or supported only by a few examples, and should not be assumed to be generally valid for each pesticide, pest, or tactic within each class.

limitations of a few examples, often weakly supported, for each tactic within each pest group.

Variation in Dose or Rate

With this tactic, resistance may be delayed or minimized by preserving a sufficient population of susceptible individuals or alleles by using low rates of a given pesticide so as not to select against heterozygotes where resistance is recessive. On the other hand, the use of high doses has also been proposed, but as a means of eliminating or reducing the frequency of heterozygotes where resistance is dominant. While laboratory studies have supported the latter approach with insecticides, there is limited evidence to confirm its success under field conditions, with the possible exception of some pests of stored grain. Using fungicides at dosage rates giving less than 100 percent control may minimize the threat of resistance, if low levels of disease can be tolerated or if a high level of resistance may occur (e.g., benomyl or metalaxyl). If resistance is linked to decreased fitness, however, or if low levels of resistance are likely to occur (e.g., dicarboximides), high dose rates might be recommended to control all individuals in the populations. Also, because of the explosive reproductive capacity of some pathogens or the high premium paid for a totally disease- and insect-free crop (e.g., apple scab and codling moth), some disease situations require virtually total control. There is no proof that herbicide-use rate has any effect on the development of resistance in weeds, although circumstantial evidence indicates that high rates may favor resistance. Because of the short generation time of rodents, any treatment that leaves significant numbers of survivors fosters selection for resistance. Both low concentrations in baits or inadequate applications fit this category. Unfortunately, specific field data are lacking.

Frequency of Application

Fewer or less frequent applications, which reduce the selection pressure over time, should reduce the rate and probability of resistance development. This tactic is assumed to be valid for management of resistance to insecticides, but it is unconfirmed. Circumstantial evidence indicates that in areas where a fungicide is used only once or twice a season, the threat of resistance development is reduced compared to full season programs. For example, in northern Europe, resistance quickly developed when metalaxyl was used full season to control late blight. Based on limited experience, it may be possible to continue cautious use of such fungicides even after resistance has developed. A specific herbicide is most commonly used only once per crop season. Postemergent herbicides or those having brief soil activity could be applied several times, especially in perennial crops, but this would tend to increase selection pressure for resistance. Paraquat-resistant weeds have occurred in a few areas following frequent applications of this herbicide. If applications of rodenticides are made monthly (as by a Pest Control Operator [PCO]), the selection pressure would be persistent and could speed selection. Treatments once or twice a year (as with urban rat control programs) would be nearly as efficient in selecting for resistance, however, because removing susceptible individuals from each generation as it reaches reproductive age speeds selection.

Local Rather Than Areawide Applications

Control of a pest with a particular pesticide in a single field or site, rather than over a large area, can leave refuges in surrounding areas to thwart resistance development; this is believed to be a useful tactic, especially with insecticides. Susceptible individuals move into previously treated areas, thus diluting the frequency of resistance. The success of this tactic may vary with insect species, refuges, and other factors. In some cases, an areawide application of the right insecticide can severely reduce a particular generation of specific insects when properly timed, thereby reducing or eliminating the need for further applications. Plants, even weeds with seeds that are easily spread, are not sufficiently mobile to allow this tactic to be very successful with herbicides; seeds probably serve more often as a way of introducing alleles conferring resistance than of moving in large enough numbers of susceptible alleles to swamp those conferring resistance. Some plant pathologists feel that this tactic is not appropriate for airborne pathogens with potential for resistance under high population pressure. For example, resistance has often occurred when metalaxyl was used to stop heavy infestations of late blight (potatoes) or blue mold (tobacco). When metalaxyl has been used over a wide area as a preventive treatment before the disease started, however, resistance has not developed in these pathogens, at least in North America. On the other hand, some experts suggest that we should "confuse" the pathogen by localized use of two or more fungicides having different mechanisms of action, together with multiple cultivars that have a number of alleles conferring resistance to the pathogen (although the latter tactic assumes a single fungicide is used in the area). To the extent that resistant rodents are considered less fit competitors (the British view), localized control would result in islands of resistance that would not readily spread. Areawide control, however, is likely to result in areawide resistance (as in Denmark) (Greaves, this volume).

Treatments Based on Economic Threshold

This tactic delays pesticide applications until the economic threshold is reached and may allow a certain level of crop damage to occur. This is a means of reducing the selection pressure for resistance. The success of this tactic in managing insecticide resistance varies with the insect pest and conditions. The establishment of valid economic thresholds and the use of pesticides only when the threshold is exceeded is a major principle of IPM. The economic threshold often varies because it depends on commodity prices. The benefit of this tactic in managing resistance to fungicides is generally unconfirmed. It may be applicable with less virulent or localized plant pathogens, when total disease control is not necessary, or when the disease occurs only occasionally. It is probably not useful for the more virulent and systemic diseases. Under current management practices, this tactic is only of marginal benefit in limiting resistance in weeds and rodents. Introduction of the newer postemergence herbicides, however, provides the potential to exploit this tactic to control weeds, based on the number of weeds that compete for resources with the cultivated plant.

Use of Less Persistent Pesticides

The selection and use of pesticides or formulations having a lower biological persistence can be a useful tactic for managing resistance. Insecticides with short residual lives tend to slow the development of resistance due to reduced exposure, but success may depend on the nature of the insect and insecticide. Persistence of a fungicide will always prolong the period of selection pressure and thus favor the build-up of resistance. It is important to point out that a less persistent fungicide applied more frequently will have the same effect on resistance (e.g., a 14-day treatment schedule of one fungicide versus a 7-day schedule of another with half the persistence). Relatively long persistence and excellent control of most weeds are believed to be mainly responsible for the numerous occurrences of triazine-resistant weeds. This tactic is not considered to be applicable to rodenticides.

Life Stages of Pest

This tactic is based on using a pesticide against the life stage of the target pest that is not so likely to develop resistance. For example, in some lepidopterous species the adults and/or very early larval stages (instars) are apparently less able to metabolize insecticides than are late instars. In theory, the rate of developing resistance would be lessened by targeting insecticides against the adults or early instars, thereby reducing the selection pressure on later instars that have a higher resistance risk due to their greater enzymatic activity for pesticide metabolism. Applying a fungicide during the sexual stage theoretically should increase the chance of selecting for a higher level of resistance in the fungus. On the other hand, when fungicides have been applied during the asexual stages (e.g., late blight and apple scab), resistance has developed very rapidly. This tactic is not applicable to herbicides or rodenticides.

Mixtures

Simultaneous use of two or more pesticides having differing mechanisms of action or target sites (Chapter 2) has been and will continue to be a very important tactic to avoid and manage resistance. Certain limitations and conditions must apply for this tactic to be successful in managing resistance in insects and other pests. The use of mixtures must start early before resistance occurs to one of the components (unless negatively correlated toxicity or enhanced susceptibility is present), each component must have similar decay rates (preferably short stability), and they must have different modes and sites of action or different resistance mechanisms (with fungicides, similar translocation). Nevertheless, resistance to two or more different insecticides can develop by the same process as with a single pesticide—it just takes longer. Mixing chemicals sometimes leads to potentiation, rather than merely additive effects, thus delaying or preventing resistance even further. In case of established resistance, potentiation may become the only means of controlling the pest (V. Dittrich, Ciba-Geigy Corporation, Basle, personal communication). Mixtures are assumed to be an important tactic in avoiding or delaying the development of resistance to single-target-site fungicides by plant pathogens. Limited laboratory data show that mixed populations of resistant and susceptible *Phytophthora infestans* shifted to the resistant populations more slowly when mixtures were used. On the other hand, some reports indicated that resistance to a specific site-inhibitor fungicide can continue to increase when one is used in combination with a multisite fungicide, due in part to the pathogen population's not being controlled by the multisite inhibitor (e.g., lack of translocation). The use of mixtures has been a major tactic in preventing both the development and spread of weed resistance to herbicides. Resistant weeds have not usually occurred where herbicide mixtures are used, but triazine-resistant weeds have often developed after 5 to 10 years where this class of herbicide has been applied alone and frequently. Once resistant weeds have developed in an area, they usually take over completely if the single-problem herbicide is used exclusively. The use of mixtures has not been a usual tactic for rodenticides. To mix an acute poison with an anticoagulant is illogical. Mixing of warfarin and vitamin D (calciferol) in England seems not to have enhanced efficacy significantly.

Alternations, Rotations, or Sequences of Pesticide Applications

The use of pesticides of differing classes or modes and sites of action in rotation, alternation, or sequence to control the same pests has been much studied and accepted to avoid resistance. It assumes that the number of generations or length of time between uses of any one material is sufficient to allow resistance to decline below a critical frequency (Georghiou, 1980; Georghiou et al., 1983). Whether this tactic is superior to pesticide mixtures and the optimum sequence, frequency, and rate of each component will likely vary according to the pest, pesticide, and other factors. It is based on the relative instability of particular resistance mechanisms and is especially viable when it is known that cross-resistance does not occur. Annual rotation or alternation is probably not a good strategy for many high-risk fungicides because resistance can develop within one growing season, but sequences of applications of different fungicides is often quite useful. Rotation of lower-risk compounds (e.g., ergosterol biosynthesis inhibitors) may be an acceptable way to prolong the life of a fungicide. As with some of the other tactics, voluntary compliance or enforceability often prevents the general use or success of this tactic in management of resistance to fungicides. Although no direct evidence documents the effectiveness of annual rotations for management of resistance to herbicides, abundant circumstantial data support the use of annual rotations or alternations of herbicides. This has not been used as a resistance-avoiding tactic, but has been used inadvertently due to the very common practice of rotating crops, mainly for other reasons, which usually requires different herbicides to avoid crop phytotoxicity and to maximize control of the different weeds. Variable sequences of different herbicides during a crop season are often used to control or manage resistant weeds once they have developed. Mixing or alternating anticoagulants is ineffective because of cross-resistance in rodents. However, the use of an acute rodenticide alternately (or periodically) in a control program with anticoagulants is thought to be the best way to prevent resistance from being selected (Greaves, Jackson, this volume).

Pesticide Formulation Technology

Although additional research is needed to substantiate this tactic, formulation technology can be used in several ways to combat pesticide resistance. It can reduce the dose or rate of pesticide applications. Synergists, adjuvants, penetrants, and materials that improve bait attractancy can be incorporated into pesticide formulations. If resistance is due to differential penetration of an insecticide, the adjuvants or penetrants used in the formulation could be useful to delay or reduce resistance. Changing the attractant in an insect bait could modify the effectiveness or potential resistance to a less effective attractant. Controlled release or longer residual type formulations might enhance the rate of resistance development due to longer selection pressure, but this has not been sufficiently tested and would depend on other factors, such as the life span of the target pest species and the effect of low levels of the insecticide on insect reproduction. No data are available, but the same factors would likely apply to fungicides and herbicides, except for bait attractants. Poorly formulated rodenticide baits could enhance the selection for anticoagulant resistance because these compounds require multiple feedings to be effective. Baits with low palatability will be insufficiently consumed, thus leaving significant numbers of survivors and fostering the selection for resistance. Other factors discussed above also would likely apply (Jackson, this volume).

Synergists

The use of pesticide synergists as a tactic for resistance management has been of special interest, but further study is required to evaluate the practical use of this tactic. It is generally based on the use of a second chemical that counteracts or inhibits the mechanism responsible for resistance to the pesticide. Insecticide synergists inhibit specific detoxification enzymes and thus can reduce or eliminate the selective advantage of individuals possessing such enzymes. Synergists as inhibitors of oxidases (e.g., piperonyl butoxide), dehydrochlorinase (e.g., chlorfenethol), esterase (e.g., DEF), and other more recent enzyme inhibitors have found some use in field applications. Their utility to inhibit the evolution of resistance would depend on the absence of an efficient, alternative mechanism of resistance in the target population. The relatively high cost of the synergist, formulation problems, the potential synergism of mammalian toxicity, and the high level of biochemical adaptation in some major insects (e.g., house fly), have militated against their use. Increased rate of metabolism by the target pathogen is not a common mechanism of fungicide resistance, but it does occur in a few cases. Furthermore, it has been shown that synergism may counteract development of resistance (e.g., a fungicide that inhibits respiration has increased the uptake of fenarimol by a fenarimol-resistant strain, thereby making the resistant strain again sensitive). The use of synergists may not be applicable with herbicides. Some synergistic interactions between herbicides (e.g., atrazine and tridiphane) have been reported due to reduced metabolism of atrazine triggered by an enzyme inhibition from tridiphane. Herbicide resistance, however, has not been due to enhanced metabolism of the herbicide by the resistant weeds. A combination of antibiotic (to reduce production of vitamin K by gut bacteria) with anticoagulant (Prolin®) appeared to give no field advantage to the formulation and would not be expected to impact on resistance development with rodenticides. Other synergist-type compounds have not been suggested.

Exploiting Unstable Resistance

Pesticide resistance often carries with it, especially during its original development, some deficiencies in fitness, vigor, behavior, or reproductive potential. These characteristics often make the resistant biotype of the target pest more susceptible to other control measures. Unstable resistance can be exploited by using other insecticides or control programs to control resistant insects preferentially or selectively until resistance diminishes. Resistant plant pathogens may be unstable at time of initial mutation or development and should be more easily controlled then. It is important to determine if resistance is stable, fit, and genetically based. By use of fungicides in which resistance is associated with a lack of fitness, the resistant mutants would not survive when the selection pressure is removed. Weed biotypes resistant to herbicides are usually less fit or competitive than the susceptible population and may be more easily controlled with alternate herbicides. In England resistant rats reportedly have higher vitamin K demand than normal rats and thus do not survive well (Greaves, this volume), although resistance in monitored English populations seems to have reached an equilibrium point rather than decreasing toward extinction.

Pesticide Selectivity

Selective insecticides often eliminate the pest species while preserving or causing less injury to the predators and beneficial insects. This is an IPM approach and will help to delay or prevent resistance development by providing additional mortality factors for resistant pests. A selective pesticide is often a specific single-target-site chemical with a higher resistance risk, but this danger might be alleviated somewhat by using less specific pesticides applied more selectively, for example in baits, as systemic insecticides in furrow, or as seed treatments. This approach is the most useful in management of resistance in insects and mites. The use of compounds with multisite action has not been a tactic to manage resistance in weeds or rodents.

New Toxophores with Alternate Sites of Action

The discovery and development of new pesticides has often been viewed as a major approach to management of resistance to earlier pesticides. Replacing older pesticides with new ones because of pest resistance has never been the primary objective of this predominately industrial activity, however. It is obvious that future priorities in pesticide development should give more attention to new or alternate target sites that will have lower risk of resistance development. While we need to encourage new discoveries, we must do everything possible to preserve all of our present pesticides. This strategy is a vital and relatively long-term solution to the control of pests resistant to current pesticides, but it can never be a permanent solution. Pests are likely to evolve various means to survive any new pesticides and other control measures. It is also becoming more difficult and expensive to make new and novel chemical discoveries. We are fortunate to have available many types of herbicides with different modes of action, but we can still benefit from breakthroughs in new chemistry to control resistant or problem weeds in certain crops. Development of new types of rodenticides has contributed to resistance management in recent years. New materials include bromethalin, vitamin D_3, and alphachlorhydrin.

Protection and Use of Natural Enemies with Pesticide Tolerance

The intentional protection of natural enemies of pests or the introduction of such predators, especially those with natural or induced tolerance to the specific pesticide, has become a tactic of much interest in resistance management. The development and release of predators with some level of resistance has shown promising results in managing insecticide resistance. Such predator resistance is usually intentionally developed by laboratory exposure during several generations. Genetic engineering offers even more potential in this area. It should be pointed out that the use of several of the previously described tactics will tend to interfere with or counteract this tactic. The use of resistant beneficials has not been used to manage fungicide resistance successfully. Laboratory data indicate that it could be a useful tactic where populations of soil antagonist strains of microorganisms (e.g., *Trichoderma* and *Gliocladium*) are used in an IPM approach. The use of resistant beneficials is not applicable to herbicides and is often not compatible with rodenticide use. With most rodents, their predators are slow breeding, are unable to match the rapid build-up of mouse and rat numbers, and are ineffective in structured urban environments.

Reintroduction of Susceptible Pests

Increasing or encouraging the immigration of susceptible pest genotypes can be effective in dealing with a small insect population with a high proportion of resistant individuals. This tactic shifts the population away from a critical frequency of resistance. The reintroduced susceptibles must be numerous enough to swamp the endemic, resistant population, thereby reducing the likelihood of mating between resistant individuals (Suckling, 1984). This tactic is often most applicable where pest control is not intensive. It is not likely to be an appropriate tactic for managing resistant fungi, weeds, or rodents.

RECOMMENDATIONS

1. Efforts should be expanded to develop IPM systems and steps taken to encourage their use as an essential feature of all programs to manage resistance.

2. Increased research and development emphasis should be directed toward laboratory and field evaluation of strategies and tactics for preventing or slowing resistance development, including efforts to:

   a. Develop models, to be tested in laboratory and field experiments, to assist in formulating hypotheses on managing resistance.

b. Develop and validate sampling and bioassay techniques for monitoring low levels of resistance.
c. Identify chemicals with negatively correlated cross-resistance and develop rotations or mixtures based on this information.
d. Determine stability of resistance in pest populations to specific pesticides.
e. Evaluate alternations and rotations of pesticides shown by research findings or field experience to have high potential as tactics to manage resistance. Design rotation schedules that will maintain acceptable levels of susceptibility.
f. Investigate—in laboratory and field—basic genetic, toxicological, and ecological factors that influence the rate of resistance development.
g. Use traditional and biotechnological genetic methods to produce pesticide-resistant biological control agents and herbicide-resistant crop plants.
h. Investigate pest migration and the factors that influence it to determine the potential for assessing the spread of resistant forms to new areas and the reinvasion of resistant populations by susceptible pests from refuges.

3. Population biologists, toxicologists, and modelers should be involved in designing and executing research and validation efforts.

4. The private sector, extension personnel, and regulatory agencies should encourage the use of promising tactics to manage resistance, while attempting to confirm or validate their usefulness (Davies, 1984).

5. As part of overall IPM strategy to manage resistance, increase efforts to understand and use components of those genetic, reproductive, behavioral, and ecological factors that may minimize the need for pesticide use and reduce resistance development.

6. The traditional method for dealing with resistance has been to switch to a new pesticide. This does not address the problem of resistance, but at best simply delays its recognition and may exacerbate it through cross-resistance. For these reasons and because further discovery and development of new and better pesticides is uncertain, greater efforts must be made to conserve existing materials as finite resources.

7. Do not depend totally or too much on any one pesticide or means to control any pest, especially with high-risk pesticides against major pests.

8. When resistance occurs, move promptly to take necessary actions and apply the best tactics to manage the resistance with all tools and technology we have available.

9. Encourage the use of crop rotations so that different herbicides will be used in successive seasons on different crops.

10. Industry should continue to search for and develop new toxophores

and, in some cases, new synergists, with emphasis on new mechanisms or approaches (e.g., behavioral-type insecticides, multisite fungicides, etc.), rather than to kill the pest by direct, immediate, and single-site action.

REFERENCES

Davies, R. A. H. 1984. Insecticide resistance: An industry viewpoint. Pp. 593–600 *in* 1984 Proc. Br. Crop Prot. Conf. Pests and Dis.

Georghiou, G. P. 1980. Insecticide resistance and prospects for its management. Residue Rev. 76:131–145.

Georghiou, G. P., A. Lagunes, and J. D. Baker. 1983. Effect of insecticide rotations on evolution of resistance. Pp. 183–189 *in* IUPAC Pesticide Chemistry, Human Welfare and the Environment, J. Miyamoto et al., eds. Oxford: Pergamon.

Georghiou, G. P., and C. E. Taylor. 1977. Operational influences in the evolution of insecticide resistance. J. Econ. Entomol. 70:653–658.

Suckling, D. M. 1984. Insecticide resistance in the light brown apple moth: A case for resistance management. Pp. 248–252 *in* Proc. 37th N.Z. Weed and Pest Control Conf.

WORKSHOP PARTICIPANTS

Tactics for Prevention and Management

HOMER M. LEBARON (*Leader*), Ciba-Geigy Corporation
DANIEL ASHTON, Bowling Green State University
AHMED NASSIR BALLA, Agricultural Research Corp., The Sudan
FAUSTO CISNEROS, The International Potato Center, Peru
R. A. H. DAVIES, ICI Plant Protection Division, Great Britain
DONALD E. DAVIS, Auburn University
JOHAN DEKKER, Agricultural University, Wageningen, The Netherlands
TIMOTHY J. DENNEHY, Cornell University
VOLKER DITTRICH, Ciba-Geigy, Ltd., Switzerland
GEORGE P. GEORGHIOU, University of California, Riverside
EDWARD H. GLASS, New York State Agricultural Experiment Station, Cornell University
KENNETH S. HAGEN, University of California, Albany
WAYNE HARNISH, FMC Corp.
WILLIAM B. JACKSON, Bowling Green State University
JOHN R. LEEPER, E. I. du Pont de Nemours and Company
JAMES V. PAROCHETTI, U.S. Department of Agriculture
FRED W. SLIFE, University of Illinois
HOWARD WEARING, DSIR, New Zealand

Pesticide Resistance: Strategies and Tactics for Management.
1986. National Academy Press, Washington, D.C.

Resistance in Weeds

FRED W. SLIFE

The evolution of herbicide development and use is presented. Weed control began around 1900, with little success. The current weed-control era began in the early 1940s and has been very successful. Problems, however, are increasing. A few weed species are becoming resistant to herbicides and are filling the ecological niches opened when herbicides are used. Soil microorganisms are increasing the rate of degradation of herbicides. Programs need to be developed to identify the problem early. Possible effective control methods (mixtures, rotations, and cultivation) are described.

INTRODUCTION

Ever since man first disturbed the natural flora to cultivate desirable plants, weeds have been a problem. The weed problem has persisted because of the great reproductive capacity of weeds, primarily seed production.

By 1900 the native weed flora in the United States had been supplemented with nearly all of the major agricultural weeds from around the world. The universal occurrence of weeds as constant components of the agricultural environment, as compared with the epidemic nature of other pests, delayed recognition of weed control in crop production.

The first attempts to use selective herbicides to control broadleaf weeds in small grains, around the year 1900, were unsuccessful. Weed-control methods consisted of crop rotation, row cultivation, fallowing, hand pulling, and hoeing. The introduction of the tractor in the early 1900s resulted in a rapid expansion of crop production and improved weed control to some degree.

During the 1930s, crop rotation was emphasized because it was considered to be the best weed-control method available. Growing crops with different life cycles, along with the variation in management associated with each crop, prevented any one type of weed from becoming dominant. Unfortunately rotations allowed a great diversity of weed species (annuals, water annuals, perennials) to persist in harmony. Most weed seeds have a high degree of dormancy that extends longer than any practical rotation (Chepil, 1946). Also, rotations and tillage practices seem to have little effect on weed populations (Dunham et al., 1958).

The current weed-control era began with the introduction of 2,4-D and MCPA in the early 1940s. The success of 2,4-D and MCPA to control broadleaf weeds in grass crops undoubtedly was the stimulus for the chemical industry to search for new herbicides. By 1960, however, problems with herbicides began surfacing, especially with 2,4-D in the U.S. Corn Belt. Although herbicides decreased, the dominant broadleaf weed complex, annual grasses, became the dominant problem. The use of 2,4-D created variations within a species and injured some corn cultivars, depending on the genetic base. Some populations of Canada thistle (*Cirsium arvense*) were severely affected by a single treatment of 2,4-D while others were not affected.

From the mid-1950s to the early 1960s a series of soil-applied herbicides became available. These compounds were so successful that soil herbicide treatments used as preplant incorporated or preemergence became the predominant mode of use. Use of combinations of herbicides increased and postemergence herbicides were used as needed.

Since the mid-1960s, weed-control programs using herbicides as the primary control measure have given the highest degree of weed control yet achieved on agricultural lands. The magnitude of the weed problem in both cultivated and noncultivated areas has been greatly reduced.

POTENTIAL PROBLEMS

Many weed scientists see the changing weed spectrum as being the greatest challenge facing chemical weed control. Herbicides have successfully opened a niche in the weed ecosystem that a few weed species tolerant to herbicides are filling. For example, broadleaf perennials are increasing in the cultivated areas, and other tolerant species have come in as seeds dispersed by air and other means.

Enhanced degradation of compounds by soil microorganisms, as is occurring with the thiocarbamate herbicide EPTC, is another potential problem. Enhanced degradation has been found only where EPTC has been used annually for 10 years. It may be that other carbamate pesticides, particularly soil-applied insecticides, are also subject to enhanced degradation. Pesticide rotation seems to modify the problem since EPTC performs well when rotated with other herbicide treatments. There is no indication that enhanced degradation is occurring with other classes of herbicides.

The development of resistance is both a blessing and a curse. It has appeared early enough that it can be corrected, and it may be possible to transfer the herbicide resistance in weeds to closely related crop species (LeBaron and Gressel, 1982). Currently it is limited to herbicides that have a specific site of action. Resistance first developed with the intensive use of the s-triazine herbicides, particularly atrazine and simazine in ornamental plantings where relatively high rates of herbicides were used annually for 10 years.

Because these herbicides have been so effective, they are used extensively in minimum-till or no-till corn production, and they have often been used at high rates to control all vegetation for prolonged periods (for example, along railroads, roadsides, and industrial sites). The triazine herbicides made it practical to grow large acreages of monoculture crops (e.g., corn and sorghum) and to depend more exclusively on these herbicides for weed control. In recent years, especially since the expiration of the triazine patents, the cost of these herbicides has decreased, providing incentive for even greater use.

It is not surprising that weed resistance appeared in these heavy triazine-use areas by 1970. Currently some 38 weed species with biotypes resistant to the *s*-triazine herbicides (atrazine) have been identified. Resistance has been identified in 25 states in the United States, 4 Canadian provinces, and 10 other countries. Perhaps the most critical areas are in Hungary and Austria, where resistance has developed extensively in monocultural corn systems. Failure to react to the problem by crop rotation or alternative controls has made atrazine an ineffective herbicide in some areas of these countries. In North America large areas of atrazine-resistant weeds have appeared in the intensive corn area of Ontario, Canada, and in the mid-Atlantic states, particularly Maryland.

Weed resistance has been confirmed in at least six classes of herbicides, the most recent being the emergence of a resistant biotype of goosegrass (*Eleusine indica*) to the dinitroaniline herbicides (Mudge et al., 1984). Weed resistance seems to be occurring rapidly in many areas where a single herbicide is used repeatedly with little or no cultivation. Resistance will develop most rapidly in orchards, nurseries, railroads, and other noncultivated areas if the potential for resistance is not recognized quickly.

IDENTIFICATION OF PROBLEM

Herbicide-resistant weeds are probably present in most weed populations. Since they may not be morphologically different than sensitive biotypes, identification can be difficult. Herbicide users know that results vary greatly with environmental weed-control conditions. Weed "escapees" are common with most treatments; in the past this has been attributed to lack of control rather than resistance.

Education programs aimed at identifying weed escapees can do a great deal in identifying the problem early. For example, in the case of atrazine use on corn, extension programs could be aimed at scouting treated fields for weed escapees, particularly when environmental conditions are ideal for herbicide performance. Additional emphasis could be placed on the appearance of *Amaranthus* spp. (pigweeds) or *Chenopodium* spp. (lambsquarters). These species, normally sensitive to atrazine, have developed resistance very rapidly with continuous atrazine use.

With any herbicide treatment the appearance of weed species normally sensitive to that treatment is cause for concern. Prompt identification and verification of resistance should be made.

CONTROL

Control techniques that can be used to prevent weeds from becoming dominant and to reduce the seriousness of resistant populations include herbicide mixtures, herbicide rotation, crop rotation, and increased cultivation. Some of these measures will be more costly than current programs. In countries where agricultural production is highly developed, weed-control programs will change to meet the challenge of weed resistance. In other countries where crop production is less productive, the use of new, more expensive control programs will be much slower.

Herbicide Mixtures

Mixtures generally give superior weed control, compared with a single herbicide treatment. Mixtures are selected for the weed spectrum each component will control; thus, high rates are avoided. This strategy delays the emergence of weed resistance, compared with using a single herbicide at higher rates. In addition herbicides known to produce weed resistance when used at high rates alone can still be used effectively in mixtures to control sensitive species.

Herbicide Rotation

Change in the herbicide program has been recommended for many years to prevent rapid changes in the weed spectrum. Shaw (1957) emphasized the need for rotational herbicide programs to present rapid changes in weed spectrum. In the countries where herbicide use is high, a wide variety of herbicide treatments are available for major crops. Preplant, preemergence, and postemergence can be utilized effectively to prevent and control resistant

TABLE 1 Soil Weed Seed Level in 10 Years of Continuous Corn as Affected by the Same Herbicide Treatment, Herbicide Rotation, and No Herbicide with Cultural Practices Only

		Seeds/kg of Soil		
		1965	1970	1975
Grass Seeds				
Corn, Corn, Corn	Same herbicide	23	129	130
	Herbicide rotation	20	5	2
	No herbicide	21	78	95
BL Seeds				
Corn, Corn, Corn	Same herbicide	102	42	4
	Herbicide rotation	98	30	5
	No herbicide	95	120	126
Total Seeds				
Corn, Corn, Corn	Same herbicide	126	171	134
	Herbicide rotation	119	35	6
	No herbicide	116	198	221

SOURCE: Agronomy Department, University of Illinois.

weeds. Recently several new herbicides have become available and more are in late development stages.

The effectiveness of herbicide rotation, compared with using the same herbicide treatment, is demonstrated in Table 1. Herbicide rotation greatly reduced the soil seed bank by preventing seed production, while the continuous use of the same herbicide allowed tolerant species to increase soil seed numbers. In this study the continuous use of atrazine over a 10-year period allowed annual grasses to increase rapidly, but greatly reduced the broadleaf population. This study also indicates that monoculture corn production is possible if herbicide treatments are changed frequently to prevent rapid shifts in the weed spectrum and to prevent or delay the emergence of resistance. Corn yields (Table 2), however, are significantly higher in a rotation than in a monoculture system.

TABLE 2 Crop Yields (1966–1977) as Affected by Weed Management Systems and Crop Rotation (bu/A)

Rotation	No Herbicide 3 Cultivations	Some Herbicide 1 Cultivation	Herbicide Rotation 1 Cultivation
Corn, Corn, Corn	104.1	127.1	131.4
Corn, Corn, Soybeans[a]	114.8	135.7	140.0
Corn, Corn, Soybeans[b]	111.6	133.3	138.3
Corn, Soybeans, Wheat	116.7	138.8	142.0

[a]First year corn yields.
[b]Second year corn yields.

SOURCE: Agronomy Department, University of Illinois.

The rapid decline in soil-weed seed in some areas is an opportunity to refine weed-control systems by effectively utilizing postemergence herbicides. For example, in cotton growing regions soil-applied grass herbicides have been used for many years. Many farmers also grow soybeans. The dinitroaniline herbicides are widely used on both crops. The grass-weed pressure in these crops is low, giving the farmer the alternative of using new postemergence grass herbicides. Row cultivation, which is a part of the weed management system, may be effective enough to eliminate the need for grass herbicides in a particular year. The strategy of using postemergence herbicides to treat the existing weed problem has high potential for delaying the emergence of resistant weeds.

Weed resistance has appeared as early as five to six years after the continuous use of the same herbicide. Most common areas are no-till corn production, nurseries, railroads, and other noncultivated areas. By stopping seed production of the herbicide-sensitive species, the resistant weeds, which are generally less fit, can dominate quickly and rapidly build up the soil seed bank.

Roberts (1968), in a series of eloquent studies over a period of years, clearly showed that the longevity of weed seeds in cultivated areas of the soil was relatively short. He demonstrated that seed loss out of the soil seed bank would occur at the rate of 50 percent per year if no new seeds were allowed to replenish the soil. Schweizer and Zimdahl (1984a,b) reported a 96 percent decline in the soil seed bank after six years in a crop-herbicide rotation study and a 98 percent decline in a continuous corn system.

Crop Rotation

Crop rotation is an important strategy that can delay the development of resistance and reduce resistant weed populations. This is true only if a variety of well-chosen herbicides are used for weed control. Unfortunately farmers are often reluctant to change their current cropping system, especially if their present program has been successful. Changing the crop is the last option most farmers will choose, since they have acquired knowledge and experience in growing, harvesting, and marketing the crop at a profit.

Crop rotation will probably receive additional emphasis because of the developing resistance problem in all pest-control disciplines. For example, in the U.S. Corn Belt alternating corn and soybeans has greatly reduced the need for corn rootworm treatments. Because crop rotation also controls the soybean cyst nematode, soybeans can be a viable crop in areas heavily infested with the nematode.

Increased Cultivation

Herbicides have been so effective that the amount of tillage previously used in crop production has been reduced. (Reduced tillage has the added benefit of decreasing soil erosion.) In no-till agriculture row cultivation is seldom used, and even in clean tillage fields row cultivation has declined.

In the U.S. Corn Belt most corn is grown in rotation, usually with soybeans. But several Corn Belt states have as much as 25 percent of the crop in continuous corn. Atrazine is generally used annually, either alone or in combination with a grass-specific herbicide in the continuous corn areas, and one- or two-row cultivations are common. Weed resistance has not been a factor in these areas thus far, possibly because of the tillage component. Weed resistance, however, is slowly evolving in areas where cultivation is a part of the management program.

In reduced tillage systems, and particularly no-till, herbicide rotation will delay the appearance of resistance. Increased rates of herbicides are required for no-till fields; thus, cross-resistance may develop. Developing tillage equipment that will control weed escapees but not destroy the surface residue would be highly desirable.

Weed control in the noncultivated areas will become more difficult and expensive as resistance becomes more widespread. Chemical rotation is critical for these areas, since there are a limited number of herbicides available for use along railroads, industrial sites, and highways. Excellent progress has been made in reducing and even eliminating grass mowing on many highways. Perennial grass competition has eliminated most annual and perennial weeds, resulting in less herbicide use. It may be possible to extend this type of weed control to other areas. Several plant growth regulators are available that inhibit foliage growth and reduce grass mowing costs. This approach to weed control is more economical than the continuous use of herbicides.

CONCLUSION

Chemical weed control used with good cultural practices has become the standard weed-control program in many parts of the world. As yet no good alternatives to these programs exist in high-production agricultural areas. Excellent progress has been made in biological control, and as more information becomes available on weed threshold levels, it may be possible to reduce herbicide use. Integrated Pest Management (IPM) programs are needed that are designed to identify more precisely and to control tolerant and resistant weed species and to enhance chemical degradation in the soil. The high degree of weed control achieved in recent years can continue if the seriousness of these problem areas is lessened.

REFERENCES

Chepil, W. S. 1946. Germination of weed seeds. Longevity, periodicity of germination and viability of seeds in cultivated sods. Sci. Agric. 26:307.

Dunham, R. S., R. G. Robinson, and R. N. Anderson. 1958. Crop rotation and associated tillage practices for controlling annual weeds in flax and reducing the weed seed population of the soil. Minn. Agric. Exp. Stn. Tech. Bull. 230.

LeBaron, H. M., and J. Gressel, eds. 1982. Herbicide Resistance in Plants. New York: John Wiley and Sons.

Mudge, L. C., B. J. Gossett, and T. R. Murphy. 1984. Resistance of goosegrass (*Eleusine indica*) to dinitroaniline herbicides. Weed Sci. 32:591.

Roberts, H. A. 1968. The changing population of viable weed seeds in arable soil. Weed Res. 8:253.

Schweizer, E. E., and R. L. Zimdahl. 1984a. Weed seed decline in irrigated soil after six years of continuous corn (*Zea mays*) and herbicides. Weed Sci. 32:76.

Schweizer, E. E., and R. L. Zimdahl. 1984b. Weed seed decline in irrigated soil after rotation of crops and herbicides. Weed Sci. 32:84.

Shaw, W. C. 1957. Basic research—the key to efficient weed control in cotton in the future. Rep. Proc. Beltwide Cotton Production-Mech. Conf. 11:5–9. Memphis, Tenn.: National Cotton Council of America.

Pesticide Resistance: Strategies and Tactics for Management.
1986. National Academy Press, Washington, D.C.

Preventing or Managing Resistance in Arthropods

JOHN R. LEEPER, RICHARD T. ROUSH, and
HAROLD T. REYNOLDS

Insecticide resistance is a widespread problem for which management tactics have been developed but have not been put into widespread practice. Genetic, reproductive, behavioral/ecological, and agronomic/control factors—over which we have varying degrees of control—influence the rate of resistance development and are key to its management. Resistance management tactics should be aimed at reducing allele frequencies, reducing dominance, and minimizing the fitness of resistance genotypes. Adequate information to confidently choose which of these tactics to use is lacking and prevents their practical use. Basic resistance research in genetics, biochemistry, physiology, and toxicology on agronomic pests is needed. The discriminate use of insecticides needs to be strengthened within integrated pest management. Improved monitoring techniques that allow for the detection of resistance at low frequencies within populations are needed.

INTRODUCTION

Many resistance management tactics have been identified over the past 40 years, but few have been put into practice; of those, most are being used to improve crop production rather than to manage resistance (e.g., economic thresholds rather than calendar spray schedules). Another problem in managing pesticide resistance is that each interest group (e.g., pesticide manufacturers, regulators, researchers, extension personnel, farmers, and public health workers, and the consumer) has a different perspective on the problem and on how it should be solved. These scientific, economic, and social/political constraints increase the complexity of the problem, because not only must we develop scientific answers to the resistance problem, we must also develop answers that meet the needs of the different interest groups.

RATE DETERMINING FACTORS

Resistance develops at different rates between species and even between populations of the same species due to genetic, reproductive, behavioral/ecological, and operational factors (Georghiou and Taylor, 1977a,b; Georghiou, 1980a,b, 1983; Wood and Bishop, 1981). The general consensus is that only the operational factors can be manipulated—everything else is beyond our control (Wood and Bishop, 1981; Georghiou, 1983). The only limitation to what is "operational," however, may be our ability to recognize how to manipulate it. For example, migration in and out of treated habitats is generally assumed to be a biological factor beyond our control. Croft and colleagues, however, have been experimenting with techniques such as pheromone lures to reintroduce susceptible genes into the treated habitats (Croft, 1984). Also, dominance of resistance was considered a genetic, nonoperational factor until Curtis et al. (1978) introduced the concept that dominance might be modified by the insecticide dose applied (effective dominance).

Directly changing pest biologies holds promise for indirectly manipulating resistance development. For example, the *Heliothis* complex, including *Heliothis zea* (Boddie) and *H. virescens* (F.), are among the most chronic, difficult to control pests in North American cotton. *H. virescens* is particularly troublesome because it has developed resistance to every major insecticide class (Sparks, 1981; Martinez-Carrillo and Reynolds, 1983). An alternative to chemical control, which can be considered an indirect resistance management tactic, is the *Heliothis* backcross hybrid (Proshold et al., 1983). Crosses of *H. virescens* with *H. subflexa* (Guenee) produce fertile daughters and sterile sons (Laster, 1972), which is perpetuated through successive generations and can reduce the rate of population increase.

Spider mites (*Tetranychus* spp.) are pests of many orchard and field crops throughout the world. Cotton seedlings can be induced to produce substances, through infestation with mites, that dramatically retard mite population growth on reinfestation (Karban and Carey, 1984). These substances also can be transported systemically within the plants and will have some degree of residual activity (up to 12 days). Although it may be some time before it is practical to inoculate cotton plants with mites to prevent mite outbreaks in the field, more immediate practical benefits from this research are possible. Plant breeders and genetic engineers could develop plant varieties with elevated intrinsic levels of the responsible substances. Or chemicals could be developed that, when applied to crops, would induce production of plant chemicals.

Influencing the reproductive rate of arthropods also offers potential for resistance management. Reducing the number of offspring per generation or the number of generations per year may reduce the need for insecticide applications. Although these tactics have not been held in high regard, because they have not been effective enough to replace pesticides, they could be used together with other pest-management practices.

These examples illustrate that some genetic, reproductive, and behavioral/ecological factors have operational components. Therefore, the term "agronomic/control" should be substituted for "operational." Agronomic refers to the various cultural practices in cropping systems, while control refers to the control and management practices in both agricultural and medical/veterinary situations. This change in terminology (1) more clearly defines the factors, (2) opens areas for consideration not traditionally thought to be within our control, and (3) encourages the further development of novel tactics less directly related to insecticide use.

TACTICS

The tactics thus far developed to prevent or manage insecticide resistance have tended to be directly related to insecticide use, which is expected, since primarily toxicologists and entomologists have addressed the problem. Insecticides, however, are only one part of resistance development. For example, the rate of change in allele frequency at any given locus in a closed population is a function of initial allele frequency, dominance, and the relative fitness of the various genotypes (Futuyma, 1979). Resistance develops more rapidly with dominance, higher gene frequencies, and a greater fitness advantage to resistant genotypes (Georghiou and Taylor, 1977a). One objective of resistance management is to maintain resistance alleles at very low frequencies. Thus, resistance management tactics should be aimed at reducing allele frequencies, reducing dominance, and minimizing the fitness of resistant genotypes.

Reducing Frequencies of Resistant Alleles

A commonly suggested method for directly reducing resistance allele frequencies is by diluting them through the mass release of susceptible insects, for example, the mass release of susceptible male mosquitoes to dilute resistance (Curtis et al., 1978). This tactic has not been put to practical field use, partly because of the cost of such a program. Another suggested method has been to eradicate resistance foci. Stringent quarantine measures and alternative controls could be used to eliminate newly established resistant foci (Sutherst and Comins, 1979). This approach requires extensive quarantine procedures and improved detection capabilities.

Decreasing Dominance of Resistance

High insecticide use rates can change the effective dominance of resistance (Curtis et al., 1978); rates that kill heterozygotes can make resistance effectively recessive. Immigration of susceptible individuals and low-resistance gene frequencies is very important to this approach (Tabashnik and Croft, 1982). The rates required to kill heterozygotes, however, might not be economically practical and might not be identified until after the heterozygotes achieve a high frequency within a population.

Minimizing Fitness of Resistant Genotypes

Most resistance management tactics involve reducing fitnesses of resistant genotypes relative to susceptible genotypes by either preserving susceptible homozygotes or eliminating heterozygotes and resistant homozygotes. Fitness can be lowered by reducing insecticide use rates, extending intervals between treatments, using short residual insecticides, and the like. Determining which tactic is most appropriate and will maintain effective control, however, is difficult. Susceptible homozygotes can be preserved by creating refugia where part of the population is not treated (Georghiou and Taylor, 1977b). Preservation may be achieved by (1) leaving areas unsprayed, (2) using higher action thresholds that tend to reduce the number of insecticide applications, (3) applying short residual compounds that reduce the effective exposure time to the remaining or immigrant subeconomic pest population (Denholm et al., 1983), (4) using selective insecticides that do not exert pressure on other species (both pest and beneficial), and (5) relying on noninsecticidal controls (biological and cultural) that may further reduce the need for pesticide applications.

Even when insecticides must be applied, reduced rates may preserve some of the susceptible homozygotes—and some beneficial arthropods—which may further reduce the need for subsequent applications (Tabashnik and Croft, 1982). The use of reduced rates, however, may not always provide economic control, and this requires more attentive scouting.

Conversely, a tactic for eliminating heterozygotes and resistant homozygotes is increased insecticide rates (Taylor and Georghiou, 1979). Tabashnik and Croft (1982) describe the conditions to determine the choice between the reduced rate (low dose) and increased rate (high dose) approaches. The information required to make an appropriate decision, including genetic data on phenotypic expression in heterozygotes and allele frequency, is generally lacking.

Other chemical approaches may be used to suppress or eliminate resistance alleles from a population. These kill heterozygotes and resistant homozygotes but often require the reintroduction of susceptible individuals, just as the increased dose tactic does. Insecticide mixtures are a common tactic, but to work most effectively the compounds must have different modes of action and metabolism, and the frequencies of resistance alleles to each insecticide must be low. Thus, individuals surviving one insecticide are likely to be killed by the other (Georghiou, 1980b). The common practice with mixtures is to use reduced rates of each insecticide, which sometimes may not be sufficient to delay resistance (Suthert and Comins, 1979). Also, using two insecticides at full rates may be less expensive than using one insecticide at the rate sufficient to kill the heterozygotes.

Materials with negative cross-resistance, those that decrease resistance to other chemicals as resistance to them increases, have a potential value in resistance management. Negative cross-resistance has been documented in both Diptera (Ogita, 1961a,b) and Homoptera (Ozaki, 1980). Although the benefits of negative cross-resistance have not been demonstrated in the field (Sawicki, 1981), they might be most efficient as mixtures.

Synergists suppress metabolic resistance mechanisms and, therefore, can prevent or overcome resistance (Ranasinghe and Georghiou, 1979). (Most resistance management tactics only delay resistance.) Unfortunately, the available synergists have undesirable characteristics, including photoinstability and phytotoxicity. Marketing and registration considerations limit the development of new synergists, and synergists cannot prevent the development of resistance through alternative means (Oppenoorth, 1976).

Where possible, insecticides conferring the lowest level of resistance are preferred, because their use reduces the selective advantages to individuals carrying resistant genotypes (Devonshire and Moores, 1982). Thus, compounds causing low levels of resistance delay its development, similar to synergists, because resistant individuals can often be killed with only a slight increase in dose.

Treating life stages where genes for metabolic mechanisms of resistance are not expressed (or only poorly expressed) is another direct tactic. For example, *Spodoptera littoralis* (Boisduval) adults and eggs are more susceptible to organophosphates than larvae, apparently due to higher microsomal cytochrome P_{450} levels in the larvae (Dittrich et al., 1980). Metabolic forms of resistance can, however, develop in adult arthropods (Plapp, 1976). This tactic would require a major change in the philosophy and mechanics in programs because control is redirected at nondamaging stages.

Although an indirect approach, insecticide rotations (alternations) can reduce resistance allele frequencies, assuming that resistant genotypes have substantially lower fitness than the susceptibles. Therefore, their frequency declines during generations between applications of the compound (Georghiou, 1980b).

Tactical Considerations in Insecticide Application

Although noninsecticidal controls that indirectly affect resistance development may become more important in suppressing populations and managing resistance, pesticides will continue to be the major control tools in the near future. Pesticide use, however, forces us to choose between mixtures, rotations, and sequences in application (Georghiou, 1980b), and adequate information to confidently choose which tactic to use is lacking. Sequences are normally forced on us by the failure of one compound and the registration of a new compound.

Keiding (1977) suggested that insecticides with simple one-factor resistance and limited cross-resistance, such as malathion, be used first in a sequence and that compounds with complicated multiple resistance or that act as selectors for resistance to other insecticides, such as dimethoate, be avoided or used last. This information, however, only became available through hindsight (Sawicki, 1975). Whether this information can be automatically extrapolated to other systems without recognizing possible metabolic differences is questionable.

A key assumption about rotations is that resistant genotypes are at a significant competitive disadvantage in the absence of selection pressure. Although resistance usually declines in the absence of a pesticide, the rates of decline may be too slow to be of much practical benefit (Curtis et al., 1978; Georghiou et al., 1983; Roush and Plapp, 1982; Emeka-Ejiofor et al., 1983). Thus, rather than significantly extending the number of times that an insecticide can be used, alternation may allow an insecticide to be used only half as often in twice as many seasons.

The use of insecticide mixtures is not without problems. Sometimes resistance to both compounds used in mixtures has developed rapidly. Some authorities on resistance feel that mixtures should never be used (Keiding, 1977). The potential utility of insecticide mixtures has been investigated experimentally since the early 1950s and has failed in some of these studies (Lagunes, 1980). Other studies have indicated that mixtures are more effective than rotations in preventing resistance development (MacDonald et al., 1983).

There are several possible explanations for these inconsistencies. Cross-resistance can occur among some of the pesticides used in the early studies of mixtures. Most field trials were conducted on such a small geographical scale, for example, within an orchard (Asquith, 1964), that resistant individuals in one plot could easily contaminate others. More important, however, most studies were conducted on "closed" laboratory populations, where there was no immigration of susceptible individuals and where the entire strain was treated in every generation. Various theoretical models (Kable and Jeffery, 1980) indicate that insecticide mixtures can significantly delay resistance development only when a portion of the population of each generation escapes selection. The theoretical models make good sense. If the entire population is treated, only those rare individuals with resistance to both pesticides can survive, and their offspring will be highly resistant. If, however, some susceptible individuals escape treatment, as usually happens, they can greatly dilute the resistance carried by the few individuals that survived the application. More research is needed to define clearly the resistance management potential of these pesticide application philosophies.

Much of the work necessary for understanding the genetics, biochemistry, physiology, and toxicology of resistance has been conducted on Diptera, primarily the house fly and mosquitoes (Georghiou, 1983). The work has also been valuable in developing a "model" of the general insect system and resistance. It would be dangerous, however, to extrapolate directly to agronomic pests what has been learned on these medically important Diptera. The metabolisms of the house fly and mosquitoes evolved under extremely different selection pressures than those of phytophagous insects (Swain, 1977; Brattsten, 1979a,b) and, therefore, may have different major detoxification pathways. With the relatively recent appreciation of the role different food sources have played in the evolutionary development of metabolic pathways, the necessity for conducting basic resistance research in genetics, biochemistry, physiology, and toxicology on agronomic pests (e.g., Lepidoptera, Coleoptera, and Acarina) has been advocated (Sawicki, 1981; Metcalf, 1983).

CONSTRAINTS ON AUGMENTING TACTICS

Implementing the above tactics will be more advantageous if the scientific, economic, and social/political constraints are recognized. The economic and social/political constraints are covered in detail in other papers in this volume (Dover and Croft, Frisbie et al., Miranowski and Carlson). Some trends appear to be eroding the advances made in integrated pest management (IPM), which has serious implications for resistance development.

Erosion of Integrated Pest Management

In the past, broad-spectrum, long-residual insecticides were applied on a calendar schedule, which continuously exposed both pest and beneficial insect populations. When lead arsenate and DDT were used, calendar spraying was thought to be inexpensive insurance for a quality crop. The first recognized cost added to this practice was the development of resistance and the loss of control within the pest populations. Farmers switched to new insecticides under development and continued on what has been aptly termed the pesticide treadmill (van den Bosch, 1978). IPM, developed in the mid 1970s, offered the farmer an opportunity to reduce pesticide applications by more critically timing and directing his sprays. The development and evolution of IPM was prompted partly by insecticide resistance. Inasmuch as IPM programs generally reduce pesticide applications, they also minimize resistance selection pressure (Brown, 1981). Although it would be difficult to document, the practice of IPM has surely slowed the development of resistance.

Pesticides are a minor portion of total production costs for many high-value crops. In these systems there is always a temptation to use pesticides as cheap insurance, particularly when farmers are in financial difficulty and as memories of past repercussions grow dim. Thus, resistance management gains made in the past may be lost as IPM programs are gradually eroded.

For example, recent cotton production practices in the United States (such as early-season insecticide use and area-wide management programs) may be eroding past IPM successes. Certain insecticides, including a pyrethroid, have recently been marketed under "yield enhancement programs"; the product is guaranteed by the manufacturer to give higher yields when applied to young cotton. Although the mechanism of yield enhancement is unclear, the insecticide seems to affect insects rather than plant physiology. Such marketing practices help form convictions among private consultants that economic thresholds do not work. Also, the risk in this practice is increased selection pressure on cotton pests. An example of an area-wide management program is that of cotton pest management, where insecticides are applied nearly simultaneously across a several square kilometer community when an economic threshold is reached on a central index field that includes less than 0.2 percent of the area (Phillips et al., 1980).

How much impact the early season and area-wide insecticide treatment programs will have on cotton pest problems and resistance management is not clear yet. They remind us, however, of the importance of socioeconomic factors on resistance management. Optimum yield for short- and long-term benefits is not always the maximum yield.

Resistance Risk Assessment

Much scientific understanding has yet to evolve concerning resistance. Until that information and support are available, social or political expediency might force the premature implementation of a program or tactic. An example of this would be to require a resistance risk assessment when registering an insecticide. Currently, appropriate information on resistance development is available only through hindsight. In addition, compounds that have had resistance develop to them tend to maintain some degree of field utility. Although the potential for resistance development should be considered when choosing an insecticide, it is premature to include risk assessment in the registration process.

Detection

To select the proper tactic for preventing or managing resistance, we must better understand resistance at the levels of the individual and the population (Sawicki, 1981). Therefore, we must develop methods for detecting resistance. Monitoring must be able to detect shifts in susceptibility early in their occurrence within a population. Current monitoring techniques (e.g., topical application, deposit-on-glass, impregnated paper) require large numbers of individuals to detect resistance alleles at low frequencies. This is frequently an impossible task because of sampling constraints, and these methods can become expensive in terms of time and resources. Therefore, techniques to detect rather than document resistance are necessary before we can act, rather than react.

Advances have been made in developing bioassays for detecting carboxylesterase and acetylcholinesterase levels in individual aphids, leafhoppers, planthoppers, and mosquitoes (Miyata et al., 1980; Saito and Miyata, 1982; Miyata, 1983). These tests, which are relatively simple and often can be used in the field, provide more effective means for detecting the frequency of a trait within a field population. They also have disadvantages. A similar test to detect the presence of the most important enzyme system in insecticide detoxification, microsomal oxidases, is currently impossible (L. B. Brattsten, du Pont, personal communication, 1984), as are similar tests for the nonmetabolic modes of resistance (e.g., target site insensitivity, penetration, sequestration, excretion). Although the presence of the enzymes can be detected, their levels cannot be determined. Further advances in test development are required if we are to begin detecting resistance at low population frequencies, which is required for the proper selection of management tactics.

CONCLUSION

Our selection of resistance tactics has been dependent on past successes and failures in the field and a great degree of luck. This is unfortunate because (1) it relies on presupposition rather than scientific fact; (2) the tactic chosen may be inappropriate for the case at hand and may lead to additional complications; and (3) tactic selection, implementation, and validation are primarily based on reaction rather than calculated action.

This realization underscores the critical need for additional basic resistance research in a diverse set of disciplines, including genetics, toxicology, biochemistry, and physiology as well as economic entomology. In addition, we need to validate and further develop—for phytophagous insects—what we have learned on the house fly and mosquitoes. We need to develop an information matrix on the biology, genetics, and modes and mechanisms of resistance to each insecticide for a broad array of species. This matrix should include species where resistance has not been a problem as well as those where it has been a serious problem. This will be no easy task, and questions of responsibility arise. Who is going to conduct the research? How is it to be funded? Who is going to coordinate it? The action taken on these points by policymakers might ultimately determine the success or futility of pesticide resistance management.

ACKNOWLEDGMENT

This paper has been approved as No. 5986 by the Director, Mississippi Agricultural and Forestry Experiment Station.

REFERENCES

Asquith, D. 1964. Resistance to acaricides in the European red mite. J. Econ. Entomol. 57:905–907.

Brattsten, L. B. 1979a. Biochemical defense mechanisms in herbivores against plant allelochemicals. Pp. 199–220 *in* Herbivores: Their Interaction with Plant Secondary Metabolites, B. A. Rosenthal and D. H. Janzen, eds. New York: Academic Press.

Brattsten, L. B. 1979b. Ecological significance of mixed function oxidations. Drug Metab. Rev. 10:35–58.

Brown, T. M. 1981. Countermeasures for insecticide resistance. Bull. Entomol. Soc. Am. 27:198–202.

Croft, B. A. 1984. Immigration as an operational factor in resistance management. Paper presented at Pac. Br. Entomol. Soc. Am. Meet., Salt Lake City, Utah, June 19–21, 1984.

Curtis, C. F., L. M. Cook, and R. J. Wood. 1978. Selection for and against insecticide resistance and possible methods for inhibiting the evolution of resistance in mosquitoes. Ecol. Entomol. 3:273–287.

Denholm, I., A. W. Farnham, K. O'Dell, and R. M. Sawicki. 1983. Factors affecting resistance to insecticides in house flies, *Musca domestica* L. (Diptera: Muscidae). I. Long-term control with bioresmethrin of flies with pyrethroid-resistance potential. Bull. Entomol. Res. 73:481–489.

Devonshire, A. L., and G. D. Moores. 1982. A carboxylesterase with broad substrate specificity causes organophosphorus, carbamate and pyrethroid resistance in peach-potato aphids (*Myzus persicae*). Pestic. Biochem. Physiol. 18:235–246.

Dittrich, V., N. Luetkemeier, and G. Voss. 1980. OP-resistance in *Spodoptera littoralis*: Inheritance, larval and imaginal expression, and consequences for control. J. Econ. Entomol. 73:356–362.

Emeka-Ejiofor, S. A. I., C. F. Curtis, and G. Davidson. 1983. Tests for effects of insecticide resistance genes in *Anopheles gambiae* on fitness in the absence of insecticides. Entomol. Exp. Appl. 34:163–168.

Futuyma, D. J. 1979. Evolutionary Biology. Sunderland, England: Sinauer Associates.

Georghiou, G. P. 1980a. Insecticide resistance and prospects for its management. Residue Rev. 76:131–145.

Georghiou, G. P. 1980b. Implications of the development of resistance to pesticides: Basic principles and consideration of countermeasures. Pp. 116–129 *in* Pest and Pesticide Management in the Caribbean. Proc. Seminar and Workshop, E. G. B. Gooding, ed. Berkeley, Calif.: Consort. Int. Crop Prot.

Georghiou, G. P. 1983. Management of resistance in arthropods. Pp. 769–792 *in* Pest Resistance to Pesticides, G. P. Georghiou and T. Saito, eds. New York: Plenum.

Georghiou, G. P., and C. E. Taylor. 1977a. Genetic and biological influences in the evolution of insecticide resistance. J. Econ. Entomol. 70:319–323.

Georghiou, G. P., and C. E. Taylor. 1977b. Operational influences in the evolution of insecticide resistance. J. Econ. Entomol. 70:653–658.

Georghiou, G. P., A. Lagunes, and J. D. Baker. 1983. Effect of insecticide rotations on evolution of resistance. Pp. 183–189 *in* IUPAC Pesticide Chemistry, Human Welfare and the Environment, J. Miyamoto, ed. New York: Pergamon.

Kable, P. F., and H. Jeffery. 1980. Selection for tolerance in organisms exposed to sprays of biocide mixtures: A theoretical model. Phytopathology 70:8–12.

Karban, R., and J. R. Carey. 1984. Induced resistance of cotton seedlings to mites. Science 225:53–54.

Keiding, J. 1977. Resistance in the housefly in Denmark and elsewhere. Pp. 261–302 *in* Pesticide Management and Insecticide Resistance, D. L. Watson and A. W. A. Brown, eds. New York: Academic Press.

Lagunes, A. 1980. Impact of the use of mixtures and sequences of insecticides in the evolution of resistance in *Culex quinquefasciatus* Say. (Diptera: Culicidae). Ph.D. disseration. University of California, Riverside.

Laster, M. L. 1972. Interspecific hybridization of *Heliothis virescens* and *H. subflexa*. Environ. Entomol. 1:682–687.

MacDonald, R. S., G. A. Surgeoner, K. R. Solomon, and C. R. Harris. 1983. Laboratory studies on the effect of four spray regimes on the development of resistance to permethrin and dichlorvos in the house fly. J. Econ. Entomol. 76:417–422.

Martinez-Carrillo, J. L., and H. T. Reynolds. 1983. Dosage-mortality studies with pyrethroids and other insecticides on the tobacco budworm (Lepidoptera: Noctuidae) from the Imperial Valley, California. J. Econ. Entomol. 76:983–986.

Metcalf, R. L. 1983. Implications and prognosis of resistance to insecticides. Pp. 703–733 *in* Pest Resistance to Pesticides, G. P. Georghiou and T. Saito, eds. New York: Plenum.

Miyata, T. 1983. Detection and monitoring for resistance in arthropods based on biochemical characteristics. Pp. 99–116 *in* Pest Resistance to Pesticides, G. P. Georghiou and T. Saito, eds. New York: Plenum.

Miyata, T., T. Saito, H. Hama, T. Iwata, and K. Ozaki. 1980. A new and simple detection method for carbamate resistance in the green rice leafhopper, *Nephotettix cincticeps* Uhler (Hemiptera: Deltocephalidae). Appl. Entomol. Zool. 15:351–352.

Ogita, Z. 1961a. An attempt to reduce and increase insecticide-resistance in *D. melanogaster* by selection pressure. Genetical and biochemical studies on negatively correlated cross-resistance in *Drosophila melanogaster*. I. Botyu-Kagaku 26:7–18.

Ogita, Z. 1961b. Genetic studies on actions of mixed insecticides with negatively correlated substances. Genetical and biochemical studies on negatively correlated cross-resistance in *Drosophila melanogaster*. III. Botyu-Kagaku 26:88–93.

Oppenoorth, F. J. 1976. Development of resistance to insecticides. Pp. 41–59 *in* The Future for Insecticides, R. L. Metcalf and J. J. McKelvey, Jr., eds. New York: John Wiley and Sons.

Ozaki, K. 1980. Resistance of rice insect pests to insecticides in Japan. Presented at the 16th Int. Congr. Entomol. Kyoto, Japan, August 3–9, 1980.

Phillips, J. R., A. P. Gutierrez, and P. L. Adkisson. 1980. General accomplishments toward better insect control in cotton. Pp. 123–153 *in* New Technology of Pest Control, L. B. Huffaker, ed. New York: John Wiley and Sons.

Plapp, F. W., Jr. 1976. Biochemical genetics of insecticide resistance. Annu. Rev. Entomol. 21:179–197.

Proshold, F. I., D. F. Martin, M.L. Laster, J. R. Raulston, and A. N. Sparks. 1983. Release of backcross insects on St. Croix to suppress the tobacco budworm (Lepidoptera: Noctuidae): Methodology and dispersal of backcross insects. J. Econ. Entomol. 76:885–891.

Ranasinghe, L. E., and G. P. Georghiou. 1979. Comparative modification of insecticide-resistant spectrum of *Culex pipiens fatigans* Wied. by selection with temephos and temephos/synergist combinations. Pestic. Sci. 10:502–508.

Roush, R. T., and F. W. Plapp, Jr. 1982. Effects of insecticide resistance on biotic potential of the house fly (Diptera: Muscidae). J. Econ. Entomol. 75:708–713.

Saito, T., and T. Miyata. 1982. Studies on insecticide resistance in *Nephotettix cincticeps*. Pp. 377–382 *in* Proc. Int. Conf. Plant Prot. in Tropics, K. L. Heong, B. S. Lee, T. M. Lim, C. H. Teoh, and Yusof Ibrahim, eds. Kuala Lumpur: Malaysian Plant Protection Society.

Sawicki, R. M. 1975. Effects of sequential resistance on pesticide management. Pp. 799–811 *in* Proc. 8th Br. Insectic. Fungic. Conf. Suffolk: Lavenham.

Sawicki, R. M. 1981. Problems in countering resistance. Philos. Trans. R. Soc. London, Ser. B 295:143–151.

Sparks, T. C. 1981. Development of insecticide resistance in *Heliothis zea* and *Heliothis virescens* in North America. Bull. Entomol. Soc. Am. 27:186–192.

Sutherst, R. W., and H. N. Comins. 1979. The management of acaricide resistance in the cattle tick, *Boophilis microplus* (Canestrini) (Acari: Ixodidae), in Australia. Bull. Entomol. Res. 69:519–537.

Swain, T. 1977. The effects of plant secondary products on insect plant co-evolution. Pp. 249–256 *in* Proc. 15th Int. Congr. of Entomol., Washington, D.C. College Park, Md.: Entomological Society of America.

Tabashnik, B. E., and B. A. Croft. 1982. Managing pesticide resistance in crop-arthropod complexes: Interactions between biological and operational factors. Environ. Entomol. 11:1137–1144.

Taylor, C. E., and G. P. Georghiou. 1979. Suppression of insecticide resistance by alteration of gene dominance and migration. J. Econ. Entomol. 72:105–109.

van den Bosch, R. 1978. The Pesticide Conspiracy. New York: Doubleday.

Wood, R. J., and J. A. Bishop. 1981. Insecticide resistance: Populations and evolution. Pp. 97–127 *in* Genetic Consequences of Man Made Change, J. A. Bishop and L. M. Cook, eds. New York: Academic Press.

Pesticide Resistance: Strategies and Tactics for Management.
1986. National Academy Press, Washington, D.C.

Preventing and Managing Fungicide Resistance

JOHAN DEKKER

Following a description of the term fungicide resistance, the factors that govern the buildup of a resistant pathogen population in the field are covered. Short-term tactics and long-term strategies to counteract the development of resistance are discussed.

INTRODUCTION

When fungicide resistance became a problem shortly after the introduction of systemic fungicides in the 1960s, the reliability of chemical disease control was at stake, especially since we did not know how to cope with the problem. Farmers who saw that a fungicide was becoming less effective often increased the dose, thus increasing the selection pressure and aggravating the problem.

The failure of some of the new, originally very effective fungicides to control disease was a surprise, and it created doubt about effective disease control using these fungicides. To better understand the resistance phenomenon, biochemical and genetic studies were conducted. These studies revealed the mechanism of action of several new fungicides and the mechanism of resistance in fungi. Greenhouse and field experiments on the ecological and population aspects of resistance have yielded considerable information about fungicide resistance.

This paper will discuss (1) the possibilities of using this information to develop tactics for preventing and managing fungicide resistance, (2) the research needed for further development and improvement of these strategies, and (3) what new approaches might offer prospects for dealing with the fungicide resistance problem.

FUNGICIDE RESISTANCE

People define fungicide resistance in different ways, depending on their interests and concerns. Fungicide resistance occurs when a fungal cell or a fungal population that originally was sensitive to a fungicide becomes less sensitive by hereditable changes after a period of exposure to the fungicide. A panel of FAO (Food and Agriculture Organization of the United Nations) experts has recommended that the word "resistance" should apply only to a hereditable decrease in sensitivity. The word "tolerance" should not be used in this sense, since it is ambiguous (FAO, 1979). Use of the word "insensitivity" in place of resistance is also not recommended, because it suggests a complete loss of sensitivity, which occurs only rarely.

The researcher first speaks of resistance after the emergence of less-sensitive cells has been observed in a petri dish and the hereditable nature of this phenomenon has been proved or seems likely. Resistance in the laboratory, however, does not mean that resistance will develop in the field. More important is a shift toward lower fungicide sensitivity in a field pathogen population, which may be called development of resistance, even if the fungicide still provides satisfactory disease control. The farmer speaks of fungicide resistance only when disease control fails. This definition is also often used by the agrochemical industry and the extension officer, who may be afraid that reports about laboratory resistance before problems in the field have occurred may confuse or even alarm the farmer. Because of the various meanings, one must define which type of resistance is being discussed: emergence of resistant cells in laboratory experiments; reduction of fungicide sensitivity in the field, but still with adequate control; or field resistance with loss of disease control.

BUILDUP OF A RESISTANT POPULATION

As Georgopoulos pointed out (this volume), the main mechanisms of resistance to fungicides are a change at the site of action in the fungal cell that decreases its affinity to the fungicide or a change in uptake of the chemical so that less of it reaches the site of action. Detoxification has rarely been reported as the cause of resistance in fungi, although it is the main mechanism of resistance in insects.

Resistant populations develop and increase in different ways, often from forces we can control. The mechanism of resistance may influence the fitness of resistant cells, as compared with sensitive cells, which is important to the buildup of a resistant pathogen population. The fitness of resistant strains appears to vary considerably for different types of fungicides. For some, resistance appears to be linked to a decrease in fitness in the absence of the fungicide. Thus, fungicides may be classified as low-risk, moderate-risk, or high-risk compounds (Dekker, 1984). With insecticides loss of fitness in resistant insects plays a lesser role. Here the strategy has been to change to an insecticide with a different mechanism of action. This strategy is possible only so long as new insecticides become available. The discovery of fungicides with a lower resistance risk offers possibilities for developing strategies to prolong their use.

The life cycle of the pathogen and the nature of the disease may influence the speed of buildup of a resistant pathogen population. For example, resistance builds more rapidly in an abundantly sporulating pathogen on aerial plant parts than in a slowly expanding soil pathogen. Environmental conditions that increase the severity of the disease may also speed the development of resistance. Another important factor in this respect is the management of pesticide application: a continuously high selection pressure by one chemical or by more chemicals with the same mechanism of action favors the buildup of a resistant pathogen or insect population.

SHORT-TERM TACTICS

New Chemicals

It would be very valuable to have information on the probability of resistance before a new chemical is used in the field. For example, experiments on an artificial medium, with or without mutagenic agents, may tell us whether emergence of resistant cells (by mutation or otherwise) is possible. If such experiments do not yield resistant cells, their emergence in the field should not be expected, but if they do, further testing should be done on the fitness of the resistant cells on the plant, compared with that of the wild-type fungus. Such experiments may give an indication of the resistance risk of the fungicide and could be used in devising strategies to minimize the chance of resistance.

Unfortunately, even when all possible laboratory and greenhouse experiments have been carried out, it is rarely possible to precisely predict what is going to happen in the field. Greenhouse conditions are never exactly the same as field conditions. Field experiments are indispensable, and they should be accompanied by careful monitoring (as outlined by Keiding, this volume). But even field experiments may not yield the results that can be obtained from large-scale application in practice, because the size of the area treated may play a role in the buildup of resistance. Thus, even if no resistance develops in laboratory, greenhouse, or field experiments, resistance could still appear in practice. For instance, with dodine and Kitazin-P, resistance problems arose only after many years. Nevertheless, experiments may indicate some of the risks involved, which is important for devising tactics to prevent or delay resistance.

High-Risk Chemicals

High-risk compounds, such as the benzimidazole and acylalanine fungicides, should not be used to control risky diseases when other, less-risky chemicals provide satisfactory control. Diseases are considered risky when they allow a rapid buildup of a resistant pathogen population, for example, by a repeating infection cycle during the growing season with an abundant spore production. Other diseases may be less risky, for example, those caused by root and foot pathogens. If a high-risk fungicide is applied to control a risky disease, stringent tactics should be used to minimize the chance for development of resistance. To avoid having such fungicides and related chemicals exert a constant selection pressure, therefore, one must remember the following factors.

- The amount of fungicide at risk applied to the crop should not exceed the minimum dose necessary for adequate disease control.
- The mode of application should be considered, for example, soil drench may allow uninterrupted uptake of the chemical and a prolonged period of selection pressure by the fungicide.
- Use of one particular fungicide or related fungicides for preharvest and postharvest application should be avoided, since the former may select for resistant cells, leading to problems in subsequent postharvest treatments.
- Treatment of a large area (e.g., all fields with a particular crop in one region or country) with the same or related fungicides should be avoided. No sensitive forms will then be available to enter the crop from outside, which limits the competition between sensitive and resistant forms during intervals of low selection pressure.
- A very thorough treatment of the crop, with little or no escape of sensitive cells, will eliminate competition between resistant and sensitive cells and thus favor the former.
- The selection pressure exerted by a chemical at risk may be reduced by rotation or combined use of chemicals or by integrating chemical control with other control measures.

Rotation or Combination

Rotations or combinations of fungicides can reduce the risk of resistance, but only when certain guidelines are followed. To reduce selection pressure the fungicides should possess different mechanisms of action. Using two risky chemicals together is not recommended: the population of fungal cells is usually so high that the chance exists for simultaneous mutations in different genes toward resistance to both chemicals. One of the fungicides in the mixture should pose little or no risk, although the use of a mixture will not

TABLE 1 Spraying Schemes and Risk for Development of Fungicide Resistance

Type	Program	Risk
a	S – S – S – S	Highest Risk
b	S – C – S – C	Alternation
c	(S+C) – (S+C) – (S+C) – (S+C)	Mixture
d	(S+C) – C – (S+C) – C	Combination b and c
e	C – C – (S+C) – C	Lowest Risk

S = Fungicide at risk.
C = Conventional fungicide.

SOURCE: Dekker (unpublished).

always stop the buildup of resistance to the compound at risk: resistance may occur in the pathogen population that is not eliminated by the nonrisky compound. Under certain conditions buildup of resistance in a mixture may occur at the same speed as when the compound is used alone (Kable and Jeffery, 1980). Theoretically this happens when there is no escape, that is, when there are no fungal cells that are not hit by the fungicide mixture and when selection pressure is not reduced during spray intervals. Such situations are rare—e.g., with postharvest treatment of citrus fruit (Eckert, 1982)—and usually do not occur in the field. In most cases the use of a mixture in the field will at least delay the buildup of resistance. If resistant strains have a reduced fitness, compared with sensitive strains, and if the interval between applications of the fungicide at risk is large enough to allow the proportion of the resistant pathogen population to drop to the preceding level, resistance problems might be avoided indefinitely.

A resistant population may build up gradually in an alternation or rotation scheme. During the period that the nonrisky compound is used, the proportion of resistant strains will not increase, and may decrease if there is reduced fitness. Resistance could then be postponed indefinitely, depending on the degree of reduction in fitness and the length of the intervals between sprays of the compound at risk. The mixture may provide longer delay of resistance at higher escape; the alternations may provide more delay at lower escape (Kable and Jeffery, 1980).

Both mixtures and alternations have some disadvantages. In mixtures the compound at risk is always present, which means constant selection pressure if the application is not interrupted. In an alternating scheme the use of the nonrisky compound is interrupted for no good reason. These disadvantages may be decreased by combining mixtures and rotations such that the nonrisky compound is constantly present and only the use of the risky compound is interrupted (Dekker, 1982). The chance for buildup of resistance may be delayed further by using the mixture only in critical situations (Table 1).

Low- or Moderate-Risk Chemicals

A fungicide may be at low risk when resistant strains show a strongly reduced fitness in greenhouse experiments or when it has been used for a few years without resistance problems. Because resistance could develop in later years, fields should be monitored. Monitoring will show whether a shift toward reduced sensitivity of the pathogen population occurs during the growing season or during consecutive seasons. Depending on the degree of this shift, measures (as discussed above for the risky compounds) can be taken to prevent or delay the buildup.

LONG-TERM STRATEGIES

To prevent or delay the buildup of a resistant pathogen population, different chemicals that are effective against a particular disease must be available. One way of increasing the number of available chemicals is to search for new site-specific inhibitors. Before the introduction of fungicides with site-specific action, little was known about the metabolic differences between the cells of a pathogen and a host, with the exception of the wall-constituent chitin, present in most fungi but not in plants. This difference is exploited by the polyoxin antibiotics, which interfere with chitin synthesis in the fungal cell wall. Examples of sites found since then include differences in tubulins constituting the spindle in fungi and plants; differences in sterols, namely ergosterol in fungi versus lanosterol in plants; and differences in the protein synthesizing apparatus, in the respiratory chain, or in enzymes involved in RNA synthesis between plants and certain fungi. More such sites will probably be discovered. Special attention should be given to site-specific inhibitors that show a low risk to development of resistance.

Further, the search should be intensified for disease-control agents that are not fungicidal in vitro, but increase the resistance of the host plant or decrease the pathogenicity of the parasite. Some of these chemicals might not or might less-readily encounter resistance.

Another concept is that of developing compounds with negatively correlated cross-resistance: a mutational change in a pathogen that confers resistance to fungicide A and, at the same time, increased sensitivity to fungicide B, and vice versa. Thus, when a combination of A and B is used, B will eliminate strains resistant to A, and A will eliminate strains resistant to B. Several combinations of such compounds have been described in the literature, but in most cases the phenomenon did not occur with all resistant strains. Moreover, occurrence of other resistance mechanisms, which do not result in negative cross-resistance to A and B, cannot be excluded. Nevertheless the phenomenon deserves further exploration.

Another phenomenon is synergism between two fungicides, A and B, especially if A is more active on strains resistant to B than on wild-type strains. The effect of respiration-inhibiting fungicides on fenarimol-resistant strains of *Aspergillus nidulans* and *Penicillium italicum* illustrates such synergism (de Waard and Dekker, 1983). Fungicide-resistant strains do not accumulate fenarimol, due to the presence of a constitutive energy-dependent efflux. Adding a respiration-inhibiting compound results in fenarimol accumulation and fungicide sensitivity. Researchers should further explore the existence of additional combinations of compounds that might counteract resistance in a similar way.

The search for integrated disease-control measures should be intensified. The use of cultivars with a certain degree of natural resistance, cultural practices, and biological control measures might be integrated with chemical control. For example, microorganisms used for biological control could be made resistant to a fungicide so that both can be used at the same time (Papavizas et al., 1982).

In addition to strengthening our research efforts, we must ensure that the number of conventional, nonsystemic fungicides is not needlessly decreased by regulatory agencies. Although these chemicals cannot perform all the tasks of systemic fungicides, they remain a reliable and invaluable tool for controlling disease when resistance to a new specific compound occurs. They can also be used as companion compounds of systemic fungicides, in mixtures, or in rotation.

The need for a varied arsenal of fungicides to cope with fungicide resistance requires that barriers for introducing new chemicals are not made higher than necessary. The risks of a new fungicide to nontarget organisms and the environment should be carefully weighed against the benefits. Not using chemicals is risky, not only for the economy and world food production but also for toxicological reasons: some fungi occurring in the crop or in the harvested product may produce mycotoxins, of which some may be carcinogenic. A senate committee appointed by President Kennedy, reporting on pesticides and public policy after the appearance of Rachel Carson's book *Silent Spring*, stressed the importance of a balanced benefit-risk equation and noted that the public lacked information concerning stringent precautions taken by the government to limit possible risks of the application of pesticides (U.S. Senate, 1966).

Finally, any long-term strategy must create possibilities for implementing tactics to prevent and manage fungicide resistance. We must develop an efficient system to communicate information among growers, extension officers, teachers, research workers, manufacturers, salesmen, the press, regulatory agencies, and the government. One example of an information distribution effort was the International Course for Southeast Asia on Fungicide Resistance in Crop Protection, held in Malaysia from October 17 to 24, 1984. The course was organized by the crop protection departments of the agricultural universities at Wageningen, the Netherlands, and at Serdang, Malaysia, in collaboration with the Food and Agricultural Organization, the Chemical Control Committee of the International Society of Plant Pathology, and the Fungicide Resistance Action Committee of the International Group of National Associations of Agrochemical Manufacturers.

CONCLUSION

Although it is not yet possible to predict the development of resistance to a new fungicide with certainty, as much information as possible should be obtained about the potential of plant pathogens to become resistant to such a fungicide. This can be done in appropriate laboratory, greenhouse, and field experiments.

If fungicides are applied that have been proved to be risky with respect to development of resistance, it is of prime importance to avoid a continuous and high selection pressure by such fungicides by using different fungicides in a mixture or in alternation. For flexible management it is also important to have a range of chemicals available. This can be achieved by the development of more fungicides with different mechanisms of action and by a reticent policy with respect to deregistration of old, conventional fungicides.

In order to alleviate the resistance problem in the future, attention should also be given to disease-control agents that increase the resistance of the host plant and to the phenomena of synergism and negatively correlated cross-resistance.

REFERENCES

Dekker, J. 1982. Counter measures for avoiding fungicide-resistance. Pp. 177–186 *in* Fungicide-Resistance in Crop Protection, J. Dekker and S. G. Georgopoulos, eds. Wageningen, Netherlands: Centre for Agricultural Publishing and Documentation.

Dekker, J. 1984. The development of resistance to fungicides. Prog. Pest. Biochem. Toxicol. 4:165–218.

de Waard, M. A., and J. Dekker. 1983. Resistance to pyrimidine fungicides which inhibit ergosterol biosynthesis. Pp. 43–49 *in* Human Welfare and the Environment, Vol. 3, Y. Miyamoto and P. C. Kearney, eds. Oxford: Pergamon.

Eckert, J. W. 1982. Penicillium decay of citrus fruits. Pp. 231–250 *in* Fungicide-Resistance in Crop Protection, J. Dekker and S. G. Georgopoulos, eds. Wageningen, Netherlands: Centre for Agricultural Publishing and Documentation.

Food and Agriculture Organization. 1979. Pest resistance to pesticides and crop loss assessment. FAO Plant Prod. and Prot. Paper, No. 6/2.

Kable, P. F., and H. Jeffery. 1980. Selection for tolerance in organisms exposed to sprays of biocide mixtures: A theoretical model. Phytopathology 70:8–12.

Papavizas, G. C., J. A. Lewis, and T. H. Abd-el Moity. 1982. New biotypes of *Trichoderma harzianum* with tolerance to benomyl and enhanced biocontrol activities. Phytopathology 72:126–132.

U.S. Senate, Committee on Government Operations. 1966. Report on Pesticides and Public Policy. Washington, D.C.: U.S. Government Printing Office.

Pesticide Resistance: Strategies and Tactics for Management.
1986. National Academy Press, Washington, D.C.

Case Histories of Anticoagulant Resistance

WILLIAM B. JACKSON and A. DANIEL ASHTON

*Genetic resistance to anticoagulant rodenticides by commensal rodents (*Rattus norvegicus, R. rattus, Mus musculus*) is widespread. Where it occurs rats and mice are not readily controlled with first-generation compounds. Second-generation anticoagulants, however, are effective in most situations. The first evidence of resistance to these compounds is now available. Management of rodent populations is important for aesthetic, economic, and public health reasons.*

INTRODUCTION

Anticoagulant rodenticides introduced in the 1950s selected for resistance within a decade, first in Norway rats (*Rattus norvegicus*) from Scotland (Boyle, 1960) then elsewhere in the United Kingdom (Rowe and Redfern, 1966; Bentley, 1969; Greaves et al., 1973); from many areas of Europe, including the roof rat (*R. rattus*) and house mouse (*Mus musculus*) (Lund, 1964, 1972); and eventually the United States (Jackson et al., 1971).

More recently resistance has been confirmed in commensal rats in Japan (Naganuma et al., 1981) and Australia (Saunders, 1978). In Malaysia the Malay wood rat (*R. tiomanicus*) (C. H. Lee, Malaysian Agriculture Research and Development Institute, personal communication, 1982) and *R. r. diardii* (Lam et al., 1982; Lam, 1984) are involved.

The World Health Organization (WHO, 1970) initially defined a resistant Norway rat as one that survived a 6–day, no-choice feeding test with 50 ppm warfarin bait. (Appropriate criteria were also specified for the roof rat and house mouse.) These criteria were verified by breeding tests (Greaves

FIGURE 1 Distribution of warfarin-resistant Norway rat populations in the United States.

and Ayres, 1976); and additional collection criteria and minimum consumption levels were added in the United States (Frantz, 1977). Field-test criteria have also been developed (Drummond and Rennison, 1973; Bishop et al., 1977).

In Europe resistant populations frequently were identified from rural sites. The first United States finding was in a rural area, but most of the identified sites in the United States have been urban (Jackson et al., 1975). A nationwide survey in the United States to determine the extent of anticoagulant resistance in rodents was facilitated under the U.S. Public Health Service Urban Rat Control Program (Jackson et al., 1973, 1975; Jackson and Kaukeinen, 1976; Jackson and Ashton, 1980; Jackson et al., 1985). Forty-five of the 98 sites sampled had resistant Norway rat populations (Figure 1). That most of the sites were urban may well be a function of the rodenticide use patterns, but more likely this results from congressional funding of *urban* rat-control programs.

The problem of mouse resistance is believed to be far greater than the rat problem, partly from differences in mouse population dynamics and feeding behavior. House mouse populations had such a high level of resistance in Europe a decade ago that conventional anticoagulants were no longer recommended for mouse control. Our knowledge of this species in the United States is meager. Ashton and Jackson (1984) reported anticoagulant resistance in mice in several areas, both urban and rural, but no comprehensive (nationwide) examination of mouse resistance has been conducted (Table 1). Canada also has reported mouse resistance (Cronin, 1979; Siddiqi and Blaine, 1982).

The following case histories illustrate the problems inherent in controlling rat and mouse populations in both rural and urban settings. They will also show how and why control programs, if done improperly, can create greater problems.

CHICAGO

The Chicago Rat Control Program began in June 1952, although rodent control efforts had begun years earlier. Developed as a tripartite program (the Departments of Health, Streets and Sanitation, and Buildings), initial treatments covered six of the city's 50 wards during the first six months of operation. Early rodenticides were cyanide gas, red squill, and 0.025 percent warfarin (1:19). The warfarin toxicant-to-bait ratio was later reduced to 1:49 (0.01 percent), when a mixer for the warfarin bait became available locally. (This level was probably insufficient to produce consistent mortality and enhanced resistance selection.) The operational pattern was to move systematically, ward by ward, block by block, through the city.

During March 1953 the effectiveness of the poisoning program was analyzed. In one block eight of nine premises that had active colonies (27 total burrows) before the poisoning operations still had colonies with evidence of activity afterward (Jackson and Evans, 1953). Operational efforts varied over the years. Our next analysis was in 1972 as part of the U.S. Public Health Service Urban Rat Control Program. Rats collected from the Lawndale target area during 1972 to 1974 were subjected to the WHO (1970) protocol for determining rodent resistance. Of the 87 rats tested, 50 (57 percent) survived (Environmental Studies Center, 1974).

By 1978 rats had been tested from four wards; 75 percent resistance to warfarin was found in three of the four wards. The trend continued at least through 1982. The Englewood area (Table 2) in 1978 had a 43 percent incidence of rat resistance, a figure similar to the Lawndale resistance pattern of 1972 to 1974. It is likely that the proportion of resistant rats in the Englewood population has since increased, but no subsequent collections have been made.

The problems of resistance in Chicago were both socioeconomic and managerial. Rodent infestations were most abundant in the lower socioeconomic

TABLE 1 Summary of Resistant House Mouse Sites in the United States

Region	Locality	Number of Sites Sampled (with resistance)	Number of Mice Tested	Number of Mice Resistant	Percent Resistant
Northeast	Hoboken (N.J.)	1 (1)	20	2[1]	10.0
	Buffalo (N.Y)	1 (1)	88+	69[1]	78.4
	Subtotal	2 (2)	108+	71	65.7
Southeast	Decatur (Al.)	1 (1)	2	2	100.0
Midwest	Decatur (Ind.)	1 (1)*	52	40	76.9
	Monroe (Ind.)	1 (1)*	9	6	66.7
	Battle Creek (Mich.)	3 (3)	23+	22	95.6
	Detroit (Mich.)	1 (1)	10	8	80.0
	Lansing (Mich.)	1 (1)	18	6	33.3
	St. Paul (Minn.)	1 (1)	4+	2	50.0
	Bowling Green (Ohio)	3 (1)*	25	2	8.0
	Cleveland (Ohio)	2 (2)*	273+	135	49.5
	Hamler (Ohio)	2 (0)*	15	0	—
	Leipsic (Ohio)	1 (1)*	19	1	5.3
	Marion (Ohio)	1 (1)	10	1	10.0
	Swanton (Ohio)	1 (1)*	49	5	10.2
	Toledo (Ohio)	2 (1)	25	3	12.0
	Whitehouse (Ohio)	1 (0)*	9	0	—
	Subtotal	21 (15)	541	231	42.7
Southwest	Fort Worth (Tex.)	1 (1)	1	1	100.0
West Coast	San Francisco (Calif.)	1 (1)	15+	4	26.7
	Total	26 (20)	667	309	46.3

* Rural sites (54/178, 30.3%).

\+ Includes F_1 individuals.

[1]Frantz, unpublished; N.Y. State Department of Health.

SOURCE: Adapted from Jackson et al. (1985).

TABLE 2 Summary of Norway Rats from Chicago, Illinois, Subjected to WHO Test for Warfarin Resistance

District	Number of Rats Tested	Number of Rats Surviving	Percent Resistant
Austin	62	44	71
Englewood	69	30	43
Garfield	11	8	73
Lawndale	212	152	72

SOURCE: Jackson and Ashton (1980).

areas, which were characterized by large transient populations, absentee landlords, abundant garbage, and vacant lots and vandalized buildings. These general conditions throughout several wards of the city provided abundant food and harborage, which supported large rat infestations. In contrast nearby areas with block clubs had overall cleaner alleys, neater back yards, and fewer rats.

Managerial problems also helped increase resistance. Warfarin baits were manufactured by city personnel and distributed throughout the neighborhoods. The quality of the bait was generally poor and inconsistent. Baitings were usually insufficient to control rodents, since warfarin baits require multiple feedings over several days, and blocks were often baited only once a month. Large dogs were present in many yards so only areas along the alleys were baited, while burrows adjacent to buildings and inside locked fences were left untouched. Inside-building infestations generally were not treated.

The systematic control program often gave way to political pressure, and rodent baitings were handled on a complaint rather than systematic basis. Thus, particular premises would be baited, but the entire block or even the adjacent premises would be left untouched. Red squill treatments, which would have killed resistant rats, were only partially effective because of induced bait shyness. Therefore, rodent control was never really achieved. Cyanide gas operations provided visible (political) evidence of rodent control but offered little overall reduction in rats, because gassings were confined to burrows in alleyways or away from structures and domestic animals. Gassing operations were also labor-intensive, requiring more manpower per unit area than baiting. Other rodenticides, including Vacor and norbormide, were used in the city, but with limited success; both compounds are no longer registered for use in rodent control in the United States. Also in 1978 the Centers for Disease Control (CDC, U.S. Public Health Service in Atlanta) restricted the use of conventional anticoagulants in federal target areas, thus leaving the city with only zinc phosphide as a toxicant tool.

Coincident with the federal program a major clean-up campaign was initiated in 1978. Vacant lots were cleared of old cars, garbage, and rubbish; lids were provided for all of the 55-gallon drums that were used as garbage cans. The program aimed at educating the public was left understaffed, however, resulting in insufficient follow-up and emphasis on long-term environmental improvement.

These practices all fostered the increasing frequency of resistant rats. The ineffective applications of anticoagulant baits meant that rats with the resistant allele survived, and enough target-area rats remained to facilitate rapid breeding. Populations quickly rebuilt, and when warfarin baits were distributed again, resistant genotypes tended to be selected. After several years of such operations, most rats in these areas were anticoagulant-resistant.

Rat Bites

The number of reported and confirmed rat bites[1] hit a low of 90 in 1968, reflecting a major clean-up campaign initiated in 1966 with the onset of support from the federal rat-control program (Table 3). Cases of rat bite increased during the 1970s, however. They peaked in 1979, a result of heavy snows hampering garbage pickup and facilitating an increase in the rodent populations, combined with ineffective rat-control efforts due to an increase in the incidence of resistance.

Second-Generation Anticoagulants

Brodifacoum, a second-generation anticoagulant, was registered for urban use in 1979. By 1984 rodent activity had decreased 75 percent in much of the area (Terry Howard, Chicago Sun Times, personal communication, 1984). Better bait materials and better placement of baits contributed to the reduction, as did the single-feeding characteristic of this rodenticide.

Many of the underlying problems that permitted the establishment of populations resistant to warfarin, however, are still present. Since no rats were collected in recent years for testing, the current incidence of resistance is not known. The potential for development of resistance to the new second-generation anticoagulants exists, and Chicago might well be the premier site in the United States for identifying this phenomenon.

DECATUR

There have been few investigations of rural rodent resistance in the United States, and most have been of mouse infestations. Although the number of

[1]Rat bites usually are not uniformly reported, and any statistics concerning them should be used cautiously. In Chicago, however, records are carefully maintained, and reported cases are investigated. Consequently, these data are indicative of trends in rat populations.

TABLE 3 Summary of Number of Confirmed Rat Bite Cases in Chicago, Illinois, from 1959 to 1982

Calendar Year	Number/Year
1959	214
1960	219
1961	191
1962	193
1963	116
1964	127
1965	155
1966	144
1967	140
1968	90
1969	131
1970	189
1971	148
1972 (to Sept. 29)	87
1973–1978 (incl.)	ca. 200
1979	323
1980	240
1981	177
1982	156

SOURCE: Chicago Department of Health.

sites and sample sizes have been small, half of the mouse infestations have had resistant mice. House mouse populations generally are confined to a single building (or a single house in a city), and treatment patterns may vary greatly. In agricultural sites such as poultry or hog farms, however, the mouse infestations may be extensive.

Usually when a problem is identified the mouse population is well established. Mouse movements tend to be vertical as well as horizontal; therefore, large groups of mice can live in a relatively small area. Mouse population dynamics can thus compound control efforts. When anticoagulants are used mice from the "visible" population only are likely to encounter the bait. Generally, the quantities of anticoagulant baits placed are insufficient to control the infestation. Even if baits are maintained continuously for the recommended 21 days, most mice will not encounter them.

In field tests mouse "waves" occur at approximately three-week intervals, and in large populations in complex environments, several waves may occur as mice are removed by poisoning (Ashton et al., 1983). Therefore, control of such populations may take months. If bait is available for only three weeks, the new immigrants may encounter only remnants of the bait, resulting in sublethal doses and enhanced selection of the resistant gene. Such was the case on a chicken farm near Decatur, Indiana.

The farm was composed of three older, deep-litter grower houses. Maintenance in the buildings was minimal, and the feeding operation was semiautomatic. Some grain was stored within the buildings. Mice had infested the building for several years, and the farmer had treated with locally produced, low-quality warfarin bait. Considering the size of the mouse population, the ample harborage and food to support a continuously breeding population, and a poor-quality bait, the selection of resistant mice was only a matter of time.

During one six-week period, 250 pounds of the poor-quality bait were consumed without any apparent reduction in the mouse infestation. Laboratory studies (conducted at the Bowling Green State University Rodent Research Laboratory) of mice collected at the site revealed that 76 percent were resistant to warfarin. Another site (Monroe, Indiana) had been treated with the same bait for several years: 70 percent of the mice tested were resistant to warfarin.

UNITED KINGDOM

Although the initial identification of warfarin resistance was in Scotland (Boyle, 1960), most resistance sites were found in England and Wales (Greaves and Rennison, 1973). Most were rural, although resistant rats apparently spread from the initial site into nearby Glasgow, and Folkestone in southeastern England had a well-identified resistant urban population.

The best-studied resistant population was at a site on the Welsh-English border near Welshpool. Initial efforts went into encircling and containing the infested area using zinc phosphide as a toxicant. As the infested area expanded 5 km/yr, however, defending the perimeter became increasingly difficult and eventually was abandoned (Drummond, 1970).

Resistance (involving all three species of commensal rodents) was documented in so many areas of the United Kingdom that field collections and studies were discontinued. Use of anticoagulants in these resistant areas was discouraged. Zinc phosphide was the prime alternative, and calciferol was recommended for house mice; later, second-generation anticoagulants became available. Some considered the resistant phenotypes to be at a reproductive disadvantage once the selective pressure of anticoagulant use was lifted (Bishop et al., 1977). In the absence of the rodenticide selection pressure, the frequency of resistance decreased from 57 to 39 percent over several years (Greaves et al., 1977).

Difenacoum (the first second-generation anticoagulant) use was initiated in 1975. By the end of the decade resistance was widespread in Hampshire in southern England: 14 percent of the rats tested (40 percent of the sites) were resistant to the compound; 85 percent were resistant to warfarin. Probably the distribution of difenacoum-resistant rats is more widespread than the survey indicated (Greaves et al., 1982).

DATA BASE

Warfarin is a vitamin K antagonist; it inhibits the reduction of vitamin K 2,3-epoxide to vitamin K (Bell and Matschiner, 1972; Bell et al., 1972), eventually blocking the vitamin K-regeneration enzyme system. This blocking inhibits the post-translational modification of prothrombin (carboxylation of glutamic acid to γ-carboxyglutamic acid) and eventual completion of the bloodclotting sequence. Ultimately the animal dies from internal hemorrhaging.

Although the exact mechanism of warfarin resistance is still under investigation, a mutation probably occurs in warfarin-resistant animals at the active site of the vitamin K-epoxide reductase, such that warfarin cannot compete as effectively with vitamin K-epoxide. The cycle continues with γ-carboxyglutamic acid being continuously supplied and subsequently producing normal prothrombin (Suttie, 1980; MacNicoll, 1981). MacNicoll (this volume) has suggested some alternatives to the vitamin K-epoxide reductase mechanism.

Most investigators have determined resistance to be related to an autosomal dominant allele (Greaves and Ayres, 1967). At one time resistance in the house mouse was thought to be polygenic, but more recently, rat and mouse resistance have been considered very similar (Wallace and MacSwiney, 1976).

In the United Kingdom, hypotheses of incomplete penetrance and modifier genes have been expressed. Strains (based on geographic origin) have been identified, and multiple allelism has been advanced to explain such variations (Greaves and Ayres, 1982). In the laboratory, maintenance of resistant animals requires diet supplementation with vitamin K. In natural populations the heterozygote is considered to have survival advantage over the resistant homozygote (Bishop and Hartley, 1976; Greaves et al., 1977). Because North American resistant animals do not seem to require this vitamin K supplementation, there may be several (or more) resistance genotypes. Research and tests to identify these genetic variations, however, have not been carried out.

Data Base—United States

In the combined efforts at Bowling Green State University and the New York State Department of Health, more than 10,000 Norway and roof rats collected from over 100 locations have been examined (Jackson et al., 1985). Nearly 50 Norway rat resistance sites have been identified, and the incidence of resistance in most samples was less than 20 percent.

Warfarin-resistant house mice have been detected at 21 sites throughout the country, but samples generally have been small. (Federal *Rat* project funds could not be used for studies on mouse populations.) About half the animals tested were resistant to warfarin, and 10 out of 11 urban samples contained resistant mice. Thus, the mouse problem seems to be widespread and critical in the United States.

Cross-Resistance

Norway rats and house mice have been tested against other first-generation anticoagulant rodenticides. With Norway rats (from 16 locations in 13 cities and 1 rural site) 165 out of 176 individuals (94 percent) resistant to warfarin were also resistant to pival. Samples of warfarin-resistant house mice and roof rats resistant to pival are small but definitive (five of five and two of two, respectively). Tests for warfarin resistance on laboratory strain, pival-resistant rats, however, produced mortality in 44 of 46 animals (Fukui, 1985).

Greaves and Ayres (1976) found a similar relationship between Welsh and Scottish strains resistant to coumatetralyl, with the Welsh strain showing a high resistance to warfarin and diphacinone. Different alleles probably are responsible for this action (Greaves and Ayres, 1982). Following this logic the pival-resistant strain should represent another allele, although this has not been confirmed.

In general an animal resistant to warfarin is likely to survive (i.e., be resistant to) feeding tests with any of the other hydroxycoumarin or indandione compounds. Consequently, alternating first-generation rodenticides for rodent control when resistance is suspected is not likely to be efficacious, but rather may enhance selection of resistant populations.

MANAGEMENT

Second-Generation Anticoagulants

The new group of anticoagulants characterized as single-feeding (but with delayed death) is a suitable tool for managing rodents. Brodifacoum and bromadiolone have been marketed in the United States and elsewhere; difenacoum, only outside the United States. These rodenticides kill rats and mice resistant to first-generation compounds. Resistance to these second-generation compounds has been found, however: in Canada, house mice resistant to bromadiolone (Siddiqi and Blaine, 1982); in England, Norway rats resistant to difenacoum (Greaves et al., 1982); and in Denmark, bromadiolone resistance in rats (Lund, 1984). Resistance to either second-generation compound in the United States has not been confirmed. There are no published data as yet indicating brodifacoum resistance in natural populations anywhere in the world, although an occasional laboratory rat has been observed to survive a test with brodifacoum; similar observations have been made in England with difenacoum-resistant strains (Lund, 1984; Cornwell, 1984; P. B. Cornwell, Rentokil, Ltd., East Grinstead, Great Britain, personal communication, 1984).

Nonanticoagulant Alternatives

Alternative rodenticides (or other control tools) must be found when resistance occurs. Acute (single-dose) rodenticides are not wholly acceptable. Red squill is currently not available. Zinc phosphide is relatively hazardous when not properly placed and quickly produces bait shyness.

A new chemical, bromethalin, was introduced commercially in December 1985 under the trade names "Vengeance" and "Assault." This rodenticide is not an anticoagulant and acts after a single feeding. Although no further feeding occurs, death is delayed several days. The compound kills anticoagulant-resistant rodents (Jackson, 1985).

Another new compound, Quintox, was introduced in 1985; a form of vitamin D (cholecalciferol), it disrupts calcium metabolism, producing hypercalcemia. It is effective against anticoagulant-resistant rodents and is similar to the calciferol formulations long-used in Europe to control resistant mice. Field-test efficacy data have not been published.

Alpha-chlorohydrin (Epibloc), a male sterilant, is toxic to resistant rats of both sexes (Andrews and Belknap, 1983; Kassa and Jackson, 1984). It was introduced as a restricted-use rodenticide, but has not been widely accepted by pest control operators (PCOs). Other compounds, still in testing and development modes by various companies, also have the potential for being effective against resistant populations. The high costs of testing and data acquisition now required for U. S. Environmental Protection Agency (EPA) registration, however, may discriminate against their development for the U. S. market.

Improving sanitation and repairing/proofing structures are long-term solutions that have long been recognized as fundamental to managing pest populations (Davis, 1953, 1972; NRC, 1980). Without them the effectiveness of toxicants is reduced, and the selection pressure for resistance is increased. With increased numbers of resistant rodents in our environment, the potential for transmission of rodent-borne diseases is also increased. Concern for resistance is important because of the public health significance, as well as the depredations of these rodents to crops and stored products. Especially when these rats and mice share our environment, the potential for transmission to humans of many rodent-borne diseases is greatly increased. The clinical forms of some of these diseases are just now being recognized (e.g., hemorrhagic fever with renal syndrome).

PREDICTIONS AND RECOMMENDATIONS

• With the continued use of first-generation anticoagulants, resistant populations will be selected with increasing frequency.

• Resistant house mouse populations especially are likely to increase in frequency because of rodenticide use both by professionals and the public. Because of the close association of this species to humans and the high potential for food destruction and contamination, a serious public health, economic, and aesthetic threat exists.

• Second-generation anticoagulants will find increasing markets. Resistance can be expected in proportion to their market penetration because of excessive placement of baits and failure to implement an integrated pest management (IPM) program. Users should be strongly encouraged (through training) in the prudent use of these rodenticides [e.g., pulsed baiting with Talon (Dubock, 1982)].

• Nonanticoagulant rodenticides (bromethalin, zinc phosphide, etc.), when not used exclusively, should be alternated with anticoagulants at least annually. Such use will mitigate against the selection and buildup of populations resistant to either first- or second-generation anticoagulants.

• Resistant populations of the commensal rodent species can be demonstrated readily. As agricultural and noncrop, nonurban uses of anticoagulants expand, resistance in native rodents can be expected following several years of persistent and careless or excessive distribution of baits.

RESEARCH NEEDS

• Commensal rodents (in the United States, these are Norway and roof rats and house mice), in the absence of effective management tools and efforts, will continue to pose serious public health problems, cause environmental destruction and deterioration, and contaminate and consume significant proportions of grains, feeds, and food products. Rodents infesting orchards and croplands can also be selected for resistance with consequences for serious economic losses. Quantification of such losses to individuals and society are needed to determine cost/benefit patterns for management programs and to provide incentives for developing new tools.

• Studies of resistance incidence in the house mouse and native rodent species have been neglected. Monitoring of resistant commensal rat populations should be continued and expanded in both agricultural and urban sites.

• The biochemical mechanism(s) of resistance need continuing study. Especially needed is support for breeding and maintenance of resistant strains for research and genetic investigations of cross-resistance and the allelic variations among different populations. (Since human anticoagulant resistance (O'Reilly et al., 1968) is an important consideration in treating vascular problems, potential health benefits accrue as well.)

- In the future if anticoagulants are labeled by EPA for use in mainland sugarcane (and other agricultural crops), the potential for selection of resistance is present. Evaluation of existing Hawaiian populations, where anticoagulants have been used peripherally, would be useful.
- Strategies for rodent management, based on IPM principles, need to be articulated and used. These will include environmental improvement, effective environmental education, and use (at least annually) of nonanticoagulant rodenticides wherever monitoring indicates anticoagulant resistance exceeds 10 percent. In lieu of monitoring, such alternation of rodenticide types should be part of the scheduled program.
- EPA registration procedures for experimental use permits, registration, and reregistration of rodenticides should be made realistic (relative to the characteristics and use patterns of the compounds), to stimulate commercial development of new products.

ACKNOWLEDGMENTS

We are pleased to acknowledge funding assistance through the Urban Rat Control Program (US PHS-CDC) for some of these studies. City of Chicago personnel greatly facilitated our monitoring studies in Chicago. Staffs at both the New York Health Department Laboratory at Troy and our own Bowling Green State University Rodent Research Laboratory carried out the resistance evaluation tests. Art Beeler was most helpful in allowing our access to his buildings and mouse populations.

REFERENCES

Andrews, R. V. and R. W. Belknap. 1983. Efficacy of alpha-chlorhydrin in sewer rat control. J. Hyg. (Camb.) 91:359–366.

Ashton, A., and W. B. Jackson. 1984. Anticoagulant resistance in the house mouse in North America. Pp. 181–188 *in* Proc. Conf. Organ. Pract. Vertebr. Pest Cont., A. C. Dubock, ed. Hampshire, England: ICI Plant Protection Division.

Ashton, A., W. B. Jackson, and J. H. McCumber. 1983. An evaluation of methods used in comparative field testing of commensal rodenticides. Pp. 138–154 *in* Vertebr. Pest Cont. Manage. Mat.: Fourth Symp., ASTM STP 817, D. E. Kaukeinen, ed. Philadelphia, Pa.: American Society for Testing and Materials.

Bell, R. G., and J. T. Matschiner. 1972. Warfarin and the inhibition of vitamin K by an oxide metabolite. Nature (London) 237:32–33.

Bell, R. G., J. A. Sadowski, and J. T. Matschiner. 1972. Mechanism of action of warfarin. Warfarin and metabolism of vitamin K. Biochem. 11:1959–1961.

Bentley, E. W. 1969. The warfarin resistance problem in England and Wales. Schriftenr. Ver. Wasser., Boden, Lufthyg. Berlin-Dahlem 32:19–25.

Bishop, J. A., and D. J. Hartley. 1976. The size and age structure of rural populations of *Rattus norvegicus* containing individuals resistant to the anticoagulant poison warfarin. J. Anim. Ecol. 45:623–646.

Bishop, J. A., D. J. Hartley, and G. G. Partridge. 1977. The population dynamics of genetically determined resistance to warfarin in *Rattus norvegicus* from mid Wales. Heredity 39:389–398.

Boyle, C. M. 1960. Case of apparent resistance of *Rattus norvegicus* Berkenhout to anticoagulant poisons. Nature (London) 188:517.

Cornwell, P. B. 1984. Meeting the demands of the food industry, William B. Jackson and Shirley S. Jackson, eds. Pi Chi Omega News. December suppl.

Cronin, D. E. 1979. Warfarin resistance in house mouse (*Mus musculus* L.) in Vancouver, British Columbia. MPM Professional Paper. Vancouver, B.C.: Simon Fraser University.

Davis, D. E. 1953. The characteristics of rat populations. Quart. Rev. Biol. 28:373–401.

Davis, D. E. 1972. Rodent control strategy. Pp. 157–171 *in* Pest Control Strategies for the Future. Washington, D.C.: National Academy of Sciences.

Drummond, D. C. 1970. Variation in rodent populations in response to control measures. Symp. Zool. Soc. Lond. 26:351–367.

Drummond, D. C., and B. D. Rennison. 1973. The detection of rodent resistance to anticoagulants. Bull. W.H.O. 48:239–242.

Dubock, A. C. 1982. Pulsed baiting—a new technique for high potency, slow acting rodenticides. Pp. 123–136 *in* Proc. 10th Vertebr. Pest Conf., Rex E. Marsh, ed. Davis: University of California.

Environmental Studies Center. 1974. Report to Chicago Health Department. Bowling Green, Ohio: Bowling Green State University.

Frantz, S. C. 1977. Procedures for collecting rats for anticoagulant resistance/urban rat control projects. Atlanta, Ga.: U.S. Department of Health, Education, and Welfare, Centers for Disease Control.

Fukui, H. H. 1985. Pivalyl-resistance in Sprague-Dawley rats (*Rattus norvegicus*). M.S. thesis, Bowling Green State University, Ohio.

Greaves, J. H., and P. B. Ayres. 1967. Heritable resistance to warfarin in rats. Nature (London) 215:877–878.

Greaves, J. H., and P. B. Ayres. 1976. Inheritance of Scottish-type resistance to warfarin in the Norway rat. Genet. Res. 28:231–239.

Greaves, J. H., and P. B. Ayres. 1982. Multiple allelism at the locus controlling warfarin resistance in the Norway rat. Genet. Res. 40:59–64.

Greaves, J. H., and B. D. Rennison. 1973. Population aspects of warfarin resistance in the brown rat, *Rattus norvegicus*. Mammal. Rev. 3:27–29.

Greaves, J. H., B. D. Rennison, and R. Redfern. 1973. Warfarin resistance in the ship rat in Liverpool. Int. Pest Control 15:17.

Greaves, J. H., D. S. Shepherd, and J. E. Gill. 1982. An investigation of difenacoum resistance in Norway rat populations in Hampshire. Ann. Appl. Biol. 100:581–587.

Greaves, J. H., R. Redfern, P. B. Ayres, and J. E. Gill. 1977. Warfarin resistance: A balanced polymorphism in the Norway rat. Genet. Res. 30:257–263.

Jackson, W. B. 1985. Single-feeding rodenticides: New chemistry, new formulations, and chemosterilants. Acta Zool. Fenn. 173:167–169.

Jackson, W. B., and A. D. Ashton. 1980. Present distribution of anticoagulant resistance in the United States. Pp. 392–397 *in* Vitamin K Metabolism and Vitamin K-dependent Proteins, J. Suttie, ed. Baltimore: University Park Press.

Jackson, W. B., A. D. Ashton, S. C. Frantz, and C. Padula. 1985. Present status of resistance to anticoagulant rodenticides in the United States. Proc. 3rd Int. Theriol. Congr., Helsinki, August 15–20, 1982. Acta Zool. Fenn. 173:163–165.

Jackson, W. B., J. E. Brooks, A. M. Bowerman, and D. E. Kaukeinen. 1973. Anticoagulant resistance in Norway rats in U.S. cities. Pest Control 41(4):56–64,81.

Jackson, W. B., J. E. Brooks, A. M. Bowerman, and D. E. Kaukeinen. 1975. Anticoagulant resistance in Norway rats as found in U.S. cities. Pest Control 43(4):12–16; 43(5):14–24.

Jackson, W. B., and R. Evans. 1953. Report on Chicago rat control program. Illinois Department of Public Health.

Jackson, W. B., and D. E. Kaukeinen. 1976. Anticoagulant resistance and rodent control. Pp. 303–308 *in* Proc. 3rd Int. Biodegradation Symp., J. Miles Sharpley and Arthur M. Kaplan, eds. London: Applied Sciences Publications, Ltd.

Jackson, W. B., P. J. Spear, and C. G. Wright. 1971. Resistance of Norway rats to anticoagulant rodenticides confirmed in the U.S. Pest Control 39(3):13.

Kassa, H., and W. B. Jackson. 1984. Bait acceptance and chemosterilant efficacy of alpha-chlorohydrin in the Norway rat, *Rattus norvegicus*. Int. Pest Contr. 26:7–11.

Lam, Y. M. 1984. Further evidence of resistance to warfarin in *Rattus rattus diardii*. MARDI Res. Bull. 12:373–379.

Lam, Y. M., A. K. Lee, Y. P. Tan, and E. Mohan. 1982. A case of warfarin resistance in *Rattus rattus diardii*. MARDI Res. Bull. 10:378–383.

Lund, M. 1964. Resistance to warfarin in the common rat. Nature (London) 203:778.

Lund, M. 1972. Rodent resistance to the anticoagulant rodenticides, with particular reference to Denmark. Bull. W.H.O. 47:611–618.

Lund, M. 1984. Resistance to the second-generation anticoagulant rodenticides. Pp. 84–94 *in* Proc. 11th Vert. Pest Conf., Dell O. Clark, ed. Davis: University of California.

MacNicoll, A. D. 1981. A review of studies on warfarin resistance in rats and mice. MAFF/ADAS Agric. Sci. Serv. Res. Develop. Rep., Pestic. Sci. Ref. Book 252(81):38–47.

Naganuma, K., A. Fujita, N. Taniguchi, and S. Takada. 1981. Warfarin susceptibility in the roof rat, *Rattus rattus*, in some locations of Tokyo. Jpn. J. Sanit. Zool. 32:243–245. (In Japanese.)

National Research Council. 1980. Urban Pest Management. Washington, D.C.: National Academy of Sciences.

O'Reilly, R. A., J. G. Pool, and P. M. Aggeler. 1968. Hereditary resistance to coumarin anticoagulant drugs in man and rat. Ann. N.Y. Acad. Sci. 151:913–931.

Rowe, F. P., and R. Redfern. 1966. The resistance of the house mouse (*Mus musculus*) to anticoagulants. Proc. Sem. Rodents and Rodent Ectoparasites. Geneva: World Health Organization (66.217:165–168).

Saunders, G. R. 1978. Resistance to warfarin in the roof rat in Sydney, N.S.W. Search 9:39–40.

Siddiqi, Z., and W. D. Blaine. 1982. Anticoagulant resistance in house mice in Toronto, Canada. Environ. Health Rev. June:49–51.

Suttie, J. W. 1980. Mechanism of action of vitamin K: Synthesis of γ-carboxyglutamic acid. CRC Critic. Rev. Biochem. July:191–223.

Wallace, M. E., and F. J. MacSwiney. 1976. A major gene controlling warfarin resistance in the house mouse. J. Hyg. 76:173–181.

World Health Organization. 1970. Provisional instructions for determining the susceptibility of rodents to anticoagulant rodenticides. W.H.O. Tech. Rep. Ser. 443:140–147.

6

Implementing Management of Resistance to Pesticides

PEST-CONTROL DECISIONS are influenced by many institutions, regulations, laws, and economics. This chapter focuses on the current status of efforts to manage development of resistance to pesticides and recommends how strategies to manage resistance might be implemented. Individuals and single companies are limited in their ability to deal with resistance primarily because: (1) resistant pests move across property boundaries, (2) information on the current and prospective pesticide susceptibility levels of pest populations is expensive to assemble, (3) information on methods for managing resistance and actions to respond to resistance are often needed at many locations at the same time, involving several related compounds made by different companies, and (4) because of potential conflicts of interest, combined with companies' needs for proprietary secrets. In responding to these difficult challenges, we will assess the roles of public agencies, groups of private firms, and the market system in managing resistance.

Once tactics for slowing pesticide resistance are developed and tested, technical progress will be achieved only if the tactics are properly, widely, and consistently applied. Four of the more important groups of organizations that affect implementation are: (1) the extension service, pest-management consultants, and farmers (considered here as a group); (2) regulatory agencies; (3) the pesticide industry; and (4) international organizations.

This report focused on the biological and genetic bases of resistance and on tactics to manage resistance. The committee recognized that few of the standard institutional mechanisms and incentives are available to bring about changes needed to encourage use of these tactics, which are often specific to particular pests and/or crops. To coordinate activities to manage resistance, it may be possible to build on existing initiatives such as the National Biological Impact Assessment Program (NBIAP).

EXTENSION, CONSULTANTS, AND PESTICIDE USERS

Education

The Cooperative Extension Service should take a leadership role in developing educational programs in the area of management of resistance to pesticides, coordinating input from state agricultural experiment stations, pest-control advisers, the pesticide industry, commodity associations, regulatory agencies, and end users. Factors to consider in a training program are

- The known toxicological, genetic, biological, and operational factors that influence selection for resistance. Published studies should be used as the primary basis for anticipating situations where resistance might occur.
- Identify and categorize pests and pesticides at high risk for developing resistance, particularly those that can develop cross-resistance.
- Review integrated pest management-compatible tactics—such as reducing selection pressure—that delay resistance development. Stress integrating pesticide use with nonpesticidal control measures. Discuss the high value of retaining, for as long as possible, low-cost pesticides that are used successfully in integrated-pest-management (IPM) programs.
- Emphasize the value of monitoring for resistance. Action thresholds should be established that determine the frequency of the population that is resistant at any time and above which it is advisable to switch to an alternative pesticide. Examples of action thresholds for development of resistance are described elsewhere in this volume (Frisbie et al.).

RECOMMENDATION 1. **The Extension Committee on Organization and Policy—IPM Task Force should conduct a feasibility study for developing an educational program on management of pesticide resistance, coordinated through the Cooperative Extension Service.**

This committee should work with representatives from the state agricultural experiment stations, U.S. Department of Agriculture-Agricultural Research Service (USDA-ARS), Economic Research Service (ERS), Animal and Plant Health Inspection Service (APHIS), professional societies, industry, consultant organizations, commodity organizations, public health agencies, and state departments of agriculture to determine whether a program thrust in this area is needed and feasible, and, if so, what the form and function of the training program should be.

Formalize Procedures for Management of Resistance to Pesticides

There may be a need to formalize and standardize procedures for dealing with resistant pest populations, as discussed by Frisbie et al. (this volume). When a control failure occurs, farmers, agricultural consultants, chemical applicators, state agricultural experiment stations, and Cooperative Extension services should work with agricultural chemical companies to determine the basis for the failure. A series of questions need to be addressed in a logical order: was the most effective pesticide applied for the specific pest and life stage; was it applied at an appropriate rate; was it applied under favorable weather conditions; did the equipment function properly; is the toxic agent active. If any of these conditions were not met, corrective actions should be taken. If the control failure persists, a bioassay should then be used to determine whether the pest population is susceptible to the pesticide. When resistance is verified at a level sufficient to justify alternative control strategies, they should be used immediately, if possible. Additional research is needed to establish action thresholds for resistance for specific pests and crops, and to develop pest-specific rapid bioassay and monitoring techniques (see Chapters 2 and 4.) When appropriate management tactics (see Chapter 5) are developed and validated, they should be implemented by the appropriate groups mentioned above.

RECOMMENDATION 2. **A formalized procedure or action plan should be developed to manage resistance to pesticides and to identify responsible individuals or agencies. The Cooperative Extension Service should take the leadership role in organizing work groups within state, regional, and national IPM programs to implement management of resistance.**

REGULATORY AGENCIES

This section concerns U.S. state and federal agencies. The committee focused on actions that could be undertaken to manage resistance without major legislative changes in state laws or in the Federal Insecticide, Fungicide and Rodenticide Act (FIFRA), the primary federal statute governing the registration and use of pesticides in the United States.

When resistance occurs, what is the appropriate role of state and federal agencies currently regulating pesticide use? Use directions and prohibitions against certain practices might be added to pesticide labels to prolong the useful life of a pesticide. As reliable methods become available, "resistance-risk" data might be required as part of the pesticide registration process and used in developing educational materials for certifying and training users and sellers of pesticides. As a first step, the mode of action should be identified. The recent experience of applying biotechnology to herbicides suggests that mode of action can be determined quite rapidly. Information on pesticide resistance often becomes available as part of other regulatory activities—reregistration, applications for emergency use exemptions, and regulatory actions taken to prohibit certain pesticide uses. Questions remain regarding what should be done with this information and how funds should be generated for regulatory activities, information collection, and research.

Pesticide Resistance in Regulatory Decisions

The committee agreed that resistance management is a legitimate activity for regulatory agencies when beneficial strategies and program opportunities arise, but recognized that there are strengths and weaknesses in each regulatory initiative considered. At the present time, the committee does not recommend major changes in state or federal regulatory responsibilities as they relate to resistance.

In the committee's judgment, there are compelling reasons why resistance is a difficult phenomenon to integrate more formally or routinely into regulatory agency decision-making. These include (1) federal agencies cannot make or implement timely decisions for local management of resistance; (2) market participants, local groups, and extension services can monitor and more effectively direct management; and (3) regulatory agencies lack funding for new management initiatives because they have higher priority pesticide regulatory objectives to pursue.

The U.S. Environmental Protection Agency (EPA) pesticide-use regulations under FIFRA do not explicitly direct the agency to consider resistance in carrying out its other responsibilities. In practice, though, the statute's basic risk-benefit balancing criterion requires consideration of pesticide efficacy in determining benefits. To the extent resistance reduces actual or anticipated efficacy—and hence benefits—EPA is already mandated to take it into account. Resistant pest populations, moreover, affect definitions of emergency use conditions, classification of pesticides, and reporting of "adverse effects"—all part of the regulatory process. Any biological factor, such as resistance, that can substantially reduce the benefits from pesticide use could affect some regulatory decisions. Still, as a practical matter, it would be difficult to encourage use of strategies to manage resistance by specifying such strategies on pesticide product labels, the major instrument available to regulatory agencies for encouraging resistance management. Because most pesticide resistance events are localized and change rapidly, use of pesticide labels to help manage resistance will rarely be feasible (Johnson, Hawkins, this volume).

Several approaches, however, do exist to encourage management of resistance. The practice of allowing market forces to reward product efficacy is well established for most pesticide products. (Johnson, this volume, discusses efficacy waivers under FIFRA.) There are possibilities, as well, for the extension service, agricultural consultants, and pesticide firms to monitor for and recommend management strategies in local areas.

Finally, in considering the regulatory approach to managing resistance to pesticides, the constraints on available resources must be recognized. For a regulatory approach to be effective, it must be timely and specific as to location, crop, and pests. Consideration of mixtures and multiple compounds so that rotations might be used would require considerable manpower, which is not currently available in the EPA. Health and environmental risks could increase if EPA and state resources were shifted from pesticide safety to managing resistance and maintaining pesticide product life.

The economic conditions under which individual pesticide firms, groups of firms, or farmers can profitably act to reduce resistance are limited by pest mobility and market structure. The following conditions are thought to favor mandatory or government programs in resistance management (Miranowski and Carlson, this volume): (1) when noncooperation by one or more chemical firms can jeopardize a regional program to manage resistance; (2) when antitrust considerations make certain cooperative efforts between firms difficult, and perhaps illegal; (3) when coordination between firms is costly, such as might occur when many companies or farmers with widely different interests in managing a pest population are involved; or (4) when a government unit is directly responsible for pest control, such as for public health pests or on public lands managed by the Forest Service.

The increasing number of pesticide resistance incidents (i.e., human health-related pests in hospitals, malaria mosquitoes, isolated rat populations, and certain agricultural pests such as the Colorado potato beetle) is frequently cited as a reason for government regulations in risk management. The increased cost of synthesizing new chemicals to replace those chemicals made obsolete by resistance also affects the availability and cost of chemical pesticides. Increased safety testing has increased pesticide registration time and costs of bringing new pesticides to market (CAST, 1981). Cross-resistance can also make it more difficult to develop new chemicals; there is concern in some quarters that we may be running out of biochemical or physiological target sites in pests that can be attacked by new chemicals. New biological techniques, however, are expected to enable us to identify additional target sites not heretofore recognized. Pesticide companies and growers are concerned about the very limited set of products—in many cases, just one or two registered compounds—available to control many major pests.

On the other hand, indicators of future scarcity of pesticides such as pesticide prices in broad classes of pesticides do not give signals of increasing rates of resistance development. Pesticide prices have been falling relative to the prices of other agricultural inputs over the past 15 years (Miranowski and Carlson, this volume). Pesticide prices are an important indicator of resource scarcity in the future even in the presence of market imperfections. One is cautioned, however, that examining changes in pesticide prices probably will not reveal and correlate with high levels of resistance on minor crops and/or pests. Also not revealed are important pest resistance episodes (rodenticides, diptericides) in poor regions or countries.

Frequently, government agencies can assist in resistance management by carrying out other regulatory and research functions. For example, the EPA has had a program supporting IPM research; as a part of IPM, efforts to manage resistance to pesticides is clearly a legitimate function of government. California and a few other states have been active in funding research to understand resistance to pesticides, with a goal of developing practical programs to reduce its buildup (Hawkins, this volume).

In cases where pesticides are approved only for experimental or emergency use, evaluation of pesticide use ideally should precede approval for commercial registration or at least be in progress while the application is being considered. Monitoring is critical to evaluation of both experimental and emergency pesticide-use programs. While accurate monitoring is expensive and virtually impossible to administer on a global scale (see Chapter 4, this volume), it should be possible on a smaller scale; it should include regular estimates of densities and distribution of pest, extent of pest-related damage, and frequencies of resistant pest genotypes. The latter should be obtained for pests in the surrounding area as well. Programs to obtain similar estimates should also be instituted for a few representative areas where pesticides are already being applied.

RECOMMENDATION 3. **Departments of agriculture within each state, in considering whether to request emergency use permits to respond to pest-control needs that have arisen because of resistance to another compound, should seek advice on whether the conditions governing the emergency use permit are consistent with validated tactics for the management of resistance. The EPA, in approving such requests, should also consider the consequences for managing resistance, especially when cross-resistance is thought to be a possibility.**

The committee agreed that the idea of requiring data on ''resistance risk'' as part of the pesticide registration requirements is currently inappropriate. Considerable research is needed before the feasibility of such data requirements can be established, and such a regulatory response would require significant resource commitments at the EPA. Whenever such data exist, however, regulatory agencies should use them in discussing potential pesticide benefits (Dover and Croft, this volume).

RECOMMENDATION 4. **Major reforms in regulatory programs do not appear justified or feasible at this time to advance the management of resistance**

to pesticides. Administrators of regulatory agencies should, however, formally enunciate policy statements that indicate awareness of, and responsiveness to, resistance management issues, to the extent that such factors can and must be considered when implementing regulatory activities. Activities to manage resistance should not be pursued to the extent that they divert program resources from high priority safety responsibilities.

Resistance Management Information in Regulatory Agencies

The committee does feel strongly that a legitimate function of regulatory agencies is the collection and dissemination of information on resistance to pesticides. Information on the efficacy of particular pesticides is critical for making informed pesticide-use decisions.

Because of their contact with farmers, extension service personnel and other local groups are in the best position to help producers formulate decisions to manage resistance. Given the frequent necessity for quick responses to resistance development by farmers, consultants, and chemical firms, however, the reporting of a resistance episode to regulatory agencies could probably not be acted on fast enough for a regulatory agency to initiate steps to foster resistance management, at least in the current production year.

It is very costly to monitor and determine the geographical boundaries of a resistant pest population. A company often may not even know that resistance exists. Sometimes, firms may be reluctant to reveal diminution of pesticide effectiveness, but it is very difficult to conceal resistance in the United States for long. On the other hand, though, the desire for repeat sales of a pesticide product, and for sales of other products in the company's product line, generally leads companies to respond quickly to assess and report the extent of resistance to pesticides.

Nevertheless, regulatory agencies can help foster solutions to pesticide resistance by compiling and disseminating accurate information. Various EPA functions, such as granting of emergency use registrations and the "adverse effects" activities, currently have important impacts on information flow. Resistance to available pesticides is reported in about 30 percent of the documents filed with the EPA by state agencies requesting emergency use registrations (section 18 of FIFRA).

Compiling information on resistance in an easy-to-access computer file and disseminating this to a repository within the USDA, such as the National Pesticide Information Retrieval System, could be useful. Because of the local nature, variable severity, and important time dimensions of resistance, it is difficult at present for state and federal programs to obtain information and respond quickly. Furthermore, there is no reliable mechanism in place to validate the accuracy of resistance information. Therefore, any steps to compile and publish resistance data should proceed only in conjunction with an effective mechanism to update and confirm its accuracy.

RECOMMENDATION 5. **Information collection and dissemination on pesticide resistance is an important function of federal agencies involved with agriculture. A new initiative in carrying out this function should be pursued. The EPA, the USDA, and state regulatory agencies should cooperate in building a permanent repository for such information, including a mechanism to confirm the accuracy of resistance data.**

Reports published by the Food and Agricultural Organization (FAO) and the World Health Organization (WHO) validate resistance episodes and could perhaps serve as a useful model in establishing a repository for resistance data.

Funding

Funding constraints must be confronted in structuring new initiatives to manage resistance. The EPA and state regulatory agencies appear to have little flexibility to reduce or adjust other program responsibilities in pesticide regulation. As stated earlier, the committee feels that safety-related pesticide regulatory activities should retain higher priority than resistance management. Therefore, public funds through regulatory channels for resistance management are limited. Resistance monitoring, information dissemination, and research activities will require both public and private programs.

One idea advanced at the convocation for raising new funds to advance resistance management is imposition of a national sales tax on pesticides. In supporting such a tax, proponents argue that the pesticide industry and users of pesticides will be the primary beneficiaries of successful resistance management and that they should defray through such a tax the costs to develop and maintain programs to manage resistance.

There is no indication that pesticide companies or farmers feel that resistance development and management needs are critical enough in the United States to justify such a tax. Pesticide firms have been surveyed and are nearly unanimously opposed to such a tax. No careful assessment of farmers is available, although the committee suspects that farmers without resistance problems will probably not be eager to bear the brunt of such a tax through higher pesticide prices since the tax would finance research and extension related to resistance for other regions or countries.

Many resistance problems have only local impacts, and these can usually be met with new pesticides or nonpesticide approaches, including wider adoption of integrated pest management. Farmers may approve localized user fees for managing resistance, as occurs with community pest-management user fees used to fund integrated pest management, pest eradication, and management of resistance to pesticides for mosquitoes. Such fees are assessed on crop acreage (for example, cotton insect eradication programs), land area (property tax), and production level (cotton bale taxes), with local growers having a direct role in controlling how the funds are used.

Pesticide manufacturers have incentives to join with other pesticide firms, university researchers, and the extension service to provide research and information on resistance. Industry is expanding voluntary efforts to report resistance. To date there have been no major financial assessments.

RECOMMENDATION 6. **Redirection of EPA funding or imposition of a national pesticide tax are not recommended. Pesticide taxes collected by local pest-control districts can be used to manage resistance and should be encouraged.**

PESTICIDE INDUSTRY

Most major pesticide firms are active in many countries, and actions taken in one country reflect its unique crops, pests, social institutions, and laws. In recent years, pesticide companies in groups and individually have increased actions to reduce the rate of resistance buildup and to prolong the market lives of pesticides. Several organizations of pesticide firms have come into existence over the past 4 or 5 years with management of resistance as their major purpose. A currently successful example of this voluntary cooperative program among companies occurred in Australia with the goal of managing development of resistance to synthetic pyrethroids by cotton bollworms (*Heliothis armigera*) (Davies, 1984).

Economically, the willingness of a particular company to take actions to reduce resistance development for a given pesticide is related to market structure, pest mobility, and cost and returns of employing resistance management tactics. The value of protecting a pesticide from resistance is affected by: number, effectiveness, and costs of existing and prospective, competitive pesticides; expense of nonchemical controls; ease of production of the compound; and effectiveness in controlling major pest(s) on major crop(s).

Groups of pesticide firms have joined together, with some successes, to prevent or forestall the emergence of resistance for certain classes of proprietary pesticides. Such action has been possible only where coordinated activities are not costly or illegal, and when coordination across pesticide products was mutually recognized as essential to prevent resistance to valuable products. With cross-resistance, or production of a single compound by several companies, cooperative efforts between firms may be used for monitoring, research, and in deployment of control tactics such as use of mixtures, recommendations for use of selective pesticides, and rotation of pesticides. The basic notion is that by joining efforts areas large enough to treat mobile pest populations effectively can be combined (Miranowski and Carlson, this volume).

Antitrust Concerns

Agreements between firms to divide up territories, customers, or time periods are usually considered illegal, especially in developed countries with mature market economies. Even if the intent of a group of firms is to regulate sales of a class of pesticides to prevent development of resistance, such group actions could be ruled anticompetitive, although the committee is aware of no such case. Companies are reticent to try collaborative activities, but individually or jointly they can take action that is not anticompetitive in order to reduce selective pressure on a pest population. Three actions that have either been tried or considered are (1) recommendations in addition to label directions by a firm on use patterns of its product to prevent resistance (such recommendations would, though, have to be consistent with labels or risk a violation of FIFRA), (2) joint recommendations to end users (farmers, public health agencies) by a group of firms on pesticide-use patterns over time and space, and (3) attempts by a firm or firms to influence or amend the registration of products by regulatory agencies (discussed below).

Recommendations on rotation or mixtures of pesticides can either be issued directly by the company or funneled through the extension service, a local government agency, or private consultants. Because such a program is only a recommendation to users, who still have the right to buy and use any registered compound at any time in a manner consistent with its label directions, recommendations to rotate or mix pesticides are not considered to be anticompetitive. Such a program requires considerable cooperation among users and sellers of pesticides.

RECOMMENDATION 7. **After consultation with the EPA; university, state, and federal researchers; and industry trade associations, the U.S. Justice Department should consider issuing a voluntary ruling that clarifies the anti-trust consequences (if any) of joint pesticide use recommendations by groups of pesticide companies offered for the purpose of reducing development of resistance to pesticides.**

Registering Pesticide Mixtures

Pesticide registration decisions can affect resistance. For example, there are occasions when mixtures are a valid management strategy, and such a strategy can be recommended with confidence for use over a wide area. Indeed, mixtures of pesticides are routinely used, consistent with EPA-approved labels on existing products, on some crops to control pest complexes that are not controllable with any single pesticide. Regulatory agencies need to be flexible and to review the merits of new label application and warning proposals quickly. Use of labels to mandate specific use patterns should be discouraged because such recommendations are not appropriate in all regions where a given pesticide is used.

At the same time, industry representatives should not attempt to use the registration process to conceal changes in pesticide efficacy from pesticide users or competitors. If a label is obtained that only allows a mixture of two compounds to be applied for a given pest, and if resistance has been developed to one of these compounds in some locales, then farmers in the areas with resistance may be induced to use an ineffective compound. In addition, a requirement that only mixtures be used could increase the amount of pesticide released in the environment.

Research on the use of pesticide mixtures to reduce development of resistance is in progress, but is not sufficiently advanced to support definitive recommendations at this time. Compatibility of pesticides as mixtures and selectivity of pesticides in controlling pests and pest complexes continue to be important topics for research.

Coordination of the research, use recommendations, and regulation of pesticide mixtures is needed. Requiring use of mixtures by only selling pesticides in this form is a severe restriction, and it should only be adopted if research demonstrates the mixture's desirability in managing resistance.

RECOMMENDATION 8. **The EPA should adopt a flexible policy on registration of pesticide mixtures. Coordination among regulatory agencies, research groups, and pesticide companies is needed when requests for registration of mixtures are proposed.**

Minor-Use Pesticides

Resistance can exacerbate shortages of pesticides in crops with small acreages and specialized pests. New pesticide introductions have been encouraged in minor-use groups by a program known as Interregional Project-4 (IR-4). The IR-4 program encourages input by universities and other groups in developing residue chemistry, efficacy, and phytotoxicity data for obtaining tolerances and product registrations. The program strives to lower the cost of registration for minor uses so that these markets are not bypassed when a chemical firm is expanding the crops and pests included on its label. In several instances, the IR-4 program has been successful in supporting minor-use registrations needed because of resistance to pesticides.

The IR-4 program can serve as a model for new efforts to encourage public sector activity in resistance management programs. If a pesticide firm has particular use instructions it wishes to include on the label, then the firm can provide authorization and financial support for university research so that a prescribed program to manage resistance is available to farmers. Efforts on the part of the USDA and universities need not be limited to residue analysis and short-term efficacy tests. Longer-term efficacy studies related to resistance management could also be included.

The IR-4 program has been effective over the past 20 years because it shares data across regions and facilitates close cooperation between the pesticide industry, grower organizations, and state and federal laboratories. The EPA has assisted by providing financial support, handling IR-4 petitions as priority actions, and promulgating specific guidelines for the registration for food and nonfood uses. Efforts to address minor-use problems faced outside of agriculture have been extremely limited. There is interest on the part of industry to include rodenticides, disinfectants, and other human health pests in this program. Resistance problems are particularly severe in some nonagricultural pests, so new approaches and funding are both needed and justified.

RECOMMENDATION 9. **Expansion is encouraged of the IR-4 activities to include pesticide registration activities related to resistance. Expanding the IR-4 concept to nonagricultural minor uses should be pursued by state and federal agencies.**

INTERNATIONAL CONSIDERATIONS

Pesticide resistance is a global problem. Resistant pest populations are not restricted by geographical boundaries, and major pesticide manufacturers operate in all parts of the world, often selling comparable products to control the same pest(s) on four continents. Large-scale movement of people and goods among countries increases the likelihood that resistant pest populations become internationally established. This is especially true in developing regions where quarantine and inspection services are often lacking or ineffective. The less-developed countries are particularly vulnerable to disease and severe economic losses from pesticide resistance. Agricultural development efforts, although often constrained by resistance in agricultural pest populations, have not generally received the attention directed toward resistance problems due to failures in disease vector control programs that affect public health programs.

Because the United States plays a large role in world agriculture, and U.S. policies are highly visible, management actions taken by the U.S. to limit resistance to pesticides can affect the global development of resistance. Global consequences should be considered in developing legislative and administrative changes in U.S. pesticide-use policy.

Roles of International Organizations

FAO and WHO These U.N. organizations are made up of member countries. Their roles in managing resistance to pesticides are to

- Encourage and assist member countries to develop and use effective, accurate monitoring systems to detect resistance.
- Provide member countries with technical assistance to analyze and interpret existing information to determine the significance of resistance episodes that are detected, and potential implications for field-level programs.
- Facilitate the collection and dissemination of information on resistance to pesticides.
- Assist all countries to carry out research on countermeasures for resistance by directly funding or by stimulating relevant research projects.
- Assist in training and education for effective management of resistance to pesticides.

The FAO regards resistance problems and related strategies as an inherent part of IPM programs. It can neither intervene nor interfere with national policies of member countries on pesticide registration or other regulatory matters. If requested, though, it can provide available guidelines and assist in securing external expertise. An international Code of Conduct on the Distribution and Use of Pesticides, now under preparation, will provide further international guidance on appropriate responses to pesticide resistance problems.

For 30 years, management of resistant disease vectors has been a high priority of WHO, which considers vector resistance the greatest technical impediment to control of these diseases. The problem has become even more critical in recent years as human pathogens have developed resistance to major drugs. Where no vaccine or effective drugs are available for mass treatment, WHO's public health programs must rely on vector control. Accordingly, WHO has developed 13 standardized tests for the susceptibility of major disease vectors to important pesticides. The WHO program for detecting and monitoring resistance involves interpreting and analyzing test results, feedback, periodic reporting, and follow-up advisement to member countries. All monitoring and detection information is stored on computer files; incidence and distribution trends are summarized and reported every 5 years at meetings of WHO's Expert Committee on Insecticide Resistance. Countries where resistance is detected receive an alert and are advised to study the epidemiological picture to see if resistance is impeding progress of disease-control efforts. This action may ultimately lead to the development of new methods or new materials for vector control. WHO's current focus on training and education is designed to alleviate a lack of trained professionals in many member countries.

The World Bank This institution may affect resistance to pesticides through its investments in agricultural development or public health projects that provide funds for pesticide purchase. World Bank officials have issued guidelines to determine whether projects with major pesticide purchases are likely to provide a positive return on investment and encourage sound safety and pest-management practices. Development of these guidelines is an important step for pest-control investments in developing countries, and the World Bank should be encouraged to apply them in addressing likely pesticide resistance problems.

CGIAR International Research Centers The international research centers are autonomous, commodity-oriented, and focus primarily on methods for increasing productivity through germ plasm research. While it is not within the mandate of these centers directly to address resistance issues, the importance of good pesticide management is recognized within many of the centers' model or better programs. IPM strategies are used in these programs partly as a measure to prevent development of resistance.

Need for U.S. Support While the concerns and programs of a number of international organizations contribute to greater global integration of management of resistance to pesticides, several problems constrain their effectiveness:

- A lack of data on pesticide use and pesticide performance in less-developed countries limits opportunities to assess the likelihood of resistance to specific materials developing in specific pest populations. Lack of such information makes it difficult to develop criteria for pesticide-dependent investments or to make wise selections of pesticide materials for use in development and public health programs.
- Information on the incidence of resistance needs to be more effectively collected and summarized, better targeted, and more broadly disseminated.
- Appropriate decision-making based on accurate information requires well-trained people. The current level of training is inadequate in many less-developed countries, and training opportunities are limited by a lack of resources.

Private and public decision-making in the United States is affected by inadequate information on global incidence of resistance to pesticides. As part of the international community, the United States depends on the maintenance of global pest susceptibility to important pesticides. The U.S. Agency for International Development (AID) provides some assistance in the areas of pesticide management, training, and improved pest management to less-developed countries and regions with which it deals. The AID policy on pesticides requires that AID projects minimize pesticide use and encourage an integrated approach to pest management, but availability of funds can limit the extent to which this policy is implemented.

RECOMMENDATION **10. The United States should support increased involvement and larger-scale organized efforts to coordinate information systems and research on resistance to pesticides. To the extent possible, the United States should provide funding and personnel to achieve increased training and education on pesticide management and pesticide resistance in less-developed countries.**

The potential benefits of these courses of action for the United States are a decreased rate of global development of resistance and an increased ability to react rapidly to accurate information about new cases of resistance in domestically important pest species. The USDA-OICD (Office of International Cooperation and Development) and AID should take a lead role in providing personnel and funding in pursuit of this goal.

Impact of U.S. Policy

No international organization or institution has regulatory authority for pesticide use. The influence of U.S. policymaking on pesticide-regulating programs in other nations is thus critical.

Both U.S. domestic pesticide use and international aid policies can affect perceptions or use of specific pesticide materials in other countries. The EPA registration process sends signals to other countries regarding U.S. judgment on benefits and risks of particular pesticides. Regulation 16 of the Code of Federal Regulations (part 216, Pesticide procedures; also codified in Section 118 of the Foreign Assistance Act) requires that any AID international assistance project involving pesticides be reviewed through an environmental impact statement (EIS). The preparation of an EIS, and its outcome, are based partly on the EPA's registration status of proposed pesticides. The process is more stringent for materials that are not registered for use in the United States.

FAO and WHO programs, and the public health and agricultural control programs in less-developed countries, rely heavily on inexpensive, practical, and effective pesticides. Actions taken by the United States on pesticide use or management of resistance can affect this reliance, especially when such actions tend to limit access in the developing nations to older, cheaper, generic chemicals that have retained efficacy for decades of use in the developed world.

- Actions that conserve the susceptibility of important pests to major pesticides aid third world goals by decreasing the global rate of resistance development.

• Policies that limit the global availability of particular pesticides can constrain less-developed countries' abilities to manage resistance.

Constraints on global availability of certain pesticides have come about because of cancellation or suspension actions taken in the United States. U.S. pesticide regulations are based on national priorities and relatively strict environmental standards. Some materials banned in the United States are considered essential for achieving public health or agricultural development goals in less-developed countries possessing different national priorities.

Additionally, the United States must be aware that less-developed countries may adopt U.S. regulations with little or no analysis or modification, even in cases where such regulations are inappropriate. Vast differences in institutions, agricultural systems, and cultural and political factors, however, can make U.S. policies on pesticide use or management of resistance to pesticides inappropriate or counterproductive for other countries.

RECOMMENDATION **11. U.S. policy recommendations or policies specific to U.S. priorities should contain qualifications that clearly limit their applicability to the current domestic situation. Assistance should be provided by AID and EPA in helping less-developed countries formulate pesticide policies.**

REFERENCES

CAST (Council for Agricultural Science and Technology). 1980. Impact of government regulations on the development of chemical pesticides for agriculture and forestry. Report No. 87. Ames, Iowa: CAST.

Davies, R. A. H. 1984. Insecticide resistance: An industry viewpoint. Pp. 593–600 *in* Proc. 1984 Br. Crop Prot. Conf., Pests and Dis.

WORKSHOP PARTICIPANTS

Implementing Management of Resistance to Pesticides

GERALD A. CARLSON (*Leader*), North Carolina State University
ARNOLD ASPELIN, U.S. Environmental Protection Agency
CHARLES M. BENBROOK, National Research Council
LUIGI CHIARAPPA, Food and Agriculture Organization of the United Nations
CHARLES J. DELP, E. I. duPont de Nemours and Company
MICHAEL J. DOVER, World Resources Institute
STAN FERTIG, U.S. Department of Agriculture
RAYMOND E. FRISBIE, Texas A&M University
NORMAN GRATZ, World Health Organization, Switzerland
LYNDON S. HAWKINS, California Department of Food and Agriculture

RICHARD HERRETT, ICI Americas, Inc.
MAUREEN HINKLE, National Audubon Society
EDWIN L. JOHNSON, U.S. Environmental Protection Agency
JOHN A. MIRANOWSKI, U.S. Department of Agriculture
KATHERINE REICHELDERFER, U.S. Department of Agriculture
PATRICK WEDDLE, Weddle, Hansen and Associates
KEN WEINSTEIN, McKenna, Conner and Cuneo

Pesticide Resistance: Strategies and Tactics for Management.
1986. National Academy Press, Washington, D.C.

Actions and Proposed Policies for Resistance Management by Agricultural Chemical Manufacturers

CHARLES J. DELP

Agricultural chemical manufacturers (industry) work independently and cooperate with each other and with academic and government institutions to study resistance and to develop and implement effective management strategies. Intra-industrial organizations facilitate cooperative resistance management activities. Industry does not support congressional legislation to broaden U.S. Environmental Protection Agency (EPA) regulatory responsibilities to include resistance management, pesticide taxes to support regulations, or a resistance research foundation created with industry assessments. Industry does support research, monitoring, and educational activities in-house and in cooperation with other organizations for resistance management.

INTRODUCTION

Agricultural chemical manufacturers (industry) are aware of the consequences of resistance to pest-control chemicals and are prepared to initiate actions to manage resistance to the chemicals they market. The efficacy of their products is a critical concern, thus, industry commits substantial resources for research, monitoring, and development of resistance management practices. In recent years intercompany cooperative actions have helped industry respond to resistance management needs worldwide.

INDUSTRY ACTIONS

Companies with long-term commitments to crop protection are making major contributions to the understanding of resistance management. They work independently and in cooperation with academic and government institutions. For example, Ciba-Geigy pioneered work in which Dittrich (1981) provided a leadership role in practical resistance research on agriculturally important arthropods, developing monitoring programs and management strategies. With herbicides, Ciba-Geigy has supported research into triazine resistance since the early 1970s (LeBaron, 1983). Its support helped determine the mode of resistance and has advanced the understanding of the photosynthetic process. This research may even result in the development of crop plants that are resistant to herbicides. Urech and Staub (1985) report on Ciba-Geigy's recent contributions on fungicide resistance.

ICI has been working to unravel the population dynamics of cereal powdery mildew strains, and Ruscoe (in press) initiated joint industry actions to deal with potential pyrethroid problems.

Du Pont has been researching benomyl resistance and monitoring methomyl sensitivity since the early 1970s. In 1981 Leeper headed an insecticide resistance management group in its research division. Monitoring is the cornerstone of the program, which also includes research into areas such as insect chemistry and toxicology, population genetics, and the potential use of synergists.

Other companies, such as Bayer, BASF, and Sumitomo, have ongoing in-house resistance research programs that at times amount to as much as 10 percent of the research program and provide grants of $10,000 to $30,000 to sponsor outside programs.

Industry has resources to facilitate practical solutions—a worldwide communications network; cooperative research and development contracts; and a broad research base in biology, genetics, biochemistry, neurophysiology, toxicology, and the like. Industry also has an excellent record of sponsoring educational activities for which speakers, teachers, and funds for symposia and training courses have been contributed. For example, 22 companies and the United Nations Food and Agriculture Organization (FAO) are supporting a series of resistance courses for the Third World. The latest, in Malaysia, was such a success that another is planned for Central America. Industry scientists are on the faculty of these courses. An increasing number of papers by industry scientists are being published, and efforts are being made for joint industry publications. There is an impressive amount of data available that could help researchers put their results into a broader context.

INTRA-INDUSTRIAL ORGANIZATIONS

Companies contributing individually to resistance management recognize the problems of insufficient or inaccurate information resulting from conflicting methods, conclusions, and management strategies. Cooperation is needed not only with academia and government but with each other. The best resistance management efforts can be nullified by the actions of one uninformed or irresponsible company or agency. Although the antitrust implications and competitive traditions make intercompany collaboration difficult, significant results have been achieved during the past five years.

In November 1979 ICI approached nine companies that develop and market photostable pyrethroid insecticides. These companies set up a technical liaison on pyrethroid resistance, the Pyrethroid Efficacy Group (PEG), which has contributed significantly to resistance management. For example, PEG helped the government of the United Kingdom deal with the problem of pyrethroid-resistant houseflies by withdrawing pyrethroids from animal houses. Industry also cooperated through PEG to prevent the use of pyrethroids on noncotton crops in Egypt, thus interrupting exposure of Spodoptera throughout the year. In Australia cooperative action by industry, government, and growers successfully implemented restrictive strategies of pyrethroid use on cotton to manage Heliothis resistance in 1984. Industry did not fully agree on the above measures, however, and some companies are reluctant to continue to cooperate if results do not support long-term economic benefits or if the scientific basis of a strategy becomes questionable (Davies, 1984).

The Fungicide Resistance Action Committee (FRAC) was developed because industry scientists knew cooperative industrial action was needed. FRAC is a steering committee organized into working groups for each fungicide type. Working groups are guided to (1) include senior scientists from companies with a related "at risk" fungicide; (2) establish trust, pool information, define problems, and assess risks; (3) agree on monitoring methods and verify field resistance; (4) verify resistance reports and potential remedies; and (5) encourage resistance research and communication. Some of the actions of the FRAC working groups follow.

Acylalanines

Four companies agreed on the risks of resistance and on a management strategy based on prepacked mixtures with fungicides having a different mode of action. The companies solicit the support of extension and advisory agencies to help restrict the use to two to four applications per season and no curative use (Urech and Staub, 1985).

Benzimidazoles

Individual companies had done much of the management work on benzimidazoles before FRAC was organized. After FRAC, research was focused on resistance in the cereal eyespot pathogen in Europe (Delp, 1984; Wade and Delp, 1985). Representatives of at least four companies sponsored meetings, research programs, and monitoring surveys and agreed with advisory officers that (1) a benzimidazole should not be used where resistant strains had caused a disease-control failure; (2) a mixture of a benzimidazole plus prochloraz would be recommended for fields with a high risk of eyespot disease or where a benzimidazole had been used for several years; and (3) a benzimidazole may provide cost-effective yield improvements in fields with resistance or poor eyespot control.

Dicarboximides

The working group of 11 companies, cooperating with officials in France, Germany, and Switzerland, conducted monitoring and research studies resulting in joint agreements to limit the recommended use of dicarboximides to control Botrytis on vines to two applications (bunch-closing and maturing of berries) in intensive disease areas.

C14–Demethylation Inhibitors (DMI)

The responsibility of this group (composed of senior scientists from five companies, with a potential of eight more) is to anticipate field resistance problems and to implement preventive strategies to avoid abuse. They cooperate with, and have commissioned many special studies through, universities and governments. Improved monitoring methods and accumulated research data are designed to lead to clear management recommendations.

The International Group of National Associations of Agrochemical Products (GIFAP), which sponsors FRAC, recently created an Insecticide Resistance Action Committee (IRAC) of which PEG is a part. The working groups of IRAC are for major crops such as cotton, fruit, rice, field crops, vegetables, animal health, and vector control. This industry committee, with objectives similar to FRAC, is conducting a worldwide industry survey of resistance problems to classify economically relevant and verified cases of field resistance according to their regional importance.

INDUSTRY POLICIES

Most companies support the following policies for managing resistance:

- Conduct research, monitoring, and education activities in support of products
- Provide financial support to outside research, monitoring, and educational facilities for pesticide resistance management
- Strengthen commitments to organizations such as FRAC, IRAC, GIFAP, and national associations
- Support special educational and management activities in the Third World

Industry does not support congressional legislation to broaden EPA regulatory responsibilities to include resistance management, pesticide taxes to support government regulations, or a resistance research foundation created with industry assessments.

CONCLUSION

As evidenced by its cooperative and voluntary programs, industry is not only concerned but it is active in resistance management, and industry is ready to work with all the groups involved (Urech, 1985). Industry can give not only products and financial support but technical resources, organizational skills, data bases, and motivation to prolong the effectiveness of pest-control agents.

REFERENCES

Davies, R. A. H. 1984. Insecticide resistance: An industry viewpoint. Pp. 593–600 *in* Proc. 1984 Br. Crop Prot. Conf., Vol. 2. Lavenham, Suffolk: Lavenham.

Delp, C. J. 1984. Industry's response to fungicide resistance. Crop Prot. 3:3–8.

Dittrich, V. 1981. The role of industry in coping with insecticide resistance. Pp. 249–253 *in* Proc. Symp. 9th Int. Congr. Plant Prot., T. Kommédahl, ed. Minneapolis, Minn.: Burgess.

LeBaron, H. M. 1983. Herbicide resistance in plants—An overview. Weeds Today 14:4–6.

Ruscoe, C. N. E. In press. Pesticide resistance: Strategies and cooperation in the agrochemical industry. *In* Rational Pesticide Use, K. J. Brent and R. K. Atkin, eds. Cambridge: Cambridge University Press.

Urech, P. A. 1985. Management of fungicide resistance in practice. Proc. EPPO Symp. Fungic. Resist. EPPO Bull. 15:571–576. Oxford: Blackwell.

Urech, P. A., and T. Staub. 1985. Resistance strategies for acylalanine fungicides. Proc. EPPO Symp. Fungic. Resist. EPPO Bull. 15:539–543. Oxford: Blackwell.

Wade, M., and C. J. Delp. 1985. Aims and activities of industry's fungicide resistance action committee (FRAC). Proc. EPPO Symp. Fungic. Resist. EPPO Bull. 15:577–583. Oxford: Blackwell.

Pesticide Resistance: Strategies and Tactics for Management.
1986. National Academy Press, Washington, D.C.

Pesticide Resistance Management: An Ex-Regulator's View

EDWIN L. JOHNSON

Regulatory officials can institute programs to deal with many aspects of pesticide resistance including information gathering, imposition of use instructions, and prohibitions designed to prolong the useful life of a pesticide. Another, probably more important consideration is whether one should undertake such programs, since they may present barriers to development and entry of new products and technology and impose additional costs on agricultural producers and industry. There is also a substantial question concerning the effectiveness of governmental regulatory intervention against countervailing incentives in the private sector once the purpose moves beyond the data and information gathering and dissemination aspects of regulatory programs. Further, one must distinguish clearly between socially valuable extensions of a pesticide's useful life and extensions that principally provide an extension of the marketability of old products for the sole benefit of the pesticide manufacturer. Technically trained individuals often see the future alternatives as limited—illustrated by the decreased rate of success in chemical screening. Economists often see the future alternatives as limitless, based only on sufficient demand, such as the past development of several generations of chemicals and the incipient development of genetically engineered pest controls. The actual situation is most likely somewhere between these extreme views. We should not a priori presume the social desirability of extending the life of our existing pest controls through programs of pesticide resistance management, particularly those of a regulatory nature. A large number of factors need to be weighed to make that determination. This paper attempts to lay out some items that need to be considered, as well as potential areas of regulatory intervention.

INTRODUCTION

Regulatory agencies can and do deal with various aspects of the pesticide resistance problem, mostly at the national level. Their participation in implementing regulatory solutions for pesticide resistance management depends on many factors (as discussed by Hawkins, Dover and Croft, and Frisbie et al., this volume). It is not an easy task to synthesize these factors into a strategy.

For example, should or can the effective life of a pesticide be extended? Continuing the market life of existing product lines has sometimes reduced industry's incentive to research and develop new product lines. Artificially extending the useful life may run contrary to a goal of encouraging the development of more effective and environmentally more desirable pest-control technologies, such as biologically derived controls.

Extension of market life by delaying the onset of resistance may mean that less of a product is used, reducing the net return to the seller, who then defends the market share of older materials to meet return on investment objectives. Industry will generally be less than enthusiastic about participating voluntarily in such a program.

The grower must also be considered. A pesticide resistance management program requires the grower to deal with complex pest-management strategies and decisions. For example, an integrated pest management (IPM) program may extend the useful life of a chemical and reduce chemical use and costs. Such programs would be in tune with the objectives of producers and environmentalists but may be complex.

INFORMATION GATHERING

The regulatory agency, as well as other governmental agencies, can play an effective central role in gathering information and transfering it to those sectors affected by the outcomes of pesticide resistance problems. An agency could choose to collect data or to require data collection by other parties, principally pesticide producers. Given the limited funding of governmental activities worldwide, the focus should be to stimulate or require others to collect and react to data on pesticide resistance. The registrant generally has the best data on product performance, and the pesticide user has the greatest interest as the recipient. The agency, then, can be an effective facilitator of data collection. Predictive information—a proactive approach—would enable an orderly approach to designing or planning resistance management decisions early. Information gathered after the fact—a reactive approach—would require pesticide resistance strategies to be developed simultaneously with pesticide resistance.

Premarketing Data

The action closest to the theoretical ideal that a regulatory agency could take would be to require regulatory data on predicted or continuing efficacy, prior to registration, with resistance buildup as a specific parameter. The United States has lowered its requirements for premarketing efficacy data to reduce regulatory intervention in areas thought to be regulated by the market. Other countries, however, routinely require efficacy data generated in that country. Requirements could perhaps be modified to obtain data relevant to forecasting the onset of resistance. Product performance over time is what is needed, not merely demonstrated efficiency at the time of registration.

But are test methods available to predict, before actual use, the time-dependent relationships of resistance development? I can only pose this question to the appropriate scientific disciplines. The answer I have received to date from within the United States and from a few foreign experts is that current methods allow us to only speculate about which pesticide chemicals are likely to create pest resistance problems. If methods can be developed does forecasting resistance development provide the necessary information for the development of resistance management strategies?

Postmarketing Data

After-the-fact resistance information is easier to obtain, but may be substantially less useful than pre-use information, which can provide the basis of an anticipatory resistance management strategy. It is easier for regulatory officials to adopt an after-the-fact surveillance system. New premarketing testing requirements will increase pesticide development costs and potentially delay marketing of new compounds. Postmarketing survey strategies, however, provide less time to adapt to observed resistance problems, particularly from a regulatory position that requires a legal change in the label or in regulations, compared with the initial approval action. By the time data become available it may be too late to introduce remedial measures in many areas already showing resistance.

I have been told that pest-control experts are aware of the places in the world where resistance may develop first because of particular environmental circumstances. Perhaps these areas could be monitored to alert the rest of the world to resistance development in time to develop effective management strategies elsewhere.

Most governments with registration programs require the periodic reregistration of products, and they reserve the right to request additional data as a part of that process. Producers of all, or of critical, products—especially those with health significance—could be required to provide new efficacy data at specified anniversaries.

A less-formal alternative might be to initiate efficacy networks of a wide range of users to exchange observed data on the current performance of registered pesticides. An example is the Environmental Protection Agency (EPA)-National Pest Control Association agreement to share information on ineffective pesticides and new products. A private-sector example is the meetings and publications of control recommendations by the National Cotton Council. International networks would be useful.

The EPA has proposed that user networks be created to provide such information for dissemination to users and pesticide producers and for reviewing use instructions and the risk/benefit posture of the pesticide. Networking, however, will require a further breakdown of the notion that certain functions are the purview of a specific organization.

Although networks provide after-the-fact information for some areas, they can provide predictive information to regions where resistance has not progressed as rapidly. Networks to date have focused on alternatives to the ineffective product and not explicitly on pesticide resistance management strategies for the affected chemical.

Other sources of information can be scanned for indications or patterns of resistance, including an organized clipping service to provide clues from scientific literature. In the United States, sources uniquely available to the regulatory program include requests for emergency exemptions based on ineffectiveness of the currently registered alternatives. Although available, in the absence of other data these approaches are hit or miss, often result in data that are hard to interpret, are late in identifying a problem, and lead to no clear solution except to remove the ineffective product from the marketplace.

EDUCATION

Education is always an important aspect of program success. A regulatory authority has a limited role in education, however, beyond making available data and resistance strategies and promulgating a clear regulatory policy. Resistance prevention strategies identifying IPM approaches are probably best conveyed through applicator certification, training, and extension activities. Labeling could perhaps communicate resistance information to users. In some developing countries the regulatory agency plays a more direct role in the training of farmers and pesticide dealers; thus it could incorporate resistance information into those programs.

REGULATORY STRATEGY

When the registration of a pesticide is governed by its risks compared with benefits, the case for registration becomes weaker as the efficacy (benefit) decreases. At some point the deterioration in benefits would dictate that the compound be removed from the market. There are two problems with this if it is the sole regulatory approach. First, rather than extending the useful life of a product, this process removes the product after it has lost its value. Second, regulatory agencies around the world share both a common shortage of resources and a universal thrust toward human health and environmental effects of a pesticide rather than efficacy. Thus, few pesticides would be banned in such a regulatory environment, unless there was a clear health or environmental hazard.

As resistance develops several things may change simultaneously. First, the pesticide becomes virtually useless against specific pests in some geographic areas, causing a shift to a presumedly more costly alternative. This reduces the total benefit of the product as well as the risk from its use. Net risk may either increase or decrease, depending on the safety of the substituted compound. Usually, newer products may be more safely used than older ones if the registration process is inducing the proper incentives into the pesticide development process.

Second, marginal benefits may be unchanged or reduced, depending on whether efficacy is reduced in areas where the pesticide is still useful. The normal scenario is that more applications or mixtures of chemicals become necessary to control a pest infestation. Such actions increase both the costs of control and the potential risk at the margin.

The combination of these two circumstances may or may not provide sufficient justification for canceling or restricting a product under U.S. pesticide law. Although there may be a shift in marginal benefits and risks, total benefits may continue to outweigh risks at sites of continuing use of the product. Cancellation would be difficult to justify, particularly if users continued to believe that further use was beneficial in their particular circumstance; users already reacting to resistance will be disinterested in the proceedings, having already shifted to alternatives.

Countries without a risk/benefit test, but that require a separate efficacy test, may be more successful in removing products when the compound's efficacy decreases below some standards of absolute performance. In the United States the courts have determined that a product may not be denied registration merely because it does not meet a predetermined level of efficacy (Cowley v. EPA, 1980); rather, it must only perform as claimed on its label. Therefore, claims such as "Aids in the control of. . ." or "Provides beneficial reduction in. . ." are enough to make it difficult to demonstrate that a compound did not meet these requirements in the absence of complete ineffectiveness. One solution is for regulatory agencies to require labels to be as unambiguous and useful to the user as possible by including resistance management information.

Any such restrictive action will probably be challenged, given the different perspectives of the affected parties. Growers dependent on the product will check crop yield (or its value) against the cost of pest control. If they are continuing to use the product in question, they probably believe it is beneficial, and they will attempt to retain it. Industry will be viewing the product from a profitability viewpoint. A product developing resistance is usually an older product that has recouped most, if not all, of its development costs. As such it will usually be worth spending money on an appeal to prolong its market life and profits. Finally, the government will be looking at the social costs of the compound, comparative production costs, returns to growers, and potential risks to the public and the environment, with available alternatives. All sides will differ on what the ultimate fate of the chemical should be. Given the uncertainty of predicting resistance, market introduction and cost-effectiveness of alternatives, uncertainty of benefit estimates, almost certain legal challenge to a proposed ban, and health and safety issues, it is unlikely that strong legal action to ban an ineffective chemical would be initiated. If it were, it would simply accelerate what nature has started: the demise of the chemical as an effective pest-control agent.

Since a ban would be unlikely and would not achieve pest resistance management, can a benefit/risk regulation be used to force use patterns that would extend the period of effectiveness? Perhaps such changes could be forced with skillful use of the threat of cancellation; a convincing case might be made that new use instructions would prevent loss by creating a favorable risk/benefit picture from an unfavorable one. A formal approach in the United States would require an extensive risk/benefit analysis equivalent to the Special Review process for cancellation, a procedure that typically takes several years to complete. The effect must first be noted and documented, a full risk/benefit analysis conducted, and a pest resistance management strategy developed and implemented on the label of the pesticide, all in the face of likely objections from manufacturers, users, and those more concerned with competing health and environmental safety priorities. How effective is this approach in dealing with progressing pest resistance?

Presuming that a regulatory resistance management strategy could be imposed and that it would be timely and effective, several issues must be explored. Is the market more effective than the government regulatory machinery? Is resistance prevention or management a valid regulatory policy objective? Can regulatory strategies with any rationality and feasibility be developed?

The regulatory mechanism is cumbersome. The EPA, for example, can regulate product labels by offering the alternative of cancellation if the label is not modified in some specific way. This can be done (1) through negotiation if the producer believes the costs of the change are not too high, (2) through a risk/benefit analysis showing that the change is necessary to prevent an unreasonable adverse effect, (3) by issuing a regulation requiring a change or defining areas where the product may not be used, or (4) by restricting use to certified applicators trained in special circumstances of use. Other than negotiation, which could get labels changed in a year or so, these procedures take from two to five years to accomplish—plenty of time for further resistance to develop.

Private-sector organizations in the United States, as well as government bodies engaged in training and extension, however, regularly review the effectiveness of current pest control by crop or region. Broadly disseminating the conclusions of these reviews should provide the timely information each farmer needs to decide whether current practices should be changed. Producers marketing alternative products are likely to reinforce information on the relative effectiveness of that company's product over the product of a competitor. Customer loyalty is important. Therefore, the producer will probably recommend ways to use the questionable product to minimize the likelihood of absolute pest-control failure. Here, the incentives appear to push government, associations, individual farmers, the producer of the product, and his competitors all in the same general direction.

Whether this would result in a resistance management approach to extend the life of the product or simply a shift to alternative controls would be a function of the alternatives, benefits, and costs of each approach as viewed by each of the operators involved. This marketplace mechanism, aided by governmental training and information, might react more quickly and more selectively than the regulatory process.

The question now becomes, Is resistance management a valid regulatory objective? The 1910 pesticide law and the 1947 Federal Insecticide, Fungicide and Rodenticide Act (FIFRA) emphasized protecting the farmer from ineffective products. Gradually, Congress modified the regulatory framework to consider potential risks of pesticide use. The 1980 amendments allowed EPA to waive efficacy data for pesticides because users, especially farmers, were believed to be highly educated and markets were sensitive enough to respond to ineffective products. It seems that regulatory objectives are not consistent with regulatory agencies expending resources on prolonging product life.

If industrial, agricultural, or any other groups bring in proposals for labeling to extend the useful life of a product, however, regulatory officials must respond and quickly review the proposals, approving them if there is no increase in hazard. Extending the useful life of a needed pesticide differs from extending its marketing life, which would only benefit the manufacturer. This distinction must be made, since it affects the risk/benefit balance for the chemical.

The question now becomes whether a pesticide management strategy can be susceptible to regulatory imposition. Regulatory schemes, to be enforceable, must be unambiguously stated on a label or in a regulation. The type of statement is difficult to visualize and even more difficult to specify. Resistance management could be implemented through a requirement that users follow IPM practices or that pesticides be mixed or rotated with other chemicals. The latter is susceptible to label instructions, and with sufficient information may be differentiated for geographical areas. As more flexible or complex options are needed to cope with resistance problems, the less practical the label becomes for conveying unambiguous, enforceable instructions.

The label is probably not suited for conveying complex strategies, and it is questionable whether it should. For example, a debate in Congress on whether a label can detail IPM approaches concluded that EPA and the U.S. Department of Agriculture (USDA) should make IPM information available to applicators through the applicator training program—but EPA should not dictate practices on labels. In countries with great diversity in cropping practices and environmental circumstances, effective and enforceable label statements are not practicable, thus the regulatory process is limited in effecting pesticide resistance strategies. One option would be to specify that compounds be used under permit. Pesticides posing resistance problems would be used only by direction of a professional pest-management consultant. California is currently the only state that has the authority to implement this approach.

A regulatory agency can require that a pesticide be used within the structure of a resistance management program. The question becomes that of timeliness of regulatory versus other approaches, the level of priority and resources that a regulatory agency ought to spend, and its ability in the United States to implement enforceable, workable strategies without additional authority.

INDUSTRY

Industry could effectively manage the useful life of its products. The incentives, however, to foster the adoption of any strategy are lacking in the United States. For example, extending useful life may imply less use, which is often counterproductive to marketing motivations and rate-of-return objectives for industry. These are generally of a short-term nature and do not accommodate giving up sales now for uncertain future, and perhaps lower, revenues. To some degree the incentives mentioned under private-sector approaches versus regulatory approaches can benefit farmers when resistance threatens. Competition and the desire to retain customer loyalty to a product line can provide growers with necessary information on what to use.

Creating strategies to extend useful life is more complex. Assuming that a company did want to express a resistance management strategy on its label, several impediments or disincentives arise. For example, the simple solution of reducing the number of applications runs against marketing and financial incentives, but could be changed by company policy. Another example is that of specifying mixtures with other products, perhaps those of competitors. Unless the competitor agrees, such a label addition can be stopped. Prior consultation with some firms, thus excluding other competitors, could lead to antitrust charges. One interesting case of collaboration to extend useful life is the Japanese fungicide experience (Delp, 1984). In countries with less vigorous antitrust enforcement and more tolerance for cartels, such joint solutions are easier to implement.

The fungicide industry has organized itself to cope with the issue of pesticide resistance. Here regulatory agencies can be receptive to industry proposals for modifying use instructions that will contribute to resistance management. Agencies can foster industry cooperative ventures, especially in strong antitrust countries, by requiring cooperation under such procedures as the 3(c)(2)(B) authority to require data to support continued registration. Regulatory agencies can also provide the necessary requirements to generate data on development of resistance to assure equitable treatment of all manufacturers for data requirements.

An organized approach by industry could include governmental and pesticide user organizations to provide real assistance in solving some of the questions. An industrial forum could be created to provide resistance research in both forecasting techniques and resistance management strategies. Such a program has the potential to minimize costs and antitrust concerns while developing feasible approaches for predicting and coping with pest resistance problems.

CONCLUSION

Regulatory agencies can help to reduce the onslaught of pesticide resistance. The most useful role is that of facilitator to foster data gathering and private-sector efforts to cope with resistance issues, including timely response to label-change requests. Regulatory mechanisms do not appear to be a primary vehicle for forcing resistance management strategies. Indeed, there are some important questions about the appropriateness of regulatory solutions and tinkering with the marketplace implicit in such an undertaking. The cost of pursuing a strong resistance management policy by regulation would be high and could hamper the development of new and more desirable pest-control tools. The conditions for embarking on such a course must be carefully thought through before regulated resistance management is adopted as a solution.

ACKNOWLEDGMENTS

I thank William Currie, Anne Lindsay, Richard Michell, Bernard Schneider, and James Touhey for their helpful ideas and comments on this paper.

REFERENCES

Cowley v. EPA, 615 Federal Reporter 2nd 1312 (1980).

Delp, C. J. 1984. Industry's response to fungicide resistance. Crop Prot. 3:3–8.

Pesticide Resistance: Strategies and Tactics for Management.
1986. National Academy Press, Washington, D.C.

The Role of Regulatory Agencies in Dealing with Pesticide Resistance

LYNDON S. HAWKINS

California has laws and regulations that place constraints on growers, pest-control advisers, and pesticide registrants as part of a resistance management program. Although the program is still in its infancy, specific pesticide-use procedures such as timing or limiting the number of pesticide applications have been established for pears and desert cotton. Expansion of a pesticide resistance management program in state governments will occur when sufficient concern is expressed by the agricultural industry. Monitoring for resistance by state governments is limited to mosquito-control programs.

INTRODUCTION

Pesticide resistance management within the regulatory framework is still in its infancy. Currently, state and local governments become involved in resistance management when a serious problem occurs in pest control and the agricultural community turns to the local or state government for assistance. Programs involved with pesticide resistance management in state or local governments are traditionally within agencies that regulate or use pesticides, such as the California Department of Food and Agriculture (CDFA)[1] and the mosquito-abatement districts. A brief review of these programs in the context of pesticide regulations that effect resistance management will

[1]The CDFA pioneered the first pesticide regulatory program and continues to evolve programs and regulations that deal with pesticide issues.

illustrate the benefits and shortcomings of government involvement in resistance management.

REGULATORY FUNCTIONS

There are two basic regulatory functions of state governments that influence resistance management: pesticide registration and pesticide use enforcement. Because pesticide labels are considered part of the law in California, growers are in violation of the law if they do not follow the label instructions. Registration personnel and pesticide-use enforcement personnel must cooperate on the registration of a product if label instructions are to be practical and enforceable.

PESTICIDE REGISTRATION

Although the U.S. Environmental Protection Agency (EPA) registers pesticides, many states also have pesticide registration programs. For example, after a product has been registered by EPA it may enter into the California pesticide registration process. California's pesticide registration process is primarily designed for the state's diverse environmental and agricultural situations that need to be considered for safe pesticide use. When a product label is submitted, the resistance management question could be considered if, for example, a resistance problem exists with particular active ingredients. A decision to review a product label in light of a resistance problem would be based primarily on local or regional pest-management problems. Whatever the outcome of the review, (1) the product could receive registration with no label changes; (2) the registrant could be asked to amend the label to better explain the product's use under California's conditions, and the label would have to be resubmitted to EPA for approval; (3) the product could be registered and a hearing held to establish the product as a restricted material with special provisions for use; or (4) the product could be denied registration.

When a resistance problem develops the state may request changes in the label to reflect the new situation. This is not likely without the support of industry or EPA. Any changes that do occur will likely be simple and will either alert the pesticide user about the problem or require the user to follow specific procedures. If the registrant wants specific procedures followed, then consideration would have to be given to making the product a restricted material. If the product were made a restricted material, government must be prepared to enforce label and, in California, permit conditions.

California may refuse to register a product that has demonstrated serious, uncontrollable, adverse effects within the agricultural environment. Should the situation evolve to a serious problem, it is too late for resistance management. An option for the state is to reevaluate the product and if necessary cancel or suspend registration. This is a time-consuming process, however. If the problem appears serious, researchers must make their results available, pest-control advisers must inform their clients, and the registrant must take appropriate action, even if it means removing the product from sale.

Requiring resistance management information for the registration process is not problem-free, especially if it involves label information. How would the label communicate a pest-management program without confusion or liability? Chemical companies are reluctant to make changes that increase liability. Furthermore, companies would have to allocate additional resources to research resistance management for a meaningful label-improvement program. Results of their research would then need to be reported to sales staff, pest-control advisers, and growers.

Assuming that the label would spell out procedures to reduce potential resistance problems and minimize overuse of the product, the label probably would not caution about the overuse of other products with the same or similar active ingredients. Even if the label statements were advisory only, the registrant probably would not provide information about a competitor's product. Therefore, a cross-resistance problem could not be adequately handled on the label. Also, the label probably could not be detailed enough to assist a grower in specific situations. A label containing resistance information would have to be quite long to explain the variables of the resistance problem, and few would take the time to read and understand it, particularly if using the product will solve an immediate pest problem.

Although there may be perceived benefits from using the pesticide registration process and the label to provide information about resistance management, it would not be the best use of government resources to focus on potential resistance problems within the registration process. If the registrant is interested in reducing the potential for resistance by anticipating the problem, alternative strategies such as crop rotation, other chemicals, or cultural control methods must be available. The pesticide label, however, is not the place for this information.

PESTICIDE USE ENFORCEMENT

California has long been recognized for its strong pesticide enforcement program. In the County Agricultural Commissioners (CAC) about 200 person years are devoted to enforcing pesticide laws at the local level by inspecting pesticide application techniques, equipment, and records; investigating accidents involving pesticides; and instructing growers, pest-control advisers, and applicators. The commissioner also issues permits for restricted-use materials. These permits can be amended to include resistance management strategies. Restricting materials on the basis of their being resistance risks and conditioning permits on resistance management, however, should be a last resort.

Regulatory agencies rely on compliance with the law by growers and others in agriculture rather than the heavy hand of law enforcement. As part of a compliance program, education is vital. Pest-control advisers are frequently the primary source of pest-management information to the grower, since they do much of the pest population monitoring and, thus, are in the best position to follow the resistance problems. To be an adviser in California, a person must (1) be licensed by the CDFA; (2) have a bachelor's degree in agriculture or related science and pass an examination in several categories, including laws and regulations, insects, weeds, vertebrates, and plant diseases; (3) renew the license every two years; and (4) have received 40 hours of continuing education. Providing the adviser with up-to-date information on resistance management will improve implementation of those practices designed to minimize the resistance problem. The best pest-control advisers are those who know about a developing resistance problem. Since advisers are frequently competitors, however, they may be reluctant to talk about specific resistance management strategy among colleagues.

University extension services can provide a balanced report on resistance to both growers and pest-control advisers. The challenges are keeping up with the numerous pest-management strategies that are being practiced and having enough information to report accurately on a potential resistance problem. A false or misleading report can be a disservice.

To make significant gains in implementing resistance management, growers and service industry must cooperate. Without guidelines and some form of governing body, a resistance management program will probably fail. The following are two examples of successful resistance management in California.

In 1978 the pear industry had no effective pesticide to control pear psylla. Perthane was being pulled off the market, and no other pesticide was registered that would adequately control pear psylla. The synthetic pyrethroids were entering the marketplace, and a few had effectively controlled pear psylla. People in the pear industry, however, were concerned about introducing the pyrethroids into a successful integrated pest-management (IPM) program in pears and about the potential for resistance. To minimize indiscriminate use a number of restrictions were placed on the use of pyrethroids. Growers were required to (1) monitor pear psylla populations, (2) treat only in the winter unless a crisis developed, (3) use winter oils as a first alternative, and (4) use lower rates of organophosphates to control the codling moth. With financial support from the CDFA, the University of California trained growers to monitor for pear psylla. The CAC also initiated a program to spot check for pear psylla populations. The program continues today with the support of growers and the agricultural community. If the program had been based on need as perceived by regulatory agencies, it would likely have failed. Cooperation among the local agricultural community proved absolutely necessary for success.

For the pear psylla program it was not necessary to establish a grower association for administrative purposes. The cotton pest problem was different. The cotton growers were faced with the problem of resistance in the tobacco budworm *Heliothus virescens* and a multitude of other cotton pests. If pest-management strategies such as pheromone traps, male confusion, area-wide pesticide application, and exchange of monitoring data were to occur, some type of formal organization had to be established, especially since funds had been collected and any decisions would affect the entire cotton-growing region. Additionally, growers had to follow requirements, and it was necessary to penalize growers for noncompliance; therefore, the operational details for the program were established by law.

With grower support legislation was passed that formed the Cotton Pest Abatement District (CPAD). Two of the key elements of CPAD were the charges to (1) eradicate, remove, or prevent the spread of any disease, insect, or other pest injurious to cotton; and (2) eradicate, eliminate, remove, or destroy any cotton plants except those that were growing under the conditions established by a valid permit. These two elements provided the needed authority to manage pests and hosts in a manner consistent with the resistance management program. Timing of pheromone releases, use of specific chemicals, monitoring procedures, and other pest-management strategies could be consistent within the district. For the first time pests could be dealt with in an area rather than on specific fields. This approach recognizes that pests migrate and that resistance management is control of a pest population, not pests within a field.

Growers can also form cooperatives. If compliance with resistance management procedures is necessary for successful control, however, a growers cooperative may not be satisfactory. Enforcement of pest-management procedures may be necessary, and penalties for noncompliance must be significant enough to achieve objectives.

NONREGULATORY FUNCTIONS

The California state government conducts several nonregulatory programs that deal with resistance management. The programs generally fall into two areas, pest monitoring and pest management. Monitoring pests for resistance has traditionally been considered research. To implement resistance management programs, however, on-the-farm monitoring must become routine. Should monitoring be conducted by a government agency, the priority will be on pesticide issues such as water, air, or soil contamination. Monitoring for resistance will become a government priority only after additional discussions and pressure from the industry bring the issue to light.

Monitoring

An important element in pesticide resistance management is monitoring pests and their tolerance to pesticides. Although researchers conduct most of the monitoring efforts, a significant exception has been the monitoring programs conducted by mosquito-abatement districts under the auspices of the Department of Health Services, Vector Biology and Control Branch.

The mosquito monitoring program is one example of a government program designed to track pesticide resistance. The program evolved from research efforts and the need to restrain the costs of controlling resistant mosquito populations. By transferring the technology gained from research into an organization responsible for stewardship of public funds, a cost-effective program will likely evolve. It also is subject to change, particularly when funds are short. In establishing a resistance monitoring program within government, consideration must be given to its duration.

Although government monitoring programs are possible, the responsibility for making day-to-day pest-management decisions is typically in the hands of the grower, with support from a pest-control adviser, chemical salesperson, or the farm adviser. Thus, government involvement must be planned carefully. Preferably, monitoring for pesticide resistance will become an accepted practice among growers and advisers. For this to occur inexpensive resistance testing procedures are necessary. California's Pest Management Analysis and Planning Program has funded research by the University of California in this area, but additional resources are needed.

Pest Management

Pest management projects in California have developed as part of the government's effort to continue to support agricultural production. The pest-management theme has become more significant in recent years, as those concerned about the adverse effects of pesticides express their feelings. Resistance management is part of the pest-management program, but long-term projects need to be developed that lead to the implementation of practical resistance management procedures on the farm. This concern stems partly from the reluctance of the chemical industry to research resistance management and to train advisers and growers in practical methods of resistance management. This reluctance by the chemical companies has placed additional pressure on government funds for research and education.

The California pest-management program allocates resources to projects that will be implemented by growers and pest-control advisers. For example, funding to develop a monitoring technique to detect resistance of mites to dicofol in cotton has been provided. The approach is to have a bioassay technique that will indicate the level of resistance in mites in a cotton field before selecting and using a miticide. Although dicofol may not be registered for use much longer, the techniques used in this research—conducted by the University of California at Davis—will be applicable to similar situations.

Although resistance is a problem in planning pest-management programs, resistance in beneficial species can be valuable. California also funded research at the University of California at Berkeley to develop resistant strains of beneficial mites. Although the project was successful, growers have been slow to implement the strategy. The major problem appears to be in assuring the grower that the beneficial mites that are being received have the level of resistance that is being claimed. Is it government's role to certify levels of resistance?

CONCLUSION

Increasingly, state agencies will find themselves facing policy questions about resistance management. Resolution will come only when growers and advisers are ready to implement resistance management procedures. Then, too, industry, universities, and governments must coordinate their efforts, because the answer is not strictly more research or more government; it is more complicated than that. Education will play as important a role as research, and maybe more so. Now is the time to begin discussions of implementing resistance management at the local, regional, and national levels.

Pesticide Resistance: Strategies and Tactics for Management.
1986. National Academy Press, Washington, D.C.

The Role of Cooperative Extension and Agricultural Consultants in Pesticide Resistance Management

RAYMOND E. FRISBIE, PATRICK WEDDLE,
and TIMOTHY J. DENNEHY

Cooperative extension and private agricultural consultants can provide educational programs and service, respectively, in managing pesticide resistance. Training programs that include a background of pesticide resistance, identifying pests with a high resistance risk, recommending tactics that are compatible with integrated pest management (IPM), and demonstrating techniques for measuring resistance frequency should be initiated by cooperative extension. Both cooperative extension and private consultants have a role to play in monitoring pest susceptibility to pesticides and establishing pesticide resistance management programs. Pesticide resistance management is an integral part of IPM. The USDA National Agricultural Pesticide Impact Assessment Program (NAPIAP) should receive increased funding to expand pesticide resistance management through the state agricultural experiment stations and cooperative extension.

INTRODUCTION

The development of pesticide resistance, specifically insecticide resistance, was one of the most significant factors resulting in the formation of pest-management programs by cooperative extension. Pesticide resistance was also elemental in American farmers more readily accepting the services of professional agricultural consultants. Farmer organizations have rallied in support of integrated pest management (IPM), not because it offers any mystical solutions but because it represents a system of pest control that is dynamic, economically sound, and offers multiple tactics that have potential for delaying the onset of pesticide resistance.

The role of cooperative extension and its responsibility in implementing agricultural IPM programs is clear; it bears primary educational responsibility within the land-grant university system to inform American farmers and ranchers on the most advanced techniques, strategies, and tactics to manage pests. Because the way pesticides are selected and used strongly influences the implementation of IPM programs (Metcalf, 1980; Georghiou, 1983), cooperative extension and private consultants must assume an active role in pesticide management.

Private agricultural consultants provide clientele with expert advice and service for the economic management of pests. For purposes of our discussion the role of cooperative extension will be viewed primarily as one of education and that of the private consultant as one of service. We fully recognize that there is no clear distinction between education and service, since both extension and consultants frequently cross these lines in meeting their educational and professional objectives.

Extension specialists and private consultants deal with pest control on a day-to-day basis, in its most practical sense. Both are practitioners of pest management. Their programs and recommendations are based on information from state, federal, and private research and are tempered by their field experience and judgment. Because of their intimate contact with agricultural production, they are usually one of the first on the scene when pesticide resistance occurs. They often shoulder the responsibility of determining the preliminary causes of pesticide control failures; this activity usually requires close cooperation between research agencies and the pesticide industry. They are frequently faced with trying to seek options to regain control in a relatively short time when a pesticide control failure occurs. In most cases this is a reactive rather than a proactive encounter with resistance.

From a survey of several states, we found that most deal with pesticide resistance on an ad hoc basis; that is, they go from one crisis to the next. The purpose of our discussion is to provide a framework for the development and application of resistance management programs by cooperative extension and private consultants. This framework emphasizes the importance of cooperative extension and agricultural consultants working with farmer organizations, the chemical industry, federal and state research agencies, and state departments of agriculture to implement a reasonable and ongoing policy of pesticide resistance management.

One has only to review the many reported cases of pesticide resistance to identify the large scope of the problem, as well as salient trends regarding the circumstances that have resulted in severe resistance problems. Metcalf (1980) recognized the importance of pesticides, but also stressed that "the only reasonable hope of delaying or avoiding pest resistance lies in IPM programs that decrease the frequency and intensity of genetic selection by reduced reliance on insecticides and alternatively rely on multiple interventions in insect population control by natural enemies, insect diseases, cultural manipulations, and host plant resistance.'' Therefore, the principle of using alternatives to chemicals for suppression of pests is well established, although this does not entirely supplant chemical controls in many IPM systems.

We have observed a positive and general trend toward increased use of biologically and ecologically sound alternatives to chemicals in many cropping systems (Huffaker, 1980; Metcalf and Luckman, 1982; Croft and Hoyt, 1983), but we feel that measures are generally not being taken to promote maintained efficacy of the diminishing number of IPM-compatible pesticides available to agriculture. Resistance management is not a major consideration of practitioners in the selection and use of pesticides. Therefore, we suggest that pesticide management, and more specifically pesticide resistance management, deserves more intensive attention from extension, private consultants, research, regulatory agencies, the chemical industry, and farmers, especially with respect to the management of resistance to IPM-compatible pesticides.

In order for pesticide resistance management to become an integral part of education and service programs, certain factors must be considered. First, we must understand the rationale or psychology of pesticide users and those who recommend the use of pesticides. Although farmers adopting IPM practices have made significant strides in the selection, timing, and application of pesticides, evidence suggests that pesticides are still used as indiscriminate mortality factors. Those who recommend or use pesticides are generally not concerned with the class of pesticide, mode of action, potential risk as a candidate for resistance, and in some cases may still not be concerned with their impact on natural enemies. Cooperative extension, therefore, must become more aware of resistance management strategies and must work to further develop such strategies into ongoing IPM programs.

The role of extension in pesticide resistance management should be one of educational leadership. Private consultants should learn about and practice pesticide resistance management and share this information with their clientele and other agricultural consultants. Chemical producers should promote product stewardship by interfacing with research and extension and augmenting and supporting resistance management. Regulatory agencies should consider resistance problems in their evaluations of benefit and risk associated with specific compounds and should consider the impact that regulations may have on resistance management programs. Farmers should strive to stay abreast of information relating to resistance and resistance prevention that could be of value in their farming operations.

TRAINING COOPERATIVE EXTENSION AND CONSULTANTS

It is imperative that cooperative extension and private consultants have a basic understanding of the various strategies in chemical management and resistance. This is a first step toward creating an awareness of the problem. Once this is accomplished, educational programs can be targeted for agricultural producers, thus creating a more informed and productive relationship with the pesticide industry as well as with research and regulatory agencies.

Cooperative extension IPM programs must stress that the choice of a pesticide is extremely important, not only in the short term but for the maintenance and use of the chemical over time. Professional improvement training programs must emphasize pesticide resistance management and provide understanding of (1) the major mechanisms of resistance, (2) the key pesticides most likely to induce resistance, (3) the factors leading to cross/multiple resistance, (4) the techniques for monitoring specific resistance problems, and (5) management strategies. Private consultants should take an active part in planning and conducting training programs. Training sessions with cooperative extension and private consultants could be held jointly or in parallel.

A professional improvement training program for cooperative extension should be identified as a priority by the Extension Committee on Organization and Policy Technical Advisory Committee on IPM. By pooling training material developed in individual states by cooperative extension, consultants, research, industry, and regulatory agencies, a general outline of the pertinent training elements can be identified. The states would then have the choice of selecting and modifying those elements that most appropriately fit their conditions for in-state training programs. We suggest that a series of pesticide resistance management workshops be established for cooperative extension training. The development of a pesticide resistance management training program, as well as the overall implementation of the program, should include experienced representatives from at least state research and extension, state regulatory agencies, private consultants, and the agricultural chemical industry. Pesticide resistance management training sessions or workshops could be the initial binding force to bring these groups together to discuss problems and learn from them in open forums.

Certain basic elements should be included in resistance management workshops.

Background of Pesticide Resistance

Training should start with an overview that stresses the known factors that influence selection for pest resistance: genetic, biological (biotic and behavioral), and operational (chemical and application technology) (Georghiou,

1983). Appropriate literature that deals with resistance in major pest classes should be included, such as Georghiou and Saito (1983), LeBaron and Gressel (1982), and Delp (1980). Case studies should be examined to develop a knowledge base and expectations based on general trends that have occurred in the development of resistance (e.g., insect resistance to chlorinated hydrocarbons, organophosphates, and synthetic pyrethroids; fungal resistance to benomyl; and weed resistance to triazine herbicides).

Identification of High-Risk Pests and Pesticides

Cooperative extension IPM specialists in each state should take the time to identify those pests and pesticides that have high probabilities for resistance development. Special attention should be given to pests that have demonstrated the ability to develop severe degrees of cross-resistance. For example, the tobacco budworm *Heliothis virescens* has developed cross-resistance. We are faced with a high risk of resistance to synthetic pyrethroids with this pest. It might be appropriate to draw upon the experience and expertise of IPM specialists when developing a training session for management of resistance to synthetic pyrethroids in *H. virescens*.

Resistance Prevention Tactics

Reducing selection pressure is one of the most obvious and IPM-compatible measures that can be used to thwart the development of resistance (Metcalf, 1980). Precise timing of insecticide applications, based on field scouting data and economic thresholds, has had a major impact in certain IPM programs by reducing the frequency and extent of treatments (Lacewell and Taylor, 1980; Frisbie and Adkisson, 1985).

The intent of most sound IPM programs is to stress nonpesticide control measures. Crop rotation, cultural practices, resistant cultivars, scouting, economic thresholds, and other management tactics that reduce the need for pesticides should be stressed. The IPM concept dictates that pesticides be resorted to as the last alternative and after nonchemical control measures have been maximized. Despite the use of nonpesticide control tactics, however, pests frequently develop populations that can cause economic loss. In such cases the choice of pesticides should be carefully made. Evaluating the relative appropriateness of pesticides for IPM programs is a very difficult and subjective task. Metcalf (1980) assigned pest management ratings to a group of insecticides, and in doing so provided generalizations regarding their IPM compatibility.

Pesticides that are short-lived and do not have a prolonged environmental persistence should be identified (Georghiou, 1983). Once these pesticides are identified they should be promoted as preferred alternatives in an IPM program. Pesticides that have long residual lives should be avoided when possible. We fully recognize that short-residual pesticides may not be available or economically feasible for all pest species. Chemicals with short residual lives may impact fewer life stages than long-residual chemicals. If possible the exposure of all life stages of a pest to pesticides should be avoided.

The resistance management training session should include all available crop-specific information concerning rotations, mixtures, or sequences of chemicals to prevent or manage resistance. This information should be developed for key pests that have a high probability of developing resistance to certain pesticides. Conscientious use of appropriate operational strategies, such as using alternative rotations or mixtures of chemicals, should become an integral part of IPM systems.

Demonstration of Available Techniques

Cooperative extension personnel should know of direct or indirect methods that are available for estimating the frequency of resistant pests at a specific location. These may involve bioassays of living subjects or biochemical assays of prepared samples of field-collected subjects. Use of such methods should be accompanied by an understanding of the frequency-dependent nature of resistance problems. Therefore, action thresholds must be discussed to emphasize that there are frequencies below which resistant pests do not seriously threaten chemical efficacy. In many cases action thresholds will need to be based on field experience, since research data in this area is often lacking.

IMPLEMENTATION

Of the pesticide resistance management programs that have been developed in the United States and around the world, extension and private consultants have been involved in one way or another, working with university and federal research or industry. The following outline is one approach in formalizing the role of cooperative extension and private consultants in pesticide resistance management.

Susceptibility Monitoring

The experience of research scientists, extension specialists, and consultants should be drawn upon to develop a list of candidate pests and pesticides for which resistance problems appear highly probable. These pests should be included in a formalized monitoring program to determine pesticide susceptibility levels (Brown, 1976). Historically, the approach has been to wait until there has been a serious economic failure due to resistance before susceptibility tests are initiated. In most cases this has amounted to postmortem descriptions of resistance. We do not propose that every key pest be included in monitoring programs—only those for which resistance poses a serious threat to the profitable production of a commodity and, therefore, where support is likely to be generated for such a program. Resistance-risk candidates can be selected for monitoring programs based on resistance information from other areas of the country and the world. Such was the case when the apple scab fungus *Venturia inaequalis* developed resistance to dodine in New York in the late 1960s (Delp, 1980). Plant pathologists in the northeastern, midwestern, and mid-Atlantic apple-producing states immediately began to monitor for resistance in their orchards. A similar scenario was experienced when apple scab developed resistance to benomyl (Hickey and Travis, 1984). In certain instances, however, resistance may occur so unexpectedly that susceptibility monitoring programs may not have had time to be initiated. This was found with the horn fly *Haematobia irritans* resistance to pyrethroid-impregnated ear tags in the southeastern United States and in Texas (Shackelford, 1984).

We recommend that pests be monitored on a regional basis and on a scale appropriate for the particular system under study. Sample collections should complement existing monitoring programs and may be obtained in the course of existing field monitoring. Cooperative extension personnel and private consultants can provide biological samples from a wide geographical and ecological range. Laboratory tests may be conducted by university, independent, and possibly chemical industry laboratories, depending on the specifics and economics of the particular cropping system. Once tests are conducted susceptibility information should be recorded and mapped.

Management Programs

When the susceptibility tests indicate a decrease in pest susceptibility to a certain pesticide(s), the following steps may be undertaken.

Pest Samples Pest samples should be collected from fields suspected of containing high frequencies of resistant pests, and laboratory bioassays should be conducted to estimate the frequency of resistant individuals and the degree of resistance (LC_{50}, LD_{50}, LT_{50}, etc.) (Teetes et al., 1975; Delp, 1980; Truelove and Hensley, 1982; Dennehy et al., 1983; El-Guindy, 1984). Routine monitoring of susceptibility to pesticides may be necessary. IPM practitioners should verify resistance problems through laboratory toxicologists. A working resistance management program depends on a close working relationship between these parties.

Field Tests Field tests should be conducted to assess the degree to which the laboratory resistance bioassay reflects loss of efficacy under typical treatment conditions in the field. Standard field trials should be performed in areas with susceptible populations and in areas with resistant populations. The laboratory bioassay must have relevance to the field such that individuals shown to be resistant in the laboratory bioassay actually do contribute to a substantial loss of efficacy under the treatment conditions in the field.

Data Reliability The reliability of field-susceptibility data must be determined. In the worst possible case one might obtain susceptibility estimates from two different locations within the same field and find that one sample contained only resistant individuals while the other contained only susceptible individuals. Although this scenario is highly unlikely, it is obvious that, given such circumstances, a single susceptibility estimate from fields would be meaningless. One way to estimate how well a single susceptibility estimate reflects average susceptibility throughout a location is to perform numerous susceptibility estimates within single locations. Dennehy and Granett (1984) did this to estimate the within-field variability in estimates of spider mite susceptibility to dicofol. Procedures should be standardized for collecting, culturing, and bioassaying test organisms.

Designation Criteria Criteria must be established for designating populations (fields) as either susceptible or resistant. Therefore, action thresholds for resistance must be developed that describe the frequency of resistant types at which a field should be designated as resistant and for which appropriate resistance management strategies should be initiated (Dennehy and Granett, 1984). Action thresholds must incorporate available information on bioassay reliability, within-field variability in susceptibility estimates, and reduction in field efficacy with increasing frequency of resistance. Dennehy and Granett (1984) established an action threshold for dicofol-resistant spider mites in cotton; when the frequency of resistant spider mites at any location was greater than or equal to 10 percent of the population, the use of alternative miticides was recommended.

Geographic Extent Once a routine susceptibility screening program discovers and validates the presence of a resistance problem, the geographic extent of resistant populations should be determined as quickly as possible using the extension IPM program network and supported, when available, by information from private consultants. Dennehy and Granett (1984) developed a wide-scale monitoring program for dicofol-resistant *Tetranychus* spp. in the cotton-producing region of the San Joaquin Valley of California. Similarly, the geographic range of the greenbug *Schizaphis graminum* resistant to disulfoton was determined in a cooperative effort between the Texas

Agricultural Experiment Station and the Texas Agricultural Extension Service IPM Program (Teetes et al., 1975). Greenbug resistance to disulfoton was at first restricted to five counties in the Texas High Plains area in 1974. It was later determined that disulfoton-resistant greenbug populations were present in the northern portion of the Texas High Plains, Oklahoma, and South Dakota. Resistance of the apple scab pathogen to benomyl was anticipated and then determined in North Carolina (Sutton, 1978). Subsequently, a three-year (1976–1979) monitoring program nicely outlined the geographical distribution of resistant fungi (Sutton, 1983).

Once the geographical distribution of resistance has been determined, cooperative extension should immediately provide this information to the growers through its educational channels. Rapid delineation of the areas with resistant populations should deter a wholesale abandonment of the use of a product based on exaggerated reports of resistance and should lengthen the life of the compound. Cooperative extension and private consultants must work closely with research laboratories and chemical producers to develop this information.

Bioassay Techniques Rapid, practitioner-assessable bioassay techniques should be developed and distributed, and cooperative extension IPM specialists and agricultural consultants should be educated in their use. A rapid bioassay method has been developed for detection of dicofol-resistant spider mites in cotton (Dennehy et al., 1983) and is currently being evaluated by field personnel in California. Techniques have also been used for benomyl- and dodine-resistant apple scab (fungal) populations (Sutton, 1978, 1983). A rapid bioassay technique was used to determine pyrethroid resistance levels of hornflies in Texas (Shakelford, 1984).

Management Strategies Management strategies that consider all useful information on the stability and inheritance of resistance (cross- and multiple resistance) and relevant information on insect ecology, biology, and toxicology must be developed. A management strategy should be recommended when the frequency of resistant individuals in a population is greater than or equal to the accepted action threshold. The most common management strategy used, once the distribution and frequency of resistant types are determined, is an immediate switch to an alternative pesticide, and often to one that possesses a different mode of toxicological action. It is the responsibility of the extension service, and indeed that of private consultants, to notify agricultural producers of effective alternative chemicals and to recommend appropriate rates and timing of applications. Growers can manipulate resistant populations or deter resistance development by rotating groups of chemicals, alternating pairs of chemicals, using mixtures of chemicals, or adding synergists to pesticides. In addition growers may influence resistance by the manner in which materials are applied (rate, volume, equipment used, etc.) or by influencing migration, overwintering, dispersion, or other aspects of pest biology/ecology. Recommendations must result from field and laboratory investigations of the system in question and should recognize the differences between cropping systems and different geographical regions of the same cropping system.

EDUCATION

It is a definite responsibility of the Cooperative Extension Service to provide educational information and training programs to make farmers and ranchers aware of resistance management. This further emphasizes the important role of pesticide resistance management within the context of ongoing IPM programs. The topics that have been previously discussed should be packaged in such a way that agricultural clientele gain a basic understanding of tactics in managing pesticide-resistant populations. Agricultural consultants who have experience in pesticide resistance management should be invited to participate in and, if possible, help develop training programs. As with most IPM training programs, close coordination should be maintained with the state agricultural experiment station, the USDA, and the state department of agriculture.

CONCLUSION

The Cooperative Extension Service should develop initiatives in the area of pesticide resistance management that would be part of ongoing state, regional, and national IPM programs in the United States. Other countries may consider these for their respective agricultural agencies. Cooperative extension should serve as the coordinating body to seek the appropriate form and function of pesticide resistance management training activities from private consultants, research and regulatory agencies, chemical industry, and IPM farmer organizations. Three specific recommendations follow:

1. Pesticide resistance management should be a high-priority area in cooperative extension and USDA federal extension IPM educational programs. The Extension Committee on Organization and Policy-Technical Advisory Committee on IPM is a logical body to address the issue.
2. The joint role of cooperative extension, private consultants, state and federal research, and the state departments of agriculture should be identified to develop techniques for sampling, bioassay, decision (action) thresholds, and appropriate management strategies for implementing pesticide resistance management programs.

To achieve this we recommend that additional funding be provided through the USDA National Agricultural Pesticide Impact Assessment Program (NAPIAP), which is responsible for determining the benefits and use of pesticides. This program is conducted through the state agricultural experiment stations and the Cooperative Extension Service and deals specifically with pesticides; therefore, it is a logical place to initiate a broader national program on pesticide resistance management. The NAPIAP has identified pesticide resistance as a project area. This area must be strengthened and expanded into pesticide resistance management. The designated NAPIAP state specialist could work closely with state agricultural experiment station scientists and Cooperative Extension Service specialists, along with the state department of agriculture, consultants, and grower groups to organize a state-based pesticide resistance management program. This program should be designated within each state as a major branch of their current IPM program and become a key area for identifying and consolidating pesticide resistance management efforts.

3. IPM research and educational programs must be emphasized. It is only through IPM programs that we can effectively reduce selection pressure leading to resistance. We believe the only reasonable course to follow is the development of alternative strategies, either chemical or nonchemical, within the context of IPM.

REFERENCES

Brown, A. W. A. 1976. Epilogue: Resistance as a factor in pesticide management. Pp. 816–824 *in* Proc. 15th Int. Congr. Entomol., Washington, D.C. College Park, Md.: Entomological Society of America.

Croft, B. A., and S. C. Hoyt, eds. 1983. Integrated Management of Insect Pests of Pome and Stone Fruits. New York: John Wiley and Sons.

Delp, C. J. 1980. Coping with resistance to plant disease control agents. Plant Dis. July:652–657.

Dennehy, T. J., and J. Granett. 1984. Monitoring dicofol resistant spider mites (Acari: Tetranychidae) in California cotton. J. Econ. Entomol. 77:1386–1392.

Dennehy, T. J., J. Granett, and T. F. Leigh. 1983. Relevance of slide-dip and residual bioassay comparison to detection of resistance in spider mites. J. Econ. Entomol. 76:1225–1230.

El-Guindy, M. A. 1984. The phenomenon of resistance to insecticides in agricultural pests. Pp. 12–13 *in* Proc. Symp. Integrated Pest Manage. Ration. Pestic. Use in Arab Countries. League of Arab States: Arab Organization for Agricultural Development.

Frisbie, R. E., and P. L. Adkisson. 1985 IPM: definitions and current status in U.S. agriculture. Pp. 41–52 *in* Biological Control in Agricultural Integrated Pest Management Systems, M. H. Hoy and D. C. Herzog, eds. New York: Academic Press.

Georghiou, G. P. 1983. Management of resistance in arthropods. Pp. 769–792 *in* Pest Resistance to Pesticides, G. P. Georghiou and T. Saito, eds. New York: Plenum.

Georghiou, G. P., and T. Saito, eds. 1983. Pest Resistance to Pesticides. New York: Plenum.

Hickey, K. D., and J. W. Travis. 1984. Management of resistant strains of orchard pathogens. P. Fruit News 63:60–62.

Huffaker, C. B, ed. 1980. New Technology of Pest Control. New York: John Wiley and Sons.

Lacewell, R. D., and C. R. Taylor. 1980. Benefit-cost analysis of integrated pest management programs. Pp. 283–301 *in* Proc. Sem. and Workshop—Pest and Pesticide Management in the Caribbean, Vol. 2, E. G. R. Gooding, ed. CICP/USAID.

LeBaron, H. M., and J. Gressel, eds. 1982. Herbicide Resistance in Plants. New York: John Wiley and Sons.

Metcalf, R. L. 1980. Changing role of insecticides in crop protection. Annu. Rev. Entomol. 25:219–256.

Metcalf, R. L., and W. H. Luckman, eds. 1982. Introduction to Insect Pest Management, 2nd ed. New York: John Wiley and Sons.

Shackelford, K. 1984. Is time running out for insecticide ear tags? The Cattleman July:59–65.

Sutton, T. B. 1978. Failure of combinations of benomyl with reduced rates of non-benzimidazole fungicides to control *Venturia inaequalis* resistant to benomyl and the spread of resistant strains in North Carolina. Plant. Dis. Rep. 62:830–834.

Sutton, T. B. 1983. Disease management strategies and techniques. Pp. 103–104 *in* Integrated Pest and Orchard Management Systems for Apples in North Carolina, G. C. Rock and J. L. Apple, eds. N. C. Agric. Res. Serv. N. C. State Univ. Tech. Bull. No. 276.

Teetes, G. W., C. A. Schaefer, J. R. Gipson, R. C. McIntyre, and E. E. Latham. 1975. Greenbug resistance to organophosphorous insecticides on the Texas High Plains. J. Econ. Entomol. 68(2):214–216.

Truelove, B., and J. R. Hensley. 1982. Methods for testing herbicide resistance. Pp. 117–131 *in* Herbicide Resistance in Plants, H. M. LeBaron and J. Gressel, eds. New York: John Wiley and Sons.

Pesticide Resistance: Strategies and Tactics for Management.
1986. National Academy Press, Washington, D.C.

Integration of Policy for Resistance Management

MICHAEL J. DOVER and BRIAN A. CROFT

An effective integrated program of resistance management raises wide-ranging policy issues addressing the need for resistance monitoring, resistance risk assessment, regulation, pesticide-use management, education, marketing, and research. This paper offers a comprehensive view of the relationship between resistance management and the various institutions that govern pesticide development and use. It also features options these institutions can take to respond to the challenge of pesticide resistance. These options embody a threefold strategy for dealing with resistance: (1) establishing joint industry/government efforts in research, monitoring, and education; (2) creating and maintaining data bases relating to resistance; and (3) developing a regulatory philosophy based on maintaining the risk/benefit balance of pesticide use.

INTRODUCTION

The vulnerability of pest-control programs to pesticide resistance appears to be growing as a result of ecological, genetic, economic, and pesticide-use factors. Contributing to the problem are

- The increase in the number of resistant species
- The industry's research focus on a relatively small number of pesticide classes
- The costs and time delays in developing new chemicals
- The increasing difficulty of finding suitable new compounds
- The intensity of pesticide use

- The lack of economical alternative pest control methods

Before resistance creates local or regional crises in agriculture or public health and the agrochemical industry becomes more concentrated as a result of resistance and other factors, a new approach must be adopted—one in which these chemicals are thought of as finite resources rather than disposable commodities.

In the next two decades chemical pesticides will probably continue as the mainstay of pest-management technologies. Given this, resistance management may become the key to continuing effective pest control. Its success will depend on how we develop, use, and regulate pesticides now and in the future. The cost and difficulty of discovering new chemicals will require placing a greater emphasis on properly managing the use of existing products rather than counting on a continuous flow of replacement compounds.

Policy issues in resistance management involve federal, state, and local governments and private-sector concerns ranging from large multinational corporations to individual farmers. Each has a significant role to play, and the interactions among these interests make the task of describing policy options a daunting one. Many of the tactics of resistance management will require that several institutions and policies be changed in concert for successful implementation.

To effect this kind of coordination, an integrated approach to policy design is needed. In this paper we present an initial approach to policy integration for resistance management. In addition to defining the roles of individual institutions, we have tried to show the linkages among institutions and policies. An understanding of these linkages is essential if resistance management is to become a reality.

RESISTANCE MANAGEMENT AS A POLICY ISSUE

Pest-control actions can resemble the depletion of any "commons." Here the commonly held resource is the susceptibility of pests to available pesticides. Individuals acting independently can deplete this resource to the detriment of all, while the benefits of conserving susceptibility may or may not exceed the cost for any individual. Thus, reliance on individual users' decisions may harm all users (Hueth and Regev, 1974; Wood, 1981). The concept of resistance as a commons issue extends as well to—

- The possibility of a domino effect in the pesticide industry from resistance
- The vulnerability of food production and public health systems to even temporary losses of effective pest control
- The reduction in the benefits of pesticide use, thus increasing the relative social cost in health or environmental effects

The future of resistance management depends on the availability of a broad range of chemical-use patterns, nonchemical tactics, and chemical pesticides (Delp, 1981; Georghiou, 1983). But can the research, regulatory, educational, and economic institutions that control pesticide production and use respond to the challenge posed by pesticide resistance? Much basic and applied research remains to be done, but resistance management is clearly feasible. Thus, although the program discussed here is predicated on the progress of a well-designed, comprehensive research effort and will probably take 10 to 15 years to implement fully, some steps can be undertaken now.

Pesticide resistance is a global problem, yet differences in national policies governing pesticides require that resistance management policy issues be addressed initially by individual countries and the specific institutions that affect pesticide production and use. Resistance management has policy implications for research, education, extension, and regulatory agencies in the public sector and for private-sector decisions on research, development, and marketing. Since the United States is a leader in pesticide development and marketing, as well as in setting the standards for evaluating, using, and managing pesticides, the focus here is on U.S. institutions and policies.

An effective integrated program of resistance management raises wide-ranging policy issues. Resistance must be detected and measured if remedial measures are to be designed and evaluated. At the same time, methods are needed for predicting the likelihood that resistance to particular pesticides in target species will develop. Constraints on the development of new chemicals, formulations, and use patterns must be reevaluated in light of the need to manage resistance. Where cooperation among pesticide manufacturers is needed and is not anticompetitive, government agencies should facilitate it. Mandatory coordination or restrictions on pesticide use, when necessary, must be enforceable. Most important, comprehensive education and research efforts are needed to support resistance management.

COMPONENTS OF A RESISTANCE MANAGEMENT PROGRAM

Resistance Monitoring

Monitoring is central to an overall resistance management program. Up-to-date information on species that exhibit resistance (Georgopoulos, 1982; Leeper, 1983) will help assess resistance risk in new products, provide a basis for initiating management action, evaluate alternative tactics, analyze product failures, assess the effectiveness of resistance management efforts, and establish priorities for education, research, and development (Staub and Sozzi, 1983). Although resistance monitoring has been a primary objective of researchers for many years, it is an almost completely new concept institutionally. No national system exists in the United States for systematically collecting and disseminating data on pesticide resistance. Monitoring by the state agricultural experiment stations is sporadic, usually done for research or in response to reports of pest-control failures. Other monitoring data, such as those collected by pesticide manufacturers in support of their products or by pesticide user groups, are often unavailable to most researchers and pest-management advisers.

The most critical constraint to fully implementing resistance monitoring is the lack of technical knowledge and suitable techniques for researchers, advisers, marketers, and users. Current methodologies are time-consuming, expensive, and of questionable validity. Only if resistance monitoring is conducted more efficiently can a national, multispecies monitoring program function. Also, techniques must be developed for detecting resistance in low proportions of pest populations, so that action can be initiated before a substantial portion of the population exhibits high levels of resistance. Thus, the development of methods for monitoring resistance must be a high priority in any resistance management program.

Also needed is a means for collecting and disseminating resistance data and related information so that advisers and users can respond rapidly to resistance problems. A technical monitoring capability must be matched with institutional capacity to monitor routinely and systematically. At issue, too, are the availability of facilities and trained personnel for monitoring and the standardization of methods for assessing resistance and interpreting results (Leeper, 1983).

Other information systems could enhance resistance management if they were available, such as—

- Pesticide usage data collected and reported within a few days of the event, coupled with data on pest infestation levels (Whalon et al., 1984)
- Data cross-referencing species names, pesticide products, active ingredients, sites of application (e.g., crops), and locations (e.g., states)
- Information on emergency outbreaks believed to be caused by product failure

As a full-fledged national resistance management program takes shape, these kinds of data will do much to support a rapid response to resistance problems.

Both private and public resources are needed to provide the technical expertise and coordination that establishing a wide-area, multispecies resistance monitoring program on a national level requires. These include the U.S. Department of Agriculture (USDA), the U.S. Environmental Protection Agency (EPA), the state experiment stations and extension services, pesticide manufacturing companies, private pest-management consultants, and pesticide users.

In addition to the critical need to conduct more research related to monitoring, a comprehensive resistance management program should also—

- Establish a national resistance monitoring program involving local, state, and federal agencies, chemical companies, and private pest-management consultants
- Link resistance monitoring data to other pesticide data, such as label information (active ingredients, pests, and sites of application), pesticide usage data, and data submitted for emergency exemptions under the pesticide law

Resistance Risk Assessment

Scientists' ability to predict resistance in a given species to a given pesticide is limited. Although several research groups have identified individual variables that affect resistance, no overall system for predicting resistance has been discovered for any major pest group. Because so little is known about how to determine the risk of resistance, gathering basic information on the mechanisms, genetics, and ecology of resistance in a wide array of target species is essential.

The key to determining the potential for resistance in a particular use is "resistance risk assessment"—a means of indicating future shifts in benefit/risk ratios for pesticide uses. The results of these assessments could be used to set priorities for monitoring, plan pesticide research and development programs, and implement specific actions for delaying or preventing the buildup of resistance.

A national resistance management program needs to begin a research effort in resistance risk assessment as an essential component of future management efforts. In addition the program should establish a historical data base on pesticide resistance, including data on species, chemicals, locales, resistance mechanisms, resistance levels, test methods used, and cross-resistance. This data base should be jointly funded by the chemical industry and the federal government.

Federal Pesticide Regulation

A resistance management program for the United States will require the involvement of pesticide regulators for three reasons. First, resistance management methods entailing innovative products or new use instructions on pesticide labels will require EPA review and approval. Second, EPA is the repository of data on pesticides, including information that may be needed for coordinating management of more than one chemical. Third, EPA is responsible for assessing risks and benefits when problems arise with pesticides. Any effort to determine or alter the risk/benefit balance as part of a resistance management effort will have to include EPA as a key participant.

Today EPA has no specific resistance policy. Although the agency can require registrants to submit "efficacy data" on their products, current policy waives these requirements for most uses. If resistance is reported in the field, a review process may be initiated, possibly leading to companies removing claims or, more often, placing warning statements on the label. The EPA has never cancelled a pesticide registration on grounds of resistance.

The EPA's most direct involvement in resistance management has been in the registration of mixtures or tank-mix requirements on labels for certain fungicides. Responding to EPA's refusal to accept resistance management as a reason for registering a pesticide mixture, the American Phytopathological Society called for recognition of "the delay or prevention of resistance as a valid registration objective" (Yoder, 1983). But EPA demurred, claiming that the problem has only recently emerged and that more scientific studies are needed to guide policy on resistance (Campt, 1983).

The EPA does favor labeling that "provides for maximum user flexibility in attempting to delay the development of resistant fungal strains while protecting the environment from unnecessary pesticidal burden" (Campt, 1983). But industry scientists and others see the issue as one of enforceability rather than flexibility. In their view resistance management requires constraining users' choices, preferably through such physical means as prepacking mixtures, so as to prevent over-reliance on any one chemical (Staub and Sozzi, 1983). The EPA's position is that fungicide mixtures do not necessarily delay or prevent resistance and that alternatives (e.g., rotation of chemicals during the season) may be just as effective. Moreover, the agency fears that if mixtures are registered or tank-mix instructions on labels are made mandatory, more pesticide may be released into the environment than is necessary. The EPA contends that deciding what users should do to counter resistance is the responsibility of users and their advisers.

These views stem in part from the local and regional nature of the onset of resistance and in part from the agency's belief that resistance management is irrelevant to the regulatory process. Resistance is seen as an aspect of the policy that waives data requirements on efficacy. In the eyes of EPA and, apparently, the majority of Congress, pesticide efficacy is expected to be regulated by the marketplace.

Unfortunately, this policy means that responses to resistance come after the fact. If a company on its own initiative determines that resistance risk is high, it may be unable to get sufficient assurance that this risk can be avoided if EPA will not accept specific label instructions or formulations designed to prevent resistance. Under these circumstances a manufacturer is unlikely to proceed with such a high-risk venture.

Pesticide use is justified on grounds that the benefits outweigh the risks. These benefits, however, can change over time due to several factors, including resistance. Thus, resistance potential represents an economic risk to the user, and if benefits are unrealized because of resistance, environmental and health risks are not offset. Given this dynamic nature of the benefit/risk balance, EPA has a responsibility to establish a specific resistance policy. But because such a policy entails some important shifts in regulatory philosophy, specific direction from Congress will be needed.

This new philosophy goes beyond the common concept of a revived efficacy requirement, which would only determine that a pesticide works when it first enters the market and that it is removed when resistance sets in. A resistance policy, by contrast, would see that a pesticide continues to work and that it is removed only as a last resort. As resistance management techniques become perfected, regulatory action could be undertaken to help restore the benefit/risk balance earlier in the course of resistance buildup. This might include expedited data review, emergency exemptions, labeling changes, or restrictions on use.

The EPA has long been criticized for its seeming inability to carry out its existing regulatory functions under the Federal Insecticide, Fungicide, and Rodenticide Act (FIFRA) (U.S. House of Representatives, 1983; Wasserstrom and Wiles, 1985). No new program or philosophy can take hold until the agency deals with the severe problems of inefficiency and ineffectiveness that have plagued it for years. This study outlines what a resistance policy for EPA should be; putting that policy in place will depend not only on the outcome of resistance management research, but also on EPA's management of itself.

Because of EPA's pivotal position in the pesticide policy area, many issues have emerged in our study (Dover and Croft, 1984). To support resistance management, regulatory policy should be modified to—

- Incorporate resistance risk into pesticide registration data requirements, once methods are available, and develop regulatory responses including label warnings, monitoring requirements, and/or use restrictions
- Allow mixtures to be registered for use in resistance management, requiring that they meet the same health and safety standards as mixtures registered for other purposes
- Establish use-by-prescription as a restricted-use category for pesticides where precision in timing, dosages, and application method are essential to resistance management
- Require certification of users and dealers for pesticides with high resistance risk
- Develop criteria for using resistance management as a basis for emergency exemption petitions, to allow for such tactics as permitting more than one alternative chemical to be made available during a resistance-caused outbreak

In a related area USDA and the Food and Drug Administration should study the effect of food-quality standards on the development of resistance to determine whether Defect Action Levels or cosmetic grading standards are excessively stringent, thus requiring greater pesticide use than necessary.

State/Local Regulation and Management

Since resistance begins at the local level, so must resistance management, of which a major aspect is controlling pesticide use. The common constraints that states will face in implementing resistance management are coordinating the actions of pesticide users and getting users to cooperate. Since few mechanisms for obtaining cooperation or enforcing coordination are in place, advisers and the industry will have to rely on persuasion and education rather than on any existing administrative structure. Success may depend on how convincing the "pitch" is rather than on the soundness of the program.

Pesticide regulation by states is one mechanism for managing use. Although some simply adhere to the minimum standards (compliance with FIFRA), others also require permits, licenses, and record-keeping systems, many of which may be useful in developing a program in resistance management. Clearly, strict regulation alone cannot prevent the buildup or spread of resistance, but an effective regulatory structure will enable a state to carry out a resistance policy, should one be established.

In addition to examining their regulatory policies, states should—

- Establish mandatory and voluntary means to coordinate pesticide use, creating pest-management districts or promoting cooperative integrated pest management (IPM) programs where resistance is a potential problem
- Provide incentives for users to adopt improved management practices, including loan or subsidy programs based on local needs and resources

Pesticide Marketing

In the highly competitive world of pesticide marketing, sales personnel face considerable pressure to sell as much of a product as possible. Where the risk of resistance exists, marketing practices encouraging overuse of a single product may work against a pesticide company's own long-range interests.

Several factors work against taking the long view. Industry marketing personnel, who seldom hold the same job for more than a few years, tend to focus on current-year sales goals rather than the longer time commitments needed to make resistance management work. Cash flow needs and the cost of production facilities may force companies into rapid production to achieve a maximum return on investment in research and development and capital equipment (Goring, 1977). Then, too, the need to hold a share of the market often leads to price cutting. If recent cuts in pyrethroid prices (Storck, 1984) trigger a major price war, pyrethroid resistance could increase as use goes up.

To increase return on research and development investment and thus spur innovation, the chemical industry has called for extension of patent protection of pesticides. Some see this as one answer to the constraints on wider acceptance of resistance management in industry. Manufacturers, however, have long augmented original product patents with patents on processes, formulations, and uses (Storck, 1984). Moreoever, long-lived chemicals and even so-called commodity pesticides can still contribute significantly to a company's profits (Lewis and Woodburn, 1984). The market lives of pesticides appear to be affected more by relative efficacy, cost, and competition from alternative chemicals than by patents (David and Unger, 1983). Patent extension may protect against price erosion, but it will not help deal with the other factors.

Pesticide manufacturers do respond when resistance is found. In the face of hard evidence, responsible companies quickly pull their product off the market in the affected area or otherwise change use practices (Delp, 1981). Some companies, however, are reluctant to press their marketing staff to sell less aggressively in the absence of definitive evidence that such moderation will, in fact, help delay or prevent resistance. Only better data acquired through research and monitoring will convince all segments of the industry of the need to restrain marketing.

U.S. companies see antitrust laws as a serious impediment to information exchanges and agreements to restrict use. A recent agreement in principle to limit sales of acylalanine fungicides to prepack mixtures (Fungicide Resistance Action Committee, 1983) could not, industry scientists argue, have been undertaken in the United States. Although agreements to limit sales would certainly be considered anticompetitive, other options, such as agreeing to limit the amount of pesticide per dose or the maximum number of applications per season, would not necessarily create antitrust problems. Moreover, pesticide manufacturers now enter into a wide variety of licensing agreements and joint ventures covering research, production, and marketing. Clearly, when companies see cooperation to be in their best interest, they cooperate. As evidence accumulates that resistance threatens whole groups of products, companies and government will have to address antitrust concerns.

In support of resistance management, chemical manufacturers should—

- Reduce employees' incentive to oversell pesticides by reviewing and revising individual company policies in compensation and promotion
- Use pricing as an incentive for resistance management, whereby companies (acting independently) might adjust their prices to discourage overuse or encourage rotation

- Limit the amount of resistance-prone pesticide that can be sold in areas of high resistance risk
- Coordinate directions and restrictions on pesticide labels, working through the USDA to obtain Department of Justice review of agreements to prevent anticompetitive activity

Education

Resistance management is a relatively new pest-management strategy. Thus, education on resistance management for users, pest-control professionals, and students aspiring to careers in pest management is essential. For students resistance management is not specifically a standard part of today's university curricula in pest management or crop production—ideal vehicles for conveying this information. Nearly two-thirds of the land-grant universities have IPM curricula at the bachelor's or master's level (J. E. Bath, Michigan State University, personal communication, 1984), although declining enrollment in these curricula presents a problem (Poe, 1983).

Many users need to know more about new pest-management strategies. Indeed, education is critical to getting users to adopt new tactics or accept necessary restrictions on pesticide use. Although the cooperative extension services have spread the principles and practices of pest management, many users still get most of their information from the retail pesticide sales force (U.S. Environmental Protection Agency, 1974), many of whom are untrained in sound pesticide management, pest identification, and problem diagnosis. Reaching both users and dealers is critical, since together they decide whether to use pesticides, which one to use, and when, where, and how much to use.

Unfortunately, the extension service lacks the time and money to undertake such a job. To assure success, sustained user education for resistance management requires permanently budgeted funds and personnel. The federal government, extension services, universities, and industry should cooperatively support an education program to—

- Produce a federal extension bulletin on resistance management and support development of state bulletins
- Develop courses on resistance management for students, professionals, and users

New Pest-Control Tactics

Properly considered, resistance management is a subset of IPM. Since using effective nonchemical control tactics contributes to resistance management, an overall resistance management policy must include a program for promoting the development and adoption of alternative pest-control methods. These include enhancement of biological control, use of crop rotation, experimentation with intercropping, and breeding for host plant resistance (Bottrell, 1979; Office of Technology Assessment, 1979).

In the public sector the major factor in developing alternative pest-control methods has been sustained support for IPM research and demonstration projects. Despite the increased emphasis in public research institutions on alternative control methods, adoption of these methods still lags behind pesticide use as the mainstay of pest control.

Considerable policy attention has been given to removing obstacles to wider adoption of alternatives (Bottrell, 1979; Office of Technology Assessment, 1979). But, save for increased research, the only substantive change has been the EPA policy on "biorational" pesticides—microbial pesticides and synthesized analogues of naturally occurring biochemicals. Registration of these products increased considerably during the late 1970s, partly because data requirements for such substances were streamlined (Chock and Dover, 1980). Beyond that, changes have been more incremental than dramatic. There are more private pest-management consultants than there were a decade ago, and more states are using computer-based information delivery systems for pest management (Croft et al., 1976), but in the absence of alternative pest control methods, these support systems are used primarily to foster improved pesticide management rather than to implement alternative tactics.

New chemicals from the pesticide industry continue to appear. Yet the fastest growing market for new products is in herbicides, relatively few of which have encountered resistance problems. As the focus of research and development narrows within the industry, the question is where the innovative chemistry for relatively smaller-market pesticides will take place. Therefore, to promote the wider variety of chemicals needed for effective resistance management, the federal government should offer incentives to develop compounds designed to manage resistance, including regulatory incentives such as expedited data review or economic incentives.

Structure of Research

The support of basic research on resistance is the most difficult to obtain over time. The chemical industry, constrained by a product orientation, cannot easily undertake long-term basic research, and few states have the resources to maintain such a program without considerable outside support. Federal basic research on an applied problem such as resistance runs the risk of "falling through the cracks" between applied research funding sources (such as USDA) and basic research funders [such as the National Science Foundation (NSF)]. Moreover, the annual budget cycle of federal agencies, highly influenced by changes in priorities in the executive or legislative branches, creates great uncertainty for planners of long-term research.

No coordinated, multidiscipline research effort for resistance management exists today in the public sector. The USDA thus far does not consider pesticide resistance a high research priority; EPA sees resistance as outside its responsibilities, and NSF has all but withdrawn from supporting pest-management research that appears to overlap with USDA's or EPA's "territory."

Meanwhile, the chemical industry's research planning remains tied primarily to the discovery, development, and defense of proprietary products. According to one research and development director, the annual industry-wide commitment to resistance research, including monitoring, is roughly $3 million. This sum—small compared with the sales of any major pesticide—indicates that most financial decision makers in the chemical companies still need to be convinced that resistance is a serious problem and that resistance management is feasible.

Until the private and public sectors can agree on their respective research roles and can decide who should pay for the research, the long list of questions about resistance and its management will go unanswered. To address these constraints, the following options should be considered.

- Create centers for the study of resistance and resistance management. Based at suitably staffed universities, these centers should be supported by federal, state, foundation, and industry funds to carry out basic and applied research in an interdisciplinary team setting.
- Establish an independent, industry-sponsored foundation to support research on resistance. This foundation could fund traditional basic science projects proposed by scientists and multidisciplinary projects sponsored by USDA or a resistance research center and sponsor annual or biennial conferences to review progress, identify promising avenues of research, and recommend future directions. Annual voluntary pledges by companies could be used to build up an endowment as well as to support ongoing research projects.

Funding a Resistance Management Program

The proposals outlined in this paper are far-ranging and potentially costly, although the savings to the chemical industry, pesticide users, health organizations, and the general public will outweigh the costs. Since the manufacturers and users of pesticides will be the principal beneficiaries of successful resistance management, they should defray the costs of developing and maintaining resistance management programs.

Accordingly, the federal government should impose an end-user tax on pesticides to finance resistance management programs. A tax of $0.02/lb of pesticide would generate over $20 million in revenue, which could be used to support monitoring, data base development, research, education, and regulatory activities. Such a tax should be phased in over five years, as resistance management programs reach the point of being able to use these funds effectively.

CONCLUSION

Now that resistance management is becoming a feasible response to resistance, a wide range of decisions must be addressed in research, regulation, education, and marketing. It is not enough simply to accelerate product development in the private sector without taking use patterns into account. Nor does resistance management justify a massive increase in regulation, even though a new regulatory philosophy is needed in the long term. What is required at this stage is a policy debate on the scope and structure of an overall resistance management program.

It is important, also, to see the linkages among the various sectors involved in pesticide policy. Our intention in this paper has not only been to be comprehensive but to show these linkages. No forward-looking resistance policy can emerge in regulatory agencies, for example, until research on monitoring and resistance risk assessment provides scientists with the tools necessary to advise decision makers on whether resistance is a potential or actual problem with a particular chemical. No management program, voluntary or mandatory, can succeed without an educated user community and pesticide sales force. In planning for the future of resistance management, we all must remember that research is not enough: the best research can only be implemented in an effective policy environment. By analyzing the current environment and the future needs as we have done, we hope that this paper will help to bring effective policies into being.

REFERENCES

Bottrell, D. E. 1979. Integrated Pest Management. Washington, D.C.: President's Council on Environmental Quality.

Campt, D. D. 1983. ...Are shared by EPA. Plant Dis. May:469.

Chock, A. K., and M. J. Dover. 1980. Current technological and conceptual impediments to widescale use of microbial control agents: Registration. Pp. 44–48 *in* Proc. Workshop on Insect Pest Management with Microbial Agents. Ithaca, N.Y.: Boyce Thompson Institute.

Croft, B. A., J. L. Howes, and S. M. Welch. 1976. A computer-based, extension pest management delivery system. Environ. Entomol. 5:20–34.

David, M. L., and S. G. Unger. 1983. The market life of agricultural insecticides. Paper presented at the Symp. Annu. Mtg. Am. Soc. Agric. Econ., Purdue University, Lafayette, Indiana, July 31–August 3, 1983.

Delp, C. J. 1981. Strategies for dealing with fungicide resistance problems. Pp. 865–871 *in* Proc. Brit. Crop Prot. Conf., Pests and Diseases, Vol. 3. Lavenham, Suffolk: Lavenham.

Dover, M., and B. Croft. 1984. Getting Tough: Public Policy and the Management of Pesticide Resistance. Washington, D.C.: World Resources Institute.

Fungicide Resistance Action Committee (FRAC). 1983. GIFAP Bull. 9:3.

Georghiou, G. P. 1983. Management of resistance in arthropods. Pp. 769–792 *in* Pest Resistance to Pesticides, G. P. Georghiou and T. Saito, eds. New York: Plenum.

Georgopoulos, S. G. 1982. Detection and measurement of fungicide resistance. Pp. 24–31 *in* Fungicide Resistance in Crop Protection, J. Dekker and S. G. Georgopoulos, eds. Wageningen, Netherlands: Centre for Agricultural Publishing and Documentation.

Goring, C. A. I. 1977. The costs of commercializing pesticides. Pp. 1-33 *in* Pesticide Management and Insecticide Resistance, D. L. Watson and A. W. A. Brown, eds. New York: Academic Press.

Hueth, D., and U. Regev. 1974. Optimal agricultural pest management with increasing pest resistance. Am. J. Agric. Econ. 56:543–552.

Leeper, J. R. 1983. Monitoring and evaluating resistance. Paper presented at the Symp. Annu. Mtg. Entomol. Soc. Am., Detroit, Mich., November 27–December 3, 1983.

Lewis, M., and A. Woodburn. 1984. Agrochemical Monitor (newsletter). 33:2–12.

Office of Technology Assessment (U.S. Congress). 1979. Pest Management Strategies in Crop Protection. Washington, D.C.: U.S. Government Printing Office.

Poe, S. L. 1983. IPM in academia: a survey of academic programs. Paper presented at the Symp. Annu. Mtg. Entomol. Soc. Am., Detroit, Mich., November 27–December 3, 1983.

Staub, T., and D. Sozzi. 1983. Fungicide resistance: A continuing challenge. Plant Disease 68(12):1024–1031.

Storck. W. J. 1984. Pesticides head for recovery. Chem. Eng. News 62:35–59.

U.S. Congress, House. 1983. EPA pesticide regulatory program study. Hearing before the Subcommittee on Department Operations, Research, and Foreign Agriculture, Committee on Agriculture, December 17, 1982. Washington, D.C.: U.S. Government Printing Office.

U.S. Environmental Protection Agency. 1974. Farmers' Pesticide Use Decisions and Attitudes on Alternate Crop Protection Methods. EPA Doc. No. EPA-540/1-74-002.

Wasserstrom, R., and R. Wiles. 1985. Field Duty: U.S. Farmworkers and Pesticide Safety. Washington, D.C.: World Resources Institute.

Whalon, M. E., S. H. Gage, and M. J. Dover. 1984. Estimation of pesticide use through the Cooperative Crop Monitoring System in Michigan apple production. J. Econ. Entomol. 77:559–564.

Wood, R. J. 1981. Strategies for conserving susceptibility to insecticides. Parasitology 82:69–80.

Yoder, K. S. 1983. APS committee's concerns on fungicide usage. Plant Dis. May:469.

Pesticide Resistance: Strategies and Tactics for Management.
1986. National Academy Press, Washington, D.C.

Economic Issues in Public and Private Approaches to Preserving Pest Susceptibility

JOHN A. MIRANOWSKI and GERALD A. CARLSON

Because of pest mobility, pesticide resistance is not easily managed by individual farmers. Pesticide manufacturer incentives in attempting to prolong the effectiveness of pesticides are influenced by the level of competition and likely new pesticide discoveries. There is little evidence in pesticide prices that pesticide companies expect rapid increases in pesticide scarcity due to resistance. Depending on market and pest mobility situations, various groups can best combat resistance development.

INTRODUCTION

Unlike most other resources used in world agriculture, synthetic pesticides are a relatively new development. Creation of social institutions to protect public safety, to encourage new pesticide discoveries, and to use effectively the stock of pesticides now available are critical for the long-term productivity of pesticides (Carlson, 1977). The private companies and individuals involved include pesticide manufacturers, formulators, retail firms, custom applicators, pest-control consultants, and farmers. Public agencies vary from country to country. The legal institutions include national and local pesticide safety regulations and pesticide use and patent protection laws.

Many resources in agriculture are allocated by the choices of farmers purchasing in unfettered markets. Because of the potential off-farm damage by pesticides, however, there are regulations concerning maximum dosages, restrictions on location of use, and other safety requirements. Economically these regulations can sometimes be justified because producers may not consider off-farm costs (health, wildlife, and environment) in their pesticide-use decisions.

When one farmer harbors or creates pests or conditions conducive to pest population growth with adverse effects for surrounding and even distant farmers, the free choices of farmers become nonoptimal (Regev, 1984). Since many insects, weed seeds, and pathogens are highly mobile, the pests are considered "common property." No single farmer owns the pests, and one farmer's pest-control decisions affect other producers.

A pesticide company may behave as the owner of a mobile pest population if the company produces the major pesticide used to control this pest population. The company will have an interest in the efficacy of the pesticide in the future and will price the product and promote its use to maximize long-term (discounted) net returns. With fewer close pesticide substitutes or longer patent protection, a pesticide manufacturer is more likely to act as a long-term manager of pest susceptibility.

Pests that are or might become resistant to one or more pesticides are the main focus of this conference. When pests lose their susceptibility to pesticide materials, they depreciate the value of these pesticides. Many types of agricultural resources depreciate in value over time: equipment wears out, land is eroded, energy stocks are drawn down, and farmers' skills and knowledge can decline. Individual farmers have long had the responsibility of deciding the rate at which they use their resources over time. When mobile pests develop resistance to currently available pesticides, however, groups of farmers, government agencies, or other social units may need to take action. A single farmer cannot decide how long the effectiveness of a given pesticide should be maintained on his farm because the pesticide-use practices of all farmers with whom he shares the particular pest population determine its susceptibility to his pesticide applications.

With mobile pests, just as with pesticide movement off the farm, there "may be" an economically superior way to organize pest control than through individual farmer decisions. The "may be" refers to the costs and special problems associated with collective pest-control actions. Actions by pesticide manufacturers, government agencies, or groups of farmers may not slow resistance development sufficiently to pay these additional costs, or the added benefits may not justify the added costs.

We plan to show how the presence of mobile, resistant pests and pesticide market structure influences pesticide companies and farmers; how one can gauge current and future scarcity of pesticides; and how to determine what factors favor various groups in managing pesticide resistance. We conclude with some implications for policy changes.

PEST SUSCEPTIBILITY: OPTIMAL USE OVER TIME

One primary economic objective in pesticide resistance management is to achieve the socially optimal amount of pest susceptibility to pesticides over time. Although the objective is relatively straightforward, the dimensions of

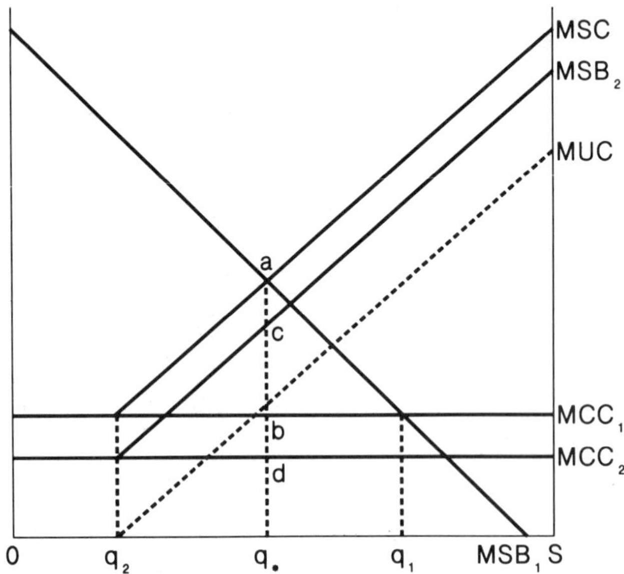

FIGURE 1 Optimal use of pest susceptibility over time: costs and benefits.

the process are complex, and the optimal solution will vary by pest and pesticide. To simplify this decision process and to help interpret specific cases, we present an analytical framework.

The analytical framework is drawn from the natural resource economics literature that helps to explain the optimal intertemporal use of scarce resources. The natural resource here is pest susceptibility to pesticides. Pest susceptibility is a renewable resource that can be harvested repeatedly, but through repeated use of specific chemicals and groups of chemicals, the stock of susceptibility gradually reduces and eventually is depleted or exhausted. For simplicity we will consider the depreciation of pest susceptibility as a function of the number of exposures or applications of pesticides. Pest susceptibility then becomes a nonrenewable stock resource that is depleted with repeated applications of a particular pesticide or pesticide group (Hueth and Regev, 1974).

Susceptibility is a common property resource of varying degree. The private versus common nature of the pest will depend on pest mobility between farms and regions. Like the commons, one individual's harvest of the stock may reduce the resource stock (susceptibility) available to other potential users. Thus, a "public" decision may increase society's welfare relative to that achieved in private optimizing decisions.

Figure 1 is a two-period graphic illustration (McInerney, 1976) of the economic dimensions of the optimal intertemporal allocation of pest susceptibility when the stock is limiting. Although a two-period analysis may not appear appropriate, it illustrates the concept without the complexity of a mathematical presentation (Hueth and Regev, 1974). In Figure 1 the fixed stock of the resource, OS, is given by the length of the horizontal axis. The curves MSB_1 and MSB_2 represent the marginal social benefits from utilizing the stock of susceptibility in periods t_1 and t_2. (Alternatively, the level of susceptibility is inversely related to the number of pesticide applications.) MSB_2 is reversed to illustrate that future resource use in period t_2 is constrained by the stock remaining after period t_1. MCC_1 and MCC_2 are the marginal pesticide costs in periods t_1 and t_2. Note that MCC_2 is lower than MCC_1. Both MSB_2 and MCC_2 have been discounted to translate all costs and benefits into present values for direct comparison.

In a static model, $MSB_1 = MCC_1$; thus, q_1 would be the optimal level of pesticide use in period t_1. But in a dynamic context, the allocation decision becomes more complex. The static solution, $MSB_1 = MCC_1$, leaves an insufficient stock of susceptibility to satisfy the optimality conditions in period t_2, that is, $MSB_2 = MCC_2$. A reallocation of the existing stock between the two periods is now necessary.

The cost of utilizing susceptibility in period t_1 in terms of foregone net benefits in t_2 is the marginal user cost, MUC, or the difference between MSB_2 and MCC_2. The marginal social cost (MSC) of using pest susceptibility in period t_1 is the sum of MUC and MCC_1.

The optimal intertemporal allocation of the stock of susceptibility is at the point where $MSB_1 = MSC$. This allocation uses $0q_*$ of the stock of susceptibility in period t_1 and q_*S in period t_2. By allocating susceptibility such that $MSB_1 = MSC$, the present value of marginal net benefits at q_* are equal in periods t_1 and t_2, or $ab = cd$.

This presentation does not include the environmental costs associated with pesticides. With some minor modifications such costs could be included. Yet even if we concentrate only on the production-related costs and benefits, private decisions can deviate from the socially desirable outcome if (1) pest susceptibility is a common property resource, or (2) the supplier of the pesticides exhibits monopolistic behavior.

To evaluate common property, the interpretation of Figure 1 has to be modified. The MSB curve becomes the marginal private benefit (i.e., marginal value product) curve for the individual producer. Likewise, the constant MCC curve applies to the private producer as well as in the aggregate social context. If each producer has sole ownership rights in the pest susceptibility on his farm and pests are immobile, then the producer's profit-maximizing behavior in allocating susceptibility between periods t_1 and t_2 will parallel the social decision sought by society. No government intervention to slow the spread of pesticide resistance can be rationalized. If each producer has open access to the common pest susceptibility, assuming that the pest is highly mobile among farms, then each producer in period t_1 will equate MCC_1 and the average private benefits, APB_1, which is above and to the right of the MSB_1 curve. With common property, $0q_0$ would be used in period t_1 and only $q0S$ would remain for t_2. The private operator has no incentive to conserve susceptibility because it will simply be extracted by other operators who are using the pesticide to control their pest problem. Thus, the common stock of susceptibility will be overexploited in the early period(s) relative to the socially desirable pattern of use. Only public intervention, such as regulations or price incentives (Regev, 1984), can move the equilibrium from q_0 toward q_*. Achieving q_*, the socially optimal allocation of susceptibility between the two periods, is both complex and administratively difficult.

The competitive structure of the pesticide industry may have a major impact on the allocation of the susceptibility stock between the two periods. Monopolistic control of a resource generally leads to underutilization of the resource in the early years relative to the later years (Dasgupta and Heal, 1979). The initial price established by the monopolist is higher than the price that would prevail in a competitive market environment.

To meaningfully analyze the pesticide market structure, often dominated by a few companies, we must confine our discussion to a single pest and the pesticides available for its control. Two cases need to be considered: (1) monopolistic marketing of a pesticide for a particular pest with no close substitutes available or easily capable of being developed, and (2) monopolistic marketing of a pesticide with close substitutes for the control of a particular pest.

In the first case, with no close substitutes, the monopolist can manage pest susceptibility over time. The pesticide firm would equate its marginal revenue and marginal cost curves to determine the profit-maximizing price and quantity combination. Because the marginal revenue curve lies below the demand curve, the optimal quantity of pesticide marketed would be less and the price would be higher than for perfect competition. This combination would tend to retard resistance development to the particular pesticide being considered. In terms of Figure 1 the optimal use of the stock of susceptibility will be less than $0q_*$ in period t_1 but the MUC curve will shift to the right, partly offsetting the incentive to overconserve the stock of pest susceptibility in period t_1. Figure 2 more clearly illustrates the monopolistic management of the stock, with $0q_m$ used in period t_1 and $q_m S$ used in period t_2. From society's viewpoint, pest susceptibility would be overconserved and pesticides would be underused in period t_1, and the stock of susceptibility may not be fully used. The monopoly situation could develop if patent protection was afforded the manufacturer and no close substitutes were available to control a specific pest.

In the second case, with the availability of close substitutes, the monopoly

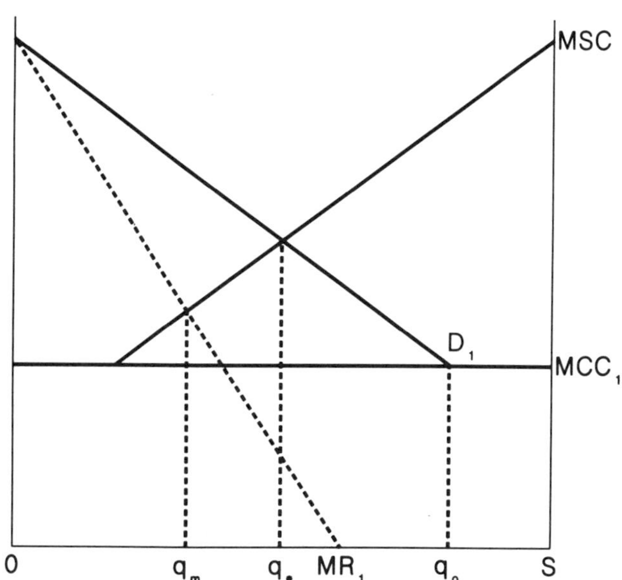

FIGURE 2 Monopolistic management of pest susceptibility over time: costs and benefits.

faces a more elastic demand and a marginal revenue curve for its particular pesticide. As the competition from substitutes increases, the monopoly advantage will disappear. But if some monopoly advantage persists and this pesticide is marketed at a higher price than would prevail in a more competitive market situation, resistance development for the monopoly product may be retarded and hastened for the close substitutes, which are competitively marketed.

RESISTANCE MANAGEMENT: INCENTIVES AND CONSTRAINTS

The only economic justification of public resistance management is that the intertemporal added social benefits outweigh the added social costs. If proposed resistance management schemes (e.g., regional pest-management cooperatives, pesticide application restrictions, education programs) cost more to implement, administer, and enforce than their potential benefits, then these proposals will reduce overall social welfare even though particular producer groups may gain.

If new pesticide products become more scarce over time, the practice of adopting a new pesticide when resistance develops becomes less viable. Alternatively, if new discoveries are adding compounds fast enough, farmers may have little or no incentive to spend time and resources in slowing resistance development. Chemical firms have economic incentives to search

TABLE 1 Indices of Prices Paid by Farmers, U.S. Average

Year	Production Items	Pesticides[a]	Fuel	Fertilizer	Tractors	Labor
			1977 = 100			
1968	50	64	50	52	44	48
1972	61	65	54	52	54	63
1976	97	111	93	102	91	93
1980	138	102	188	134	136	126
1981	148	111	213	144	152	137
1982	150	119	210	144	165	143
1983	153	125	202	137	174	147

[a]Composite index of agricultural chemicals.

SOURCE: U.S. Department of Agriculture (1984a).

for compounds in areas where resistance to currently used compounds is developing, since a new chemical will become more valuable as pests become resistant to old compounds.

Over the past 40 years commercial agriculture, with minor exceptions, has had moderately priced replacement pesticides available when resistance has developed. Therefore, why should farmers expect that the pesticide industry will fail to develop new economic alternatives in the future? If new pesticides are not expected to be readily available in the future, we should see higher relative pesticide prices to reflect the increasing scarcity. Although agricultural chemical prices were increasing between the years 1968 to 1983, the rate of increase was substantially behind all other production input prices (Table 1). A more detailed breakdown of price increases paid by major pesticide categories is provided in Table 2. The statistical series may be shorter, and variation in the number of chemicals included over time may reduce comparability, but the implications are very similar to those contained in Table 1. Additionally, the price increases that did occur for pesticides may have been more related to current market features, such as higher energy prices, more stringent environmental regulations, and general inflation as opposed to the increasing future pesticide scarcity. Thus, based on expectations of future pesticide scarcity, the aggregate market evidence does not indicate the need for overwhelming concern over future pesticide availability and pest susceptibility.

Pesticide companies have incentives to monitor resistance and deploy products to maintain product life (Delp, 1984). Although higher pesticide prices, especially by firms with large market shares, may encourage lower pesticide-use rates and lower resistance development rates, most chemical companies are striving to maintain and expand market shares. Thus, conflicts with business objectives are not uncommon when pursuing joint actions to prolong the effectiveness of pesticides (Delp, 1984).

Decisions to develop new chemical-control alternatives and to price existing products are influenced by patents, the ability to maintain trade secrets, and safety test requirements. The first two influences encourage the development of new compounds; the last factor decreases effective patent lives and reduces incentives to invest in additional pesticide research. These factors should lead to higher pesticide prices, thus discouraging use and reducing resistance buildup.

Pesticide firms may also use marketing practices to reduce resistance buildup. For example, resistance has not developed with pesticide mixtures as opposed to a single compound formulation. Selling mixtures is not without costs, especially to the firm and the environment.

Integrated pest management (IPM) activities, community pest-control organizations, and other attempts to regulate entire pest populations could have major impacts on pesticide resistance development. Yet such actions have problems. These actions are usually costly to organize, administer, and enforce (Rook and Carlson, in press). Enforcement may be difficult to ensure. Such actions may be difficult to expand to many crop-pest situations, and they may be unnecessary if the pest species has limited mobility. For example, resistance in corn rootworm control may be largely endogenized into the individual farmer's decision-making process with few external impacts.

National, state, and local laws may restrict the frequency of pesticide use and the actual chemicals selected. Australia has initiated a program to voluntarily limit pyrethroid use on cotton to a particular period of cotton growth. Although this cooperative program between farmers and industry has been successful, it is difficult to agree on such restrictions until resistance problems become quite serious. Unless the rates and consequences of resistance de-

TABLE 2 Estimated Increases in Average Prices Paid by Farmers for Selected Pesticides

Year	Herbicides	Insecticides	Fungicides
		(Percent)	
1975	43	31	40
1976	− 1	4	11
1977	− 9	− 4	5
1978	− 6	4	6
1979	3	5	7
1980	7	10	22
1981	15	8	7
1982	2	6	4
1983	− 4	8	3
1984	− 6	1	NA
1985*	0	− 3	− 1

*Projected.

SOURCE: U.S. Department of Agriculture, 1975, 1977, 1979, 1980, 1982, 1983, 1984b.

velopment are known, opposition from farmers and pesticide firms may be strong, especially if the pest is not highly mobile between farms.

ORGANIZATIONS TO MANAGE RESISTANCE

It is in the self-interest of various economic agents to manage pests and protect the future productivity of pesticides. Because of the differences in pest mobility, detection of pests, farm sizes, and arrays of chemical and nonchemical controls, there may be several social institutions for managing resistance. The simple scheme outlined below considers six different management configurations: none, farmers, groups of farmers, chemical company, groups of chemical companies, and government units and laws.

These units range from small to large spatial areas of influence. The importance of resistance management for a particular pesticide will increase when few substitute chemical or nonchemical controls are available. The extra resources expended for maintaining susceptible pest populations may be considerable. Monitoring resistance development, switching compounds, rotating compounds, changing sales efforts, and other actions call for high levels of scientific and managerial manpower. Coordination costs between groups of farmers or several chemical firms will increase as the spatial area increases. As benefits from resistance management diverge, the ability to reach agreements will decline. Thus, government agencies may need to expand their role or governments may need to enforce mandatory laws or regulations. The effectiveness of such laws and regulations will be limited by their ability to be enforced.

Conditions Favoring No-Resistance Management

- Little evidence of or very slow resistance development.
- Substitute pesticide and nonpesticide controls are available and competitive with current pesticide control.
- New substitute compounds are readily developed.

Conditions Favoring Resistance Management by Farmers

- Very low pest mobility, with the farmer ''raising'' and ''owning'' his pests.
- Substitute pesticide and cultural controls are far more costly, and competitive replacement pesticides are not forthcoming, that is, specialized crop with limited pesticide market; unique pest lacking susceptibility to other compounds.
- High-value crop subject to large pest damage in absence of control.

Conditions Favoring Voluntary, Multifarm Resistance Management (Cooperative, Community, Consultants' Clients)

• Pests are sufficiently mobile within a confined region so that multifarm coordination of resistance management is beneficial.

• Economics of size can be realized in resistance management while maintaining low coordination costs.

• Benefits and costs are proportional to level of participation.

• Minor benefits accrue to free-riders; these individuals impose minor costs on the coordinated effort.

Conditions Favoring Resistance Management by a Single Chemical Company

• Possession of a highly profitable compound with no potential or actual close substitutes.

• Monitoring of resistance is not costly.

• Strong monopoly position permits the company to market the compound so as to manage resistance.

• Pest is sufficiently mobile so that incentives for voluntary management by growers are uneconomic.

Conditions Favoring Resistance Management by Multiple Chemical Firms (Contracts, Informal Agreements)

• Resistance can be easily managed by mixtures of compounds owned by several firms.

• Resistance monitoring information is valuable for several products sold by several firms, either from potential cross-resistance or by multiple producers of the same compound.

• Coordination of selective pressure on a pest population can be achieved by joint action such as rotation of compounds over time and space.

Conditions Favoring Use of Government Pest Control, Regulatory Agency, or Laws for Resistance Management

• Free-ridership by one chemical firm or a few farmers can jeopardize a regional coordination effort, typically characterized by a highly competitive pesticide market and a highly mobile pest.

• Coordination between firms is costly, such as might occur when many companies or farmers with widely different interests manage resistance for a given pest population.

• Governmental unit that is responsible for pest control in an area (e.g., public health, Forest Service) can be responsible for resistance management.

• Government agencies can use existing regulations to respond to resistance by quickly making new compounds available. Section 18 of the Federal Insecticide, Fungicide and Rodenticide Act (FIFRA) allows the U.S. Environmental Protection Agency (EPA) to speed the clearance process for new pesticides when resistance is a serious problem.

These approaches are not mutually exclusive, and it may be helpful to have several groups attempting resistance management. The main limitation of having several approaches is that each approach will require scarce resources to preserve susceptible pests. The overlapping efforts may require more total resources to accomplish a given end relative to a coordinated effort.

POLICY IMPLICATIONS

What policy implications can be drawn from the above analytical framework with respect to resistance management?

• If the pest being considered is rather immobile and if farmers are informed on pesticide resistance development, then the optimal allocation of pest susceptibility over time can be achieved through farmers maximizing their long-run returns. Public information (education) programs may be needed to create an awareness among farmers of pesticide resistance development.

• If the pest being considered is mobile, optimal management of pest susceptibility may require some form of organization or regulation, given the "common property" nature of pest susceptibility. Yet some form of intervention can be justified only if the added benefits outweigh the added costs. Lazarus and Dixon (1984) evaluated a regional program for control of corn rootworm pesticide resistance and concluded that the potential gains were very small relative to farm incomes. Such a result can be expected when the pest exhibits limited mobility and when viable substitute controls such as crop rotations are available. Additionally, the magnitude of control program costs should be an important factor in the decision to organize or regulate.

• The attractiveness of resistance management to the pesticide industry will depend on market structure. In a relatively competitive pest-control situation with many close substitutes, there would be little incentive for resistance management. If there were few substitutes and few firms competing in the marketplace to control a particular pest, industry pesticide resistance management would be more viable given the lower transaction costs involved. But oligopoly theory would imply that an equilibrium management strategy may be difficult to achieve in a dynamic market situation. Government intervention, including a tax solution, to allocate susceptibility is possible, but it would be extremely difficult to achieve an optimal allocation even if pesticide resistance could be retarded.

• A monopolistic market situation can lead to an industry solution to resistance management if no close substitutes are available. Where the monopolist provides the only efficacious control alternative, pest susceptibility will only be allocated optimally by coincidence. More likely, susceptibility will be overconserved. Yet the costs of public intervention may easily outweigh the benefits in this situation, especially if external environmental costs are considered. If close substitutes are available the monopoly-controlled compound will incur slower rates of resistance development, but resistance development may be hastened for the close substitutes produced more competitively. DDT resistance development in the 1960s and early 1970s may be a good illustration of such competitive market consequences.

CONCLUSION

In summary the development of pesticide resistance is not an argument for resistance management in and of itself. The best group to implement resistance will depend on market and pest mobility conditions.

REFERENCES

Carlson, G. A. 1977. Long run productivity of insecticides. Am. J. Agric. Econ. 59:543–548.

Dasgupta, P. S., and G. M. Heal. 1979. Economic Theory and Exhaustible Resources. Welwyn, England: Nisbet/Cambridge.

Delp, C. J. 1984. Industry's response to fungicide resistance. Crop Prot. 3:3–8.

Hueth, D., and U. Regev. 1974. Optimal agricultural pest management with increasing pest resistance. Am. J. Agric. Econ. 56:543–552.

Lazarus, W. F., and B. L. Dixon. 1984. Agricultural pests as common property: Control of the corn rootworm. Am. J. Agric. Econ. 66:456–465.

McInerney, J. 1976. Natural resource economics: The basic analytical principles. J. Agric. Econ. 27:31–53.

Regev, U. 1984. An economic analysis of man's addiction to pesticides. Pp. 444–453 *in* Pest and Pathogen Control: Strategy, Tactical, and Policy Models, Gordon R. Conway, ed. New York: John Wiley and Sons.

Rook, S. P., and G. A. Carlson. In press. Farmer participation in pest management groups. Am. J. Agric. Econ.

U.S. Department of Agriculture. 1975. Evaluation of Pesticide Supplies and Demand for 1974, 1975, and 1976. Agric. Econ. Rep. No. 300.

U.S. Department of Agriculture. 1977. Evaluation of Pesticide Supplies and Demand for 1977. Agric. Econ. Rep. No. 366.

U.S. Department of Agriculture. 1979. Evaluation of Pesticide Supplies and Demand for 1979. Agric. Econ. Rep. No. 422.

U.S. Department of Agriculture. 1980. Evaluation of Pesticide Supplies and Demand for 1980. Agric. Econ. Rep. No. 454.

U.S. Department of Agriculture. 1982. Farm Pesticide Supply and Demand Trends, 1982. Agric. Econ. Rep. No. 485.

U.S. Department of Agriculture, Economic Research Service. 1983. Inputs Outlook and Situation. June.

U.S. Department of Agriculture. 1984a. Agricultural Statistics 1984. Washington, D.C.: U.S. Government Printing Office.

U.S. Department of Agriculture, Economic Research Service. 1984b. Inputs Outlook and Situation. August.

Glossary

allele: any of several particular forms of a gene.

adverse effects: any change in the information on known risks of pesticide use; must be reported to EPA as part of FIFRA requirements.

antitrust: a body of laws making illegal various actions that have the effect of changing prices, allocating sales, or otherwise restraining trade.

competition: the interaction between populations in which there is mutual inhibition of each other's growth due to the sharing of common resource(s).

critical frequency: the frequency of resistance within a population at which specific strategies should be enacted in order to manage resistance successfully. It is important to recognize that critical frequencies have not been established for the vast majority of pest/pesticide situations.

deme: a local population of closely related organisms.

density dependence: situations in which the rate of growth of populations or relative fitness of individuals varies with the standing density of the population.

density independence: situations in which the rate of growth of populations or relative fitness of individuals is independent of the density of the population.

diploid: individual organisms or cells with two separate sets of genes (chromosomes).

dominance: situations in which the expression of one allelic form of a gene determines the phenotype of heterozygous individuals and obliterates the expression of recessive alleles of that gene.

ecology: the study of the distribution and abundance of organisms and their interactions with their physical and biotic environment.

economic efficacy: "acceptable" pest control from the user's perspective, under field conditions. Many variables influence the user's impression of what is "acceptable." They include effectiveness, cost of the pesticide and its alternatives, commodity value, perception of pest severity, injury or public health standards, etc. The fact that a pest population has a verified level and frequency of resistance does not necessarily mean that there has been a loss of economic efficacy, or that use of the pesticide should be discontinued. Conversely, it is possible to have a loss of economic efficacy without resistance (i.e., the introduction of a more effective pesticide, cultivar, or cultural practice; the microbial degradation of the pesticide before it reaches the pest).

economic threshold: that pest population density or damage level at which control measures should be taken to prevent economic injury from occurring.

efflux: passing out, flow out, or pumping out from a cell.

epistasis: the nonadditive interactions between genes where the phenotypic expression of alleles of one gene affects the expression of alleles of other gene(s).

emergency use permits: pesticide-use permits granted under section 18 of FIFRA for specific locations and time periods; granted if unusual pest or pesticide availability conditions arise.

genomic shock: an environmental effect that causes movement of genetic elements (e.g., a transposition of genes within a chromosome).

genotype: the combination of genes borne by an individual organism.

genetically effective component (of migration): the contribution of immigrants to the genetic makeup of the population in the succeeding generation.

fitness: the relative probability of survival and reproductive yield of individuals of a particular genotype.

haploid: having only one complete set of chromosomes.

heterokaryotic: containing genetically different nuclei (in cells).

IR-4: an interregional project supported by USDA and the land-grant colleges to provide efficacy and safety data necessary for receiving a tolerance and registration of a pesticide for a minor-use market.

linkage disequilibrium: a nonrandom association of alleles at two or more loci.

locus: a synonym for gene; the position of a gene on a chromosome (pl. loci).

market structure: a description of a group of buyers and sellers in a market that emphasizes number of participants, numbers and types of substitute products, information exchange, and other features affecting level of competition.

minor-use: a pesticide market that is small in sales volume because the pesticide is only used on infrequently occurring pests or on pests of a crop that has a small number of acre-applications per year.

multinucleate: having two or more nuclei per cell.

neolamarckian: the process by which the genotype of individual organisms is changed due to the directed action of the environment. After the French zoologist, J. G. Lamarck (1744–1829), who proposed a theory of evolution operating through the inheritance of acquired characters.

phenotype: the physical manifestation of the genes borne by an individual organism.

plasmid: short, circular segment of nonchromosomal DNA.

plastid: a cytoplasmic organelle concerned with photosynthesis and/or storage of food.

plastome: the complex of genes in the plastid (also: plastid genome).

pleiotropy: multiple phenotypic expressions of a single gene.

population biology: the study of the genetic and ecological behavior of populations of organisms.

population genetics: the study of genetic diversity and the mechanisms of genetic change (evolution) in populations of organisms.

refuge: a place or period of time in which organisms are free from the action of predators or substances that inhibit their growth and reproduction.

relative reproductive rate: the numbers of progeny produced by an individual of a particular genotype in the course of a generation compared to those produced by other individuals in the population.

resistance: the inherited ability in a strain of pest to tolerate doses of toxicant that would prove lethal to a majority of individuals in a normal population of that species. This definition implies a statistically significant shift in LC_x (or LD_x) values that are normally established through laboratory bioassays. Laboratory documentation of resistance, however, does not necessarily indicate a current or impending loss of economic efficacy in the field.

selection: changes in the genetic composition of populations resulting from the differential survival or reproduction of specific genotypes.

selective advantage: the extent to which the relative fitness of individuals of a particular genotype exceeds the mean fitness of all genotypes in the population.

selective disadvantage: the extent to which the relative fitness of individuals of a particular genotype is less than the mean fitness of all genotypes in the population.

thylakoid: flattened membrane sacs within chloroplastids in which chlorophyll molecules are incorporated.

tubulin: a protein that is a subunit of microtubules, which are found in structures such as the mitotic spindle.

user fees: taxes or fees that are proportional to use level and are charged to users of public services.

variance: the average value of the squared deviations of observations from their mean. It is a measure of the magnitude of variation in a character. In the study of inheritance of continuously distributed characters (Quantitative Genetics), primary concern is the proportion of the variance in the phenotype ("phenotypic variance") that is due to underlying variation in the genotypes of individuals in the population.

additive genetic variance: the proportion of the phenotypic variance that is due to cumulative expression of alleles of the same gene or different genes that are acting independently in the determination of the value of a continuously distributed character, e.g., height.

nonadditive genetic variance: the proportion of the phenotypic variance that is due to cumulative expression of alleles of the same gene or different genes that are not independent in their determination of the phenotype, e.g., the contributions of dominance and epistasis.

(genetic) covariance: the average value of the product of the differences between an array of observations taken in pairs and their means. It is used as a measure of the direction and extent a character varies among pairs of individuals and as a measure of how two different characters vary within individuals. As used in Chapter 3, the genetic covariance is the direction and extent of variation in the expression of the same set of genes on two different characters, e.g., tolerance to two different pesticides.

Volterra principle: The prediction (and observation) that in cases where a prey population is held in check by a predator, the killing of predators and prey (e.g., by the use of pesticides) is likely to result in an increase in the numbers of prey (pests).

xenobiotic: foreign chemical (such as a pesticide).

Index

A

D

E

F

G

H

I

N

O

P

Q

R